The Cultural Landscape -
Past, Present and Future

The Cultural Landscape -

Past, Present and Future

Edited by

Hilary H. Birks

H.J.B. Birks

Peter Emil Kaland

Dagfinn Moe

Botanical Institute
University of Bergen
Bergen

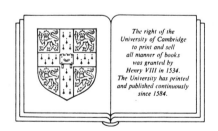

*The right of the
University of Cambridge
to print and sell
all manner of books
was granted by
Henry VIII in 1534.
The University has printed
and published continuously
since 1584.*

CAMBRIDGE UNIVERSITY PRESS
Cambridge
New York   Port Chester
Melbourne   Sydney

Published by the Press Syndicate of the University of Cambridge
The Pitt Building, Trumpington Street, Cambridge CB2 1RP
40 West 20th Street, New York, NY 10011, USA
10 Stamford Road, Oakleigh, Melbourne 3166, Australia

First published 1988
Reprinted 1990

Printed in Great Britain at the University Press, Cambridge

British Library cataloguing in publication data

The cultural landscape: past, present and future.
1. Landscape. Effects of civilisation, to 1986
I. Birks, Hilary H.
719'.09

Library of Congress cataloguing in publication data available

ISBN 0 521 34435 2 hardback

Contents

Part One - Present and Future

Part Two - Past

Part Three - Abstracts

Contents

Contents ix

List of Contributors

B. Aaby — Geological Survey of Denmark, Copenhagen, Denmark

J. O. Adejuwon — Department of Geography, University of Ife, Ile-Ife, Nigeria

B. Ammann — Systematische-Geobotanisches Institut, University of Bern, Bern, Switzerland

S. Th. Andersen — Geological Survey of Denmark, Copenhagen, Denmark

I. Austad — Sogn og Fjordane College, Sogndal, Norway

K.-E. Behre — Niedersächsisches Landesinstitut für Marschen- und Wurtenforschung, Wilhelmshaven, West Germany

B. E. Berglund — Department of Quaternary Geology, University of Lund, Lund, Sweden

H. J. B. Birks — Botanical Institute, University of Bergen, Bergen, Norway

S. J. P. Bohncke — Institute of Earth Sciences, Free University, Amsterdam, The Netherlands

J. M. Bos — Albert Egges van Giffen Instituut voor Prae- en Protohistorie, University of Amsterdam, Amsterdam, The Netherlands

M. Bowler — Department of Botany, University College, Galway, Ireland

E. D. Bozilova — Department of Botany, University of Sofia, Sofia, Bulgaria

A. Brande — Institute of Ecology, Technical University of Berlin, Berlin, West Germany

F. M. Chambers — Department of Geography, University of Keele, Keele, U.K.

R. A. Dodgshon — Department of Geography, University College of Wales, Aberystwyth, U.K.

K. J. Edwards — Department of Geography, University of Birmingham, Birmingham, U.K.

J. Eilart	Nature Conservation Society of the ESSR and Institute of Geology, Tallinn, USSR
O. Ekanade	Department of Geography, University of Ife, Ile-Ife, Nigeria
U. Emanuelsson	Department of Ecology, Plant Ecology, University of Lund, Lund, Sweden
J. Empett	Department of Biology, King's College, London, U.K.
A. E. Evans	Department of Biology, King's College, London, U.K.
K. Fægri	Botanical Institute, University of Bergen, Bergen, Norway
A. Fasteland	Historical Museum, University of Bergen, Bergen, Norway
A. Fleming	Department of Archaeology & Prehistory, University of Sheffield, Sheffield, U.K.
B. Fredskild	Greenland Botanical Survey, University of Copenhagen, Copenhagen, Denmark
M.-J. Gaillard	Department of Quaternary Geology, University of Lund, Lund, Sweden
B. van Geel	Hugo de Vries-Laboratorium, University of Amsterdam, Amsterdam, The Netherlands
L. Hauge	Sogn og Fjordane College, Sogndal, Norway
S. Hicks	Department of Geology, University of Oulu, Oulu, Finland
J. Hughes	Department of Botany, University of Durham, Durham, U.K
B. Huntley	Department of Botany, University of Durham, Durham, U.K.
M. Ihse	Department of Physical Geography, Remote Sensing Lab, University of Stockholm, Stockholm, Sweden
S. Indrelid	Historical Museum, University of Bergen, Bergen, Norway

M. Kabailiené	Geological and Mineralogical Department, Vilnius State University, Vilnius, USSR
P. E. Kaland	Botanical Institute, University of Bergen, Bergen, Norway
A. J. Kalis	Seminar für Vor- und Frühgeschichte, J. W. Goethe-Universität, Frankfurt, West Germany
R. S. Kelly	Gwynedd Archaeological Trust Ltd., Bangor, U.K.
K. J. Kirby	Nature Conservancy Council, Northminster House, Peterborough, U.K.
A. Kreuz	Seminar für Vor- und Frühgeschichte, J. W. Goethe-Universität, Frankfurt, West Germany
M. Kvamme	Botanical Institute, University of Bergen, Bergen, Norway
H. Küster	Institut für Vor- und Fruhgeschichte, University of Munich, Munich, West Germany
D. Lóczy	Geographical Research Institute, Hungarian Academy of Sciences, Budapest, Hungary
J. Lüning	Seminar für Vor- und Frühgeschichte, J. W. Goethe-Universität, Frankfurt, West Germany
J. M. Line	Computer Laboratory, University of Cambridge, Cambridge, U.K.
E. Long	Department of Biology, King's College, London, U.K.
D. Moe	Botanical Institute, University of Bergen, Bergen, Norway
K. Molloy	Department of Botany, University College, Galway, Ireland
P. D. Moore	Department of Biology, King's College, London, U.K.
S. E. Moseley	Department of Biology, King's College, London, U.K.

E. J. Nilssen	Institute of Biology and Geology, University of Tromsø, Tromsø, Norway
A. Norderhaug	Kjønnerød, Borgheim, Norway
M. O'Connell	Department of Botany, University College, Galway, Ireland
B. V. Odgaard	Geological Survey of Denmark, Copenhagen, Denmark
G. Olsson	Department of Ecology, Plant Ecology, University of Lund, Lund, Sweden
J. P. Pals	Albert Egges van Giffen Instituut voor Prae- en Protohistorie, University of Amsterdam, Amsterdam, The Netherlands
I. Perry	Department of Biology, King's College, London, U.K.
T. Persson	Department of Quaternary Geology, University of Lund, Lund, Sweden
S.-M. Price	36 Fair Green, Diss, Norfolk, U.K.
O. Rackham	Corpus Christi College, Cambridge, U.K.
M. Ralska-Jasiewiczowa	Institute of Botany, Polish Academy of Sciences, Cracow, Poland
G. Regnéll	Department of Ecology, Plant Ecology, University of Lund, Lund, Sweden
J. Regnéll	Department of Quaternary Geology, University of Lund, Lund, Sweden
D. E. Robinson	Department of Natural Sciences, National Museum, Copenhagen, Denmark
L. Saarse	Nature Conservation Society of the ESSR and Institute of Geology, Tallinn, USSR
J. A. Schmit	Schmit & Company, 80 Cedar Grove Road, Media, Pennsylvania 19063, USA
M. Tolonen	Department of Botany, University of Helsinki, Helsinki, Finland
S. B. Tonkov	Department of Botany, University of Sofia, Sofia, Bulgaria
Y. Vasari	Department of Botany, University of Helsinki, Helsinki, Finland

J. A. J. Vervloet   Netherlands Soil Survey Institute,
                              Wageningen, The Netherlands

I. Vuorela   Department of Geology, University of
                              Helsinki, Helsinki, Finland

Foreword

This book is a permanent record of the international symposium held during the field excursion entitled The Cultural Landscape - Past, Present and Future, which was held in July 1986 in western Norway. It was primarily a celebration of the 100 year annniversary of the Botanical Insititute of the University of Bergen, which was founded as The Botanical Museum in 1886.

After leaving Bergen, the excursion visited the west Norwegian coastal heath landscape and saw demonstrations of how heathland was formerly maintained and utilized. It continued to the forested fjord area along Sognefjord and saw examples and demonstrations of lowland farms and the utilization of trees. It concluded in the mountains in the region of the summer farms, thus completing the three major cultural landscapes of western Norway.

At Sogndal, the symposium was held during two days, with lectures and poster sessions. The chapters of this book are contributions from these lectures and posters, and the abstracts summarize other contributions given at the symposium. The principal symposium and excursion organizers were H. J. B. Birks, P. E. Kaland, M. Kvamme and D. Moe, and they received help from many other individuals, in particular, I. Austad. The symposium and the excursion were assisted financially by Clara Lächmans Fond, INQUA - EuroSiberian Subcommission for the Study of the Holocene, Letterstedtska föreningen, Olaf Grolle Olsens Legat, Unesco-IUGS and University of Bergen. They were also a part of the International Geological Correlation Programme Project 158b. This book is a contribution to this project. The production of the book has been supported financially by Olaf Grolle Olsens Legat.

As editors of this book we want to thank many people. It would not have been possible without the contributors, and we thank them for their contributions and their patience during the editing process. The following have reviewed one or more manuscripts: B. Aaby, B. Ammann, S. Th. Andersen, K.-E. Behre, H. H. Birks, H. J. B. Birks, K. J. Edwards, D. R. Foster, B. Fredskild, K. Fægri, S. P. Hicks, B. Huntley, U. Miller, D. Moe, P. D. Moore, M. O'Connell, B. Odgaard, G. Régnell, D. Robinson, T. Taylor, B. van Geel, W. van Zeist.

This book has been produced as a camera-ready copy at the Botanical Institute. It has been produced with the NORTEXT typesetting program by Norsk Data A/S, and we wish to thank Norsk Data A/S for their cooperation. The typesetting has been done by Gerhard Datasats A/S, and we express our gratitude to Leif Rødder for all his advice and practical assistance.

The whole book has been typed in to the Norsk Data word processing program by Annechen Ree. She has undertaken this massive task skillfully and cheerfully in a language which is not her mother tongue. The preparation of the diagrams has been done by Siri Herland, who has calmly and efficiently coped with some rather difficult material. She has also had the delicate job of laying out the pages and arranging the text and figures. To them both we express our most grateful thanks.

In addition to the normal scientific editing of a book of this type, the copy editing, text production, and the printing of the camera-ready original of a book of this size have presented many unforseen problems for the editors. We attempted to overcome them by dividing the responsibility for different tasks: HJBB and HHB have been responsible for the scientific and copy editing and linguistic revisions, HJBB for the index and the bibliography, PEK and HHB for the text production, PEK for the printing, and DM for the diagrams.

HHB, HJBB, PEK, DM

We wish to acknowledge the special contribution of Peter Emil Kaland, who has undertaken the difficult and arduous task of producing and printing the text as a camera-ready copy. He has devoted several months to mastering the intricacies of computerized type-setting and throughout he has remained cheerful and enthusiastic. The appearance of this volume owes much to his labours.

HHB, HJBB, DM
Bergen, May 1988

Preface

Knut Fægri

Like H. G. Wells' conception of peace ("Joan and Peter"), the cultural landscape can only be understood by its antithesis: untouched, unspoilt nature. The realization of this dualism is the basis for understanding and appreciating either.

To Medieval man nature was hostile, forbidding. His world was the patch of friendly, cultivated ground, isolated in a fearful, dangerous matrix. On the whole, this sentiment also prevailed during the following centuries, only to be changed by Rousseau and his romanticist followers. And we should not forget a contemporary of the great Jean Jacques: Linnaeus, to whom Nature certainly was neither foreign nor hostile, even though, as a good child of his century, he relished the sight of well-cultivated land.

The realization that nature was a positive factor, indeed so much so that it should be protected against the depredations of man, most remarkably did not originate in densely built-up Europe, where nature really was endangered, but in the USA which at that time had plenty of it. Nature became something to be cherished and protected. Later, the idea was also adopted in Europe, not least in Linnaeus' home country, where the Linnaean philosophic tradition is so important even in today's intellectual life, and where Rutger Sernander was one of the original banner-carriers. Under his enthusiastic aegis (e.g. Sernander 1919), 'natural' Swedish landscapes were protected as memories of an idyllic pastoral period past. Les temps perdus.

Only when natural successions started their own course inside the carefully protected and fenced-off areas and turned them into impenetrable wildernesses did the realization gradually develop that the borderline between cultivation and nature had, perhaps, been wrongly placed. Man's influence was wider and subtler than originally thought, and the landscape in general was far from being unadulterated nature. The concept of the cultivation landscape was born as a landscape in apparent equilibrium, where man's influence was only one of several equally strong influences.

In the end, it was realized that, even in Scandinavia, a virgin landscape was a fiction. With some small and doubtful exceptions all vegetation types were created or modified by man. The degree of modification varies, from the open arable field with its heavy machinery to the grazing areas in the mountains and the deep forests, although just there one might perhaps find areas that had never seen an ax or a (man-made) fire. Instead of conceiving of the Scandinavian landscape as a vast 'natural' area with small cultivated patches, the idea developed that the whole fjord or the whole valley is one great cultivation effect. The emphasis shifted from the alternative cultivated/ uncultivated concept to the idea of a gradient of human impact. The task of defining and preserving this gradient is a challenge to modern nature conservation (Fægri 1954a, 1962).

Long before this had been realized, there was a desire to reconstruct the history of the cultivation landscape, which in the thinking of those days meant the open fields and pastures in contrast to the closed forest.

When and how did what happen? Based on Firbas' pioneering papers (1935, 1937) we started in a primitive fashion to play with 'cultivation pollen' and forest density in our pollen diagrams (e.g. Fægri 1940), but it was Iversen's landnam paper (1941) that gave the answer, or at least a preliminary answer. During the war years there were few opportunities to test his conclusions against uncontaminated or undisturbed nature - a scarce commodity in Iversen's native Denmark. However, during the last 40 years, pollen analysis, plant ecology and - not least - prehistoric archaeology have contributed to the recognition of the borderline, or rather the transition zone, between uncontaminated nature and what eventually became known as the cultural landscape.

Exactly when the term cultural landscape was first coined I do not know, nor is it very important: it came by itself. It certainly was not common before the Second World War, but it is found in, for example, a Swedish dictionary of 1939 as a term with predominant scientific use. In post-war dictionaries it is frequent. According to information from Norsk språkråd (the Norwegian language council) it was first recognized as a Norwegian word in 1960. It should be kept in mind that *kultur-* in Teutonic languages is very versatile and easily used as the first element in compounds with native words. A pre-war Danish dictionary lists 71 such compoundings and that list would be far from complete today. The term *kulturlandskap* or *Kulturlandschaft* would therefore immediately present itself on the day the concept had become so well defined as to need and justify a term of its own. The difficulty - if any - is that it is formally so commonplace as to be taken for granted, like *heath* or *forest*.

The 'natural' landscapes of preceding generations are now understood for what they really are: relics of earlier types of land-use, which were maintained by extensive methods demanding little machinery and much manpower and which therefore became uneconomical. At one time they were essential for extracting a livelihood from a not too cooperative nature. By abandoning these methods and discontinuing traditional land-use, the landscape was left to regenerate in response to other uses or non-use. A landscape is dynamic, it does not remain the same when conditions change. Therefore the remains of the old landscape types are important as records of what once constituted major parts of the land, and in the best Sibylline traditions, such records become more precious the scarcer they are. The remains still with us carry some of the information necessary for understanding the composition and structure of yesteryear's vegetation and for understanding the successions that occurred after the old techniques became obsolete.

The theme of the symposium on which this volume is based is the cultural landscape in general. The theme of the field trip associated with the symposium was the western Norwegian landscape, showing how people have, before industrialization, worked the whole landscape with the primary object of keeping body and soul together. In evaluating the methods and the economics of this management one should not forget that a variable, but always important part of the economic landscape of western Norway, is situated under the surface of the sea, and further in from the coast also in the lakes and streams.

The very strong west-east gradient of physical factors - topography, soil, climate, vegetation - has its natural counterpart in the change in the relative importance of fishing and sea-hunting from the outer coast to the heads of the fjords, where lakes and rivers are more important fishing grounds than the sea. Although horizontal distances are not very great, some 150 km, the transect covered by the trip traversed completely different landscapes with correspondingly different (pre-industrialization) economics. The vertical dimensions add to the complexity of understanding the dynamics of these landscape types. Lack of intimate ecologic understanding has lead to an unwarranted negative evaluation of many landscapes and their vegetation. The coastal ericaceous heath is a case in point. It has been almost unequivocally interpreted in the negative: only the investigations of the last decades have shown that, properly managed, the heath is a highly productive plant community. The snag is that, in the present economic climate, heath utilization does not pay: the yield/work ratio is too unfavourable.

Hopefully, the Cultural Landscape field trip (Birks 1986c) amply demonstrated the danger of evaluating phenomena in too simple a manner. Even within western Norway - and even more pronounced if eastern Norway is also considered - the difference between west and east is so great that one has to think in quite different ways. Schematicism is bound to block understanding. No text-book solution is available for western Norway because (1) there is no single solution, and (2) none of us who have worked in this landscape during the last 50 years, and are still continuing to do so, feel competent to sit down and write a precept for the ecological management of the western Norwegian landscape. The only thing that is certain is that one must approach the landscape with humility and an open mind. To come along with a ready precept, concocted in another area and hopefully eminently suited for that, is to beg for disaster. The primary demand is to acquire an ecological feeling for the landscape - to which the mechanical solutions must be adapted.

Since the days of my predecessor Jens Holmboe (1906-1925) the Bergen Museum/University of Bergen Botanical Department has had western Norway as its major research area, trying to understand this very complicated landscape. Realizing that western Norway still, or until quite recently, could show modern examples of many old cultivation techniques, it was natural to invite colleagues to visit these sites as part of the 100 years jubilee of our institute, and to make their visit a combination of the general symposium and a demonstration of some field areas of possible interest. This would also give us an opportunity to demonstrate to and discuss with colleagues from other parts of Norway and from neighbouring countries the special problems confronting us - and them. In our rugged, varied landscape the effects of cultivation, as expressed in pollen diagrams, are much subtler than the almost brutal large-scale effects registered in more uniform landscapes elsewhere. The contrast between our presumed natural landscapes and the patches of cultivation was - and is - so striking that, 50 years ago, we never saw the need for a definition of what constituted the cultivated *vs.* 'natural' landscape. Today, the botanical, phytosociological and ethnobotanical definition of what constitutes the cultural landscape has been considerably sharpened and is still under active scrutiny. We hoped that by demonstrating the rather extreme western Norwegian landscape we may help to give that scrutiny more depth. Some of the problems met with and discussed are presented in this volume.

The old landscape of ecological economy cannot be saved entirely. Only small parts can be maintained, preserved and restored. To achieve this we must, through a thorough study of this economy, understand how the landscape functions and know the interactions of the multiple techniques in the day-to-day work in the field. They may look simple, but they embody millenia of experience.

Many scientists have contributed to the modern views of the cultural *vs.* natural landscape. In ending this introduction I should like to pay tribute to one botanist who has, perhaps indirectly, done more than anybody else to dispel the old belief in 'natural' landscapes: Lars-Gunnar Romell, a professed anti-romanticist and anti-Sernanderian, who nevertheless in innumerable papers and articles defined and defended "the landscape of the muzzle and the scythe" against unnecessary despoliation by the harsh machinery-harvesting and machinery-thinking of so-called modern agriculture.

Part One

The Present and Future

Introduction

H. J. B. Birks

Much of the traditional cultural landscape of north-west Europe is rapidy disappearing as a result of changes towards intensive modern agriculture or silviculture. Before any preservation of traditional cultural landscapes can be attempted, it is essential to document the present-day cultural landscape and to study present-day processes that influence the floristic, vegetational and ecological patterns that comprise it.

Landscapes can be studied at many scales, ranging from the single element of the landscape mosaic such as individual meadows or woodland stands to the entire landscape mosaic consisting of different patches, corridors and matrix elements (Forman & Godron 1986). Different ecological approaches are appropriate and useful at different scales, and the range of methodologies represented by the chapters in this part reflect this diversity of scale. Descriptive plant sociology of the vegetation of a particular landscape element is one of the most used techniques in European vegetation science. Its use as a means of primary documentation is illustrated by *Hughes & Huntley's* extensive survey of upland hay meadows in Britain. They also use the newly developed and powerful multivariate technique of canonical correspondence analysis (ter Braak 1986) to explore hay meadows/management relationships and suggest that meadow/management patterns vary from area to area, a conclusion of considerable importance in devising management plans. Canonical correspondence analysis has very considerable potential in studies of past and present cultural landscape ecology because it allows vegetation/land-use relationships to be studied directly and simultaneously (Jongman *et al.* 1987).

Hauge similarly uses a descriptive phytosociological approach as a basis for vegetation mapping, monitoring and restoration of a small, typically steep meadow-forest area in western Norway. He combines these phytosociological data with historical information about past agricultural use to formulate a management plan for the area, involving traditional land-use, restoration of buildings, terraces and irrigation ditches, and tree pollarding. *Austad* describes and illustrates the diverse pollarding practices within different woodland elements of the cultural landscape mosaic in an area of western Norway where pollarding still continues, but as a rapidly dwindling part of the agricultural economy. She draws on her extensive field observations, discussions with farmers and documentary records to present insights into how almost all trees were harvested to their fullest as a source of winter fodder for animals and of wood products. Such intensive exploitation must have considerable ecological impact at the landscape scale.

Rackham reviews the traditional methods of woodland and tree management in the British Isles and highlights the strong differences in tree utilization between the British Isles and, for example, western Norway presented by Austad and Hauge. By utilizing ecological, palaeo-ecological, archaeological, documentary and dendrochronological evidence, Rackham traces these fundamental differences back to important historical, ecological and climatic differences. The latter, for example, result in the overriding importance of collecting adequate winter animal-fodder in western Norway compared with the British Isles. The traditional uses of trees and woodland in Britain - woodland, wood-pasture and non-woodland trees in hedges - appear to date back to Anglo-Saxon times at least, if not to Celtic times. The British tradition of being "a nation of shopkeepers rather than peasants" has resulted not in local self-sufficiency but in trade, regional specialization and communication, in contrast to the remarkable self-sufficiency of isolated areas in western Norway.

A recurrent theme of all these chapters is the problem of preservation and maintenance of traditional cultural landscapes in the future, whether the landscape be woodlands, meadows or a mosaic of both. As Hughes & Huntley emphasize, conservation of the cultural landscape is not solely nature conservation, as it requires preservation of traditional land-use practices, buildings, walls and other components of the landscape mosaic, and of traditional ways of life. *Kirby*, a professional conservationist, discusses the very real problems of preserving traditional woodland management in Britain. He emphasizes that social and economic conditions are not the same today as when many traditional practices developed. He raises the key question - why do we wish to maintain traditional practices? Is it for purely historical or cultural reasons (e.g. Hauge's ecological museum and Hughes and Huntley's museum) or because of what traditional management produces in way of landscape features, including biota, aesthetics, etc.? If the latter, Kirby suggests that similar end-results could be achieved more easily and more economically with new management techniques than by re-introducing traditional practices. These are fundamental philosophical questions facing everyone concerned about cultural landscape preservation, and answers to them are required before embarking on management and preservation plans. In all probability, the answer will vary from place to place and from person to person!

Moving from the fine-scale of landscape elements to the cultural landscape as a functional whole, *Indrelid* discusses a single beautiful fjord-valley in western Norway. He shows how the farms in this valley are situated in 3 distinct topographical settings, each with its own agricultural advantages and disadvantages. He describes the basis for subsistence on farms in such an extreme topography as a steep west-Norwegian fjord and discusses the critical importance of infield and outfield, and summer farms at or above the altitudinal tree limit. By analyzing historical records for farm economics in the last century, Indrelid shows how essential mountain summer farms were in providing fertile pastures for the period when the next winter's fodder was being grown in the infields and outfields. These records also show the importance of leaf- and shoot-gathering as winter animal-fodder. An important lesson for landscape ecologists from Indrelid's chapter is the need to consider the entire functional culture landscape rather than parts of the mosaic if the ecology of the landscape is to be understood.

Spectacular rates and extent of change in cultural landscapes are reconstructed from aerial photographs and computer cartography for southern Sweden by *Ihse* and from LANDSAT imagery for two areas in Hungary by *Lóczy*. Both chapters illustrate the complexity and challenge of documenting and interpreting changes in cultural landscapes. Ecology at the landscape scale involves interactions with descriptive community ecology, historical ecology and palaeoecology, geomorphology, archaeology, population dynamics, historical geography, pedology, sociology and ethnology.

There is more to cultural landscape ecology than description and documentation of patterns in time and space. There is consideration of underlying causal processes. Several chapters consider, in part, the interaction between patterns and processes (e.g. Austad, Hughes & Huntley). However, the 3 chapters by Emanuelsson, Olsson, and Dodgshon are explicitly process-orientated, *Emanuelsson* presents a model that relates human exploitation of the landscape to population size and soil nutrients, and emphasizes the critical importance of in-organic nutrients from manure and fertilizers in food production at different technological levels and of ecological control. The latter concept suggests that it is easier to maintain the *status quo* than it is to regain it, a lesson all concerned with landscape management should not forget. The interaction of nutrients, ecological control and climate can result in important geographical and temporal variations in land-use within a given technological level. This model requires careful testing both in space and time by means of quantitative ecosystem and palaeo-ecological studies, respectively. It has the potential of providing a much-needed conceptual basis for interpreting spatial and temporal patterns in cultural landscape development (e.g. regression periods) in simple ecological terms.

By combining the approaches of ecosystem ecology, historical ecology and historical geography, *Olsson* attempts to derive nutrient budgets and estimates of nutrient-use efficiency in the 18th century for two contrasting villages in southern Sweden. She shows that annual yield was greater with three-course rotation than with two-course rotation and that in terms of required manure and available space, the former was considerably more efficient. When these local-scale estimates are considered within the regional environment, geological and edaphic factors appear as the major controls on agricultural development and patterns in the 18th century. *Dodgshon's* chapter reminds us that the cultural landscape consists not only of plants, vegetation, soils and manure but also of humans who similarly can have an idiosyncratic behaviour and complex ecology. He links ecology, agricultural history, sociology and economics to explain the patterns of farming in extreme marginal habitats in the Scottish Highlands in the 16th-19th centuries. These patterns with the development of outfields in the 16th century can only be interpreted against a back-cloth of population growth, a closed society and intensive use of labour. This chapter highlights the problems of interpreting changes in cultural landscape patterns solely in terms of conventional ecological factors.

The 11 chapters in this Part reflect the diverse approaches currently being used in the description, documentation and interpretation of present-day cultural landscape patterns at a variety of scales. They also illustrate the problems of understanding underlying processes over periods of 100-400 years. Such understandings are required today not only for achieving effective preservation of cultural landscapes in the future but also as analogues for interpreting changes in land-use practices over the long, palaeoecological time scales of 1000-5000 years that are considered in Part 2.

Tree Pollarding in Western Norway

Ingvild Austad

Introduction

In a strongly seasonal climate with a non-productive winter season, populations must store provisions for the lean periods. When people started to keep animals for domestic use (meat and milk production), winter fodder had to be provided for cattle. In almost all Norway, the winter is too cold and too long to allow cattle to graze outside throughout the year. Large quantities of fodder are required for the winter. The practice of collecting twigs and leaves for domestic animals is probably the oldest form of fodder harvesting. In contrast to collecting hay, which requires sharp sickles or scythes and a highly developed metallurgical technique, leaf fodder can be collected without tools (Reinton 1976). However, iron tools make such harvesting more efficient. Special knives, very similar to those used today in western Norway, have been found in archaeological deposits of Late Viking Age (Fig. 1). Material finds and illustrations in books, manuscripts and paintings from Medieval times back to the Iron Age show that this practice was common in much of Europe (Andersen 1985, this volume; Troels-Smith 1960), although perhaps not in Britain (Rackham this volume).

Information from preliminary work on new taxes in Norway from 1863 suggest that this management type was then common and widespread. This practice affected both the ecology and scenery of the cultural landscape.

Tree utilization was a result of experience handed down over generations. Combined utilization of different tree products (not only fodder), or the use of trees in particular ways throughout a season or life span can be found. All species of deciduous trees were used for animal fodder, also *Pinus sylvestris*, *Juniperus communis* and *Calluna vulgaris*. Although cutting trees for collecting animal fodder was widely practised in Norway, the choice of species, technique and utilization varied from area to area, as did the special names given to tree management. This chapter concentrates on pollarding in the fjord-districts of western Norway, mainly Sogn og Fjordane county.

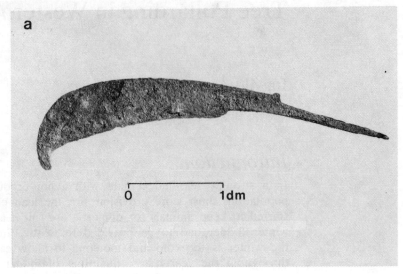

Fig. 1. 'Snidil', the special knife used for cutting small twigs from deciduous trees. Fig. 1a shows a very old knife, dating back to the Late Viking Period (800-1050 A.D.). Fig. 1b illustrates a common knife with a wooden handle still used today (lower) and a new commercial knife (1986) with a plastic handle (upper). Photo 1a: Historical Museum, University of Bergen. Photo 1b: L. Hauge 1986.

The pollarding process

In this section I describe and illustrate the major types of pollarding.

Styving (naving): Low pollarding/topping

This is the process of topping trees, i.e. cutting back branches at a height of 2-3 m, above the reach of grazing animals. The trees were first cut at 10-15 years age or when the stem diameter exceeded 15 cm. Usually men climbed the trees using a short ladder. Hard cutting stimulated the growth of resting buds near the top of the stumps and the trees developed a candelabra form (Fig. 2). Cutting mainly took place in July and August. Trees were cut every 5-7 years. Trees most commonly treated this way were *Betula pendula* and *B. pubescens*. The latter is more common, but in practice they are used indiscriminately. The English term "low pollarding" is probably appropriate (Rackham 1976).

Lauving: Lopping

Whereas styving refers to what happens to the tree, the actual fodder-collecting is called lauving. The cut branches were often very coarse. They were usually gathered by women and children who cut them into shorter pieces (0.75-1 m). The twigs were bunched and tied together with a twisted twig of *Sorbus aucuparia* or *Betula*. The English term 'lopping' defines this process (Hæggström 1983). Lauving was carried out every year. The number of bunches collected depended

on the number of livestock to be fed through the winter and the hay crop for the year.

Young shoots were sometimes cut directly from the tree-bases, so-called stubbelauving. *Salix caprea, Sorbus aucuparia* and *Alnus glutinosa* were usually cut in this way (Fig. 3). Rackham (1976) describes this as "coppicing stool above ground".

Young shoots were also cut directly from suckers, so-called rotlauving. This practice was often done when clearing grazing-land. *Populus tremula* was the most common species used in this way. *Alnus incana* was treated in a way between these two techniques. Some farmers set aside areas that were cut frequently, every 4-5 years, rather like a small plantation.

Lauvkjerv: Bunch of twigs

Twigs collected for animal fodder were gathered in bunches. The size of the bunch varied, but usually corresponded to the armspan size of the worker, namely the amount of twigs she could put under her arm while bunching them. According to the type of tree, growth for the year, and thickness and stiffness of twigs and twiglets and their mutual angle, the weight of fresh bunches varied from about 4 to 6 kg. *Betula* was usually the heaviest material, whereas *Fraxinus* and *Ulmus* were the lightest.

Lauvrak: Stack of bunches

The bunches of twigs were usually stacked together in a special way.

The tops were stacked to the centre, while the bases faced outwards. One stack contained about 200 bunches.

Rispelauv: Plucking-leaves

In some places leaves were collected for fodder by plucking and then sacking them. *Ulmus* and *Fraxinus* were most commonly used. This

Fig. 2. Low pollarding of *Betula*. Balestrand, Sogn. Photo: L. Hauge 1984.

Fig. 3. Demonstration of coppicing *Alnus incana* for collecting animal fodder. Lærdal, Sogn. Photo: L. Hauge 1984.

practice is documented from Havråtunet, Osterøy, near Bergen as late as 1951 (Gjerdåker 1951).

Rakelauv: Raking-leaves

Raking up autumn leaf-fall, mainly *Populus* and *Fraxinus*, was practised in some places, mostly for use as bedding in stalls to help absorb liquid dung during the winter.

Ris: High pollarding/Shredding

Lauving (collecting leaf fodder) was late-summer work. In addition, branches were sometimes collected during winter for twigs and bark and later fed to the animals. *Ulmus* and also *Fraxinus* were especially suitable and were therefore kept as a reserve if sufficient ordinary leaf-fodder could be collected from other trees (*Betula*, *Alnus*). *Ulmus* was also valuable for its possible use in human nutrition (cf. Nordhagen 1954). *Pinus* was the alternative.

It was very important to have as high a production of twigs as possible, and often *Ulmus* was allowed to grow extremely tall, resulting in many layers of branches (Fig. 4). This was very hard work, but in winter farmers had more time to spare than in summer.

Fig. 4. An old, high-pollarded tree used for animal fodder in early spring. There are cutting marks as high as 15 m above the ground. The productivity of these trees was very high, but the work was dangerous and difficult. Luster, Sogn. Photo: I. Austad 1984.

This could be called "high pollarding" or "shredding" (Rackham 1976), but it must not be confused with common pollarding carried out in summer. Rougher material (e.g. branches) was also cut and fed to animals for bark. If the spring was late and the farmer had insufficient fodder, other trees were also used in this way, most commonly *Salix caprea* and *Betula* (Fig. 5).

Skav: Peeling

In connection with 'rising' and tree felling, bark was peeled with a special tool, cut into small pieces, mixed with water and given as fodder in winter and early spring (Ropeid 1960).

The importance of tree pollarding in earlier times

To investigate how common and important pollarding was in earlier times, information about 9 farms in the Hafslo district of Sogn in 1863 is presented in Table 1. This information comes from the preliminary work for new taxes.

Only *Betula*, *Populus* and *Alnus* are mentioned as bunches of twigs collected in summer, while the ris and skav for feeding cattle probably relate to fodder collected in winter or early spring (*Ulmus*).

In 1863 Hafslo had 210 farms of which only 3 had no possibility for pollarding. The county covered 557 km², and 56 km² were more-or-less wooded. With an average fodder collection as presented in Table 1, tree-fodder harvesting must have greatly influenced the landscape ecology.

Bunches of twigs, weight and utilization

In old documents only the number of bunches of twigs is mentioned. The information varies greatly, and in some counties it is lacking or incomplete.

To get an idea of the impact of pollarding on landscape ecology, we require an estimate of the total quantity of material collected. We need to know the weight of fresh bunches plus the amount of waste material.

Some farmers still pollarding participated in a survey in 1985. Table 2 shows average values of (1) fresh bunches, (2) the same bunches dried, (3) waste material from these bunches, and (4) weight of utilized material (fodder). These figures are somewhat higher than in Lunde (1917). In his investigation, weight of dried leaves per bunch was estimated to 800 g.

Feeding

The potential for collecting and using tree fodder for domestic animals depended on natural conditions. People used whatever they could find.

Fig. 5. Spring high pollarding. In early spring it is common in some districts of western Norway to cut off branches, especially of *Salix caprea*, to feed sheep. Lærdal, Sogn. Photo: L. Hauge 1986.

Farm name	Land register number	Bunches of twigs with leaves, ris and skav
Lad	2.4	1200 bunches of *Betula* 200 bunches of *Alnus*
Prestegaard	3.5	400 bunches of *Betula* 200 bunches of *Alnus*
Eikum	35.68	Ris for 4 milking cows, 3000 bunches of *Betula* 2600 bunches of *Alnus*
Kinsedal	39b.108	2000 bunches of *Betula* 200 bunches of *Populus*
Urnæs	40b.109	4000 bunches of *Betula* 600 bunches of *Populus* 1000 bunches of *Alnus*
Cotters under the farm Urnæs		10000 bunches of twigs
Skjeggestad	5.7	400 bunches of *Betula* Ris and skav for 10 milking cows
Kjærlingnes	3.4	Skav and ris for 5 milking cows, 6000 bunches of *Betula*
Heggestad	5.8	2500 Bunches of *Betula* Ris and skav for 6 milking cows
Kvam nedre	10.23	No facilities for twigs, ris and skav

Table 1. The importance of tree pollarding in 1863 in Hafslo county, Sogn og Fjordane. Data from 9 typical farms are presented. (Martrikkelforarbeidet til skattematrikkelen, 1863, etter lov av 6.6.1863. "Herreds-beskrivelse" Hafslo.)

Type of leaves	Number of bunches	Fresh bunches	Dried bunches	Waste material	Utilized material
Betula 1	1	10.2	6.7	3.6	3.1
Betula 2	1	8.6	5.7	3.4	2.3
Betula 3	6	-	3.3	1.8	1.5
Betula 4	3	-	3.5	1.6	1.9
Alnus 1	1	8.5	4.9	3.0	1.9
Alnus 2	1	7.2	4.0	2.7	1.3
Alnus 3	1	7.9	4.5	3.0	1.5
Fraxinus 1	6	-	2.9	-	-
Fraxinus 2	3	-	3.0	0.75	2.25
Salix caprea 1	6	-	2.2	-	-
Ulmus 1	3	-	3.5	1.0	2.5

Table 2. Weight of bunches of twigs with leaves' and their utilization by sheep in Sogn in 1985. The figures represent average values in kg for one bunch.

Betula, Salix caprea, Alnus and *Sorbus aucuparia* are common over much of Norway, and were thus the trees that offered the highest potential. Leaves (twigs) of *Ulmus* and *Fraxinus* together with *Salix caprea* and *Sorbus aucuparia* were used preferentially for milking cows and calves. *Betula* and *Alnus* were considered to have a lower nutritional value. *Betula* had an acrid effect on milk products and was used only for sheep and goats. Horses were fed on *Populus tremula*. Domestic animals received at least one feeding with leaves (twigs) a day.

Leaves (twigs) of *Corylus avellana, Tilia cordata, Prunus padus,* *Quercus robur* and *Q. petraea* were also used, especially for sheep, but to a smaller extent. This was partly because they are less suitable (dry *Corylus* leaves are extremely brittle), and partly because these trees were required for other important purposes that were incompatible with fodder production (Høeg 1974). *Corylus* sticks were used for barrel hoops, *Quercus* for tanning bark and *Tilia* for ropes. *Tilia* trees with a typical pollard form can still be found. The need or obtaining many young shoots for rope production may be the reason for cutting *Tilia* in this way (Fig. 6). In connection with peeling, the discarded material (e.g. branches) was often used for hay-sticks (Gjerdåker 1951).

Fig. 6. Pollarded *Tilia* trees. The reason for cutting *Tilia* may be for bast production and fodder. Such trees represent cultural history and form an aesthetic element in the landscape. Balestrand, Sogn.
Photo: L. Hauge 1984.

Fig. 7. Bunches of *Fraxinus* twigs drying on a fence. These leaves are regarded as especially friable, and the distance to the barn must be short. Final drying most frequently took place inside the barn. Vik, Sogn. Photo: I. Austad 1980.

Drying and storing

The collected bunches had to be dried. In outfields the bunches were commonly placed on pollarded trees beyond the reach of animals or on special structures. A common type in Sogn consists of 3 fairly large branches (including those cut in styving) built into a tripod. The secondary branches were cut back to make short pegs for suspending the bunches. These constructions were protected against grazing cattle.

In infields the bunches were commonly put along the stone walls or against the barns, or hung on hay-sticks or fences (Fig. 7). It was also common to place bunches on hay-racks (Figs. 8a, 8b). The time required for drying depended on the weather.

Dried leaves of *Ulmus*, *Fraxinus* and *Alnus* are very friable. Bunches of these often had to be transported from the outfields to the barn almost fresh, and be dried in or near the barn. *Betula* is more robust and could be transported long distances after drying.

The quantity of twigs collected on each farm was great (Table 1). The bunches were not only heavy (Table 2) but also bulky. On some farms there are special tall, ventilated barns for storing bunches, often located in the outfields. Usually the normal barn had a special loft for storing the bunches.

Bunches of *Betula* were often stored and stacked in outfields (Fig. 9). The farmers had special, permanent places for stacking. They also placed small semi-permanent stacks in the pollarded trees, beyond the reach of animals.

Fig. 8. Drying bunches on hay-racks. This was common in earlier days (Fig. 8a) and special racks were used. It can still be seen in the 1980's. Photo 8a: from Fægri *et al*. (1981).
Photo 8b: Luster, Sogn. A. Ryhl 1980.

Pollarding in western Norway today

Because of the hard work involved in pollarding, very few farmers still continue this practice. An investigation in 1981 showed that in middle Sogn, 15 farmers still collected twigs for feeding their animals (Austad 1985b; Fig. 10). The number of bunches collected was small, only 200 on average per farm. The reason for still using this kind of fodder, according to the farmers, is to improve the health of their sheep by controlling parasites.

Pollarding activity was concentrated on trees in the infields near to the farm. Trees in the outfields or woodland were cut very rarely. *Fraxinus* and *Ulmus* were frequently cut, as were *Salix caprea* and *Betula*. Pollarding was done in August after harvesting of raspberries, a common crop in this area. Farmers usually cut twigs from old pollarded trees (bollings). Trees were topped very rarely. Some of the material collected came from coppicing and clearing grazed areas. Drying of the bunches was confined to fences and hay-sticks, and the material was stored in barns.

Fig. 9. Bunches collected and stacked for drying and storing are unusual sights to-day. *Betula* was commonly kept in stacks in the outfields through winter. The picture shows the last stack observed in the area. Luster, Sogn. Photo: O. Balle 1980.

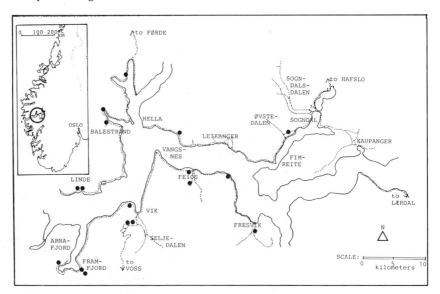

Fig. 10. Distribution of farms in Sogn where pollarding is still practised (from Austad 1985).

Only sheep were fed with twigs. Farmers usually started feeding with *Alnus* and *Betula*, then *Salix caprea*, and ended in spring with *Ulmus* and *Fraxinus*. The latter is regarded as preferred fodder for sheep and therefore was kept to the end of winter-feeding.

Restoration and management

The discontinuation of pollarding in western Norway today, especially in outfields, results in a considerable loss to the Norwegian heritage because the practice ceases and pollarded trees disappear from the landscape. Its high historical, aesthetic, recreational and ecological value combines to provide strong arguments for preserving and managing special areas.

Rejuvenating old, intensively managed trees is a difficult task in the protection and management of semi-natural deciduous woodlands. Without a severe reduction of crown-size, those trees whose branches were heavily cropped in the past become top-heavy and vulnerable to windthrow. Therefore, it is essential that existing trees in special areas are rejuvenated and that traditional cutting techniques are reintroduced for their long-term management.

An experimental area in a deciduous forest in Luster county of Sogn, with high-pollarded elm trees shows that restoration of such trees is possible. One or two trained people can restore 4-5 trees per day. The work has been done in October and April from 1983 to 1986 (Austad *et al.* 1985; Fig. 11).

Fig. 11. Restoration of *Ulmus* trees. Luster, Sogn. Photo: K. Brinkmann 1985.

Discussion and conclusions

Comparisons with descriptions from other areas suggest that pollarding in Sogn is not only representative for other districts in Norway, but also for earlier periods in other European countries (Andersen this volume; Emanuelsson this volume; Rackham this volume).

Sogn is one of the last areas in Europe where pollarding is still carried out as an agricultural practice, and where, most readily, people can give information about the techniques involved. However, there are few farmers still pollarding trees for animal fodder and they are usually elderly. Trees managed in this way are therefore disappearing rapidly from the landscape, mainly because of the ageing trees, lack of pollarding and felling of trees for timber and subsequent replacement by *Picea* plantations (Fig. 12). In wooded pastures, pollarded *Betula* is especially at risk. High pollarding/rising, peeling and techniques of constructing drying and storage stacks are now lost. Very often we also find a secondary use of trees, e.g. pollarding of *Ulmus* and *Fraxinus*, in areas where these trees were earlier only used during the winter.

Despite tree pollarding for animal fodder being so widespread and intensive and being carried out until recently (1945) in Norway, very few people know of the practice or realize its importance, not only for the economy of small farms, but also as an important factor in the ecology of the cultural landscape (Fig. 13). Moreover, we know little of the importance of these trees for the ecology of birds and insects. In areas where pollarded trees still occur as farm-yard trees and in hedgerows, pastures and deciduous woodlands, they provide attractive recreation areas (Fig. 14).

We know that leaves represent valuable fodder for both nutrition and digestibility. Lunde (1917) compared the nutritional value of *Ulmus, Fraxinus, Betula* and *Alnus incana* with hay and *Trifolium pratense* (Table 3). These results indicate that leaves (bunches of twigs) are as good as hay. Brelin *et al.* (1979) confirmed the relatively high nutritional level of leaves in Sweden.

In 1983 analyses of nutritional value and digestibility of *Ulmus, Fraxinus, Tilia, Betula, Salix caprea, Populus tremula* and *Sorbus aucuparia* leaves were conducted to compare with results of earlier investigations (Nedkvitne & Garmo 1986). They showed differences in nutritional value and digestibility of leaves from different woods. Trees also exhibit varying nutritional values during the year. June seems to be the best month for collecting fodder. *Ulmus* is in a class by itself. Trees can also be a source of protein, and again early summer is optimal.

Fig. 12. The effect of afforestation in the old cultural landscape, with a *Picea* plantation for seed production. Kaupanger, Sogn. Photo: I. Austad 1980.

Tree	water %	ash %	fat %	material without N (sugar) %	material with N (protein) %	fibre %
Ulmus glabra	12.6	9.9	2.9	49.2	13.2	12.3
Sorbus aucuparia	11.9	5.9	6.5	50.4	9.9	15.4
Salix caprea	11.5	6.1	3.8	50.3	11.6	16.7
Populus tremula	10.8	5.5	6.0	43.5	13.3	20.9
Fraxinus excelsior	11.6	6.3	3.0	50.4	12.0	16.7
Alnus incana	11.9	3.9	5.9	43.6	17.6	17.4
Betula spp.	11.7	3.9	7.0	49.2	12.0	16.2
Ordinary hay	14.96	5.42	2.2	44.43	8.51	24.56
Trifolium pratense	15.65	5.17	1.88	36.76	10.98	28.56

Table 3. Nutritional value of leaves of different trees (from Lunde 1917).

It is interesting to see that traditional stock-feeding agrees with these results. It is also relevant to note that these new analyses have confirmed farmers' intuition and the old analyses. However, little is known about the value of using tree fodder for controlling sheep parasites, for example concentrated in the grazing-period in late autumn when parasite infection often occurs.

Protection of historic buildings and farms should be coupled with protection and management of their associated cultural landscape in which pollarded trees are essential elements. In other Scandinavian countries, restoration and management of cultural landscapes and their elements have been carried out. For example, on Åland, Finland, restoration experiments in wooded grazing areas and overgrown hay-meadows have been attempted. The results give valuable information about the methods required to produce particular field-layers and to reconstruct aspects of the old cultural landscape, including topping and lopping of trees (Hæggström 1983).

In Sweden there are areas where restoration and management have been carried out to preserve the most interesting features of the old cultural landscape (Nilsson 1970; Fig. 15).

Pollarded trees and the pollarding practice represent important facets

Fig. 13. Tree pollarding and peeling enhanced the value of a farm. Some deciduous woodlands in western Norway are instructive examples of former utilization. Luster, Sogn.
Photo: I. Austad 1982.

Fig. 14. Low pollarded *Betula* trees in a grove or park-like woodland. This is an attractive recreation area with managed trees and a diversity of herbs and shrubs. The ecological importance and cultural significance of this habitat add to its conservation value. Balestrand, Sogn.
Photo: L. Hauge 1984.

of Norwegian cultural history and agricultural tradition. It is vital to understand and appreciate the value of pollarded trees in the cultural landscape not just as a source of creative thought or as museum specimens, but as essential elements of the working countryside.

Because of the hard work involved in pollarding, the Norwegian Government has a special responsibility to encourage pollarding by offering farmers subsidies or assistance with labour.

Summary

(1) In earlier times pollarding was widespread in Europe. From the Iron Age until the Second World War, this form of tree management was common in much of Norway as a means of obtaining winter animal-fodder.

(2) Pollarding practices in Sogn og Fjordane county, western Norway are described. Remnants of the old, cultural landscape in marginal agricultural land can still be found. The county is one of the few areas in western Europe where traditional agricultural methods are still practised.

(3) In some pastures and deciduous forests, regular pollarding strongly affects the composition of plant communities. The effect on landscape ecology equals the effect of hay-making or grazing.

(4) Pollarded trees are still very common in Sogn og Fjordane, especially pollarded *Betula*. They are today an important part of the cultural heritage and the aesthetics of the landscape.

(5) A few farmers still cut trees for animal fodder, but trees managed by this practice are rapidly disappearing from the cultural landscape.

Fig. 15. Restored, pollarded trees at Steneryd, Blekinge, south-east Sweden.
Photo: I. Austad 1984.

In some areas, the lack of active pollarding has made it necessary to restore and manage pollarded trees as a means of preserving a valuable aspect of the cultural heritage.

Acknowledgements

I express my special thanks to Anders Timberlid for extracting information from the preliminary work for the new taxes from 1863. I am also indebted to Knut Fægri and Lars Walløe for valuable comments and discussions on the manuscript, and Gary Fry for revising the English.

Nomenclature

Plant nomenclature follows J. Lid (1985) *Norsk-svensk-finsk Flora*, Det norske Samlaget, Oslo.

Galdane, Lærdal, Western Norway - Management and Restoration of the Cultural Landscape

Leif Hauge

Introduction

The cotter's farm at Galdane is a well preserved component of the former fine-scale agricultural landscape. Managed vegetation types in combination with land-use patterns, terraces, stone-walls, paths, ditches and buildings reflect a dynamic landscape with a wide range of variation. These elements occur in their natural surroundings, close to several old roads and buildings in typical west Norwegian terrain.

The survival of this cultural landscape is under threat. As the farm is abandoned, typical landscape-elements such as pollarded trees, hay-meadows, birch groves and juniper-fields are rapidly disappearing (Austad this volume). The major threat is encroachment by shrubs and trees.

The objective of the Galdane restoration and management plan is to preserve a part of the old pastoral landscape that was common at the beginning of the century. Permanent plots have been established so that management effects can be monitored and compared with plots where natural vegetation is allowed to develop. A further objective is to establish models for economic, administrative and practical management for protecting other important cultural landscape areas.

The aim of my research at Galdane is to study the ecological processes in semi-natural vegetation as an aid to protecting and managing traditional agricultural landscape-units. Special effort is directed at identifying the former agricultural use of the various fields. Future trends in vegetation-structure and composition can be predicted by analysing changes since the farm was abandoned.

The work is a part of an Økoforsk research programme.

The Galdane area

Galdane is located in Lærdal in the innermost part of the Sognefjord, western Norway (Fig. 1). The total farm area is 26.8 ha, mostly forest, bare rock and scree. About 0.5 ha is cultivated land. The cotter's

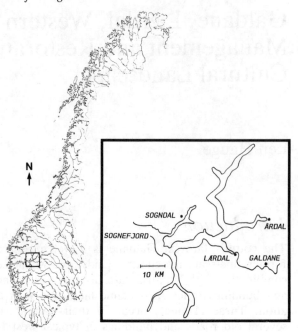

Fig. 1. Location of the study area.

farm at Galdane is part of the main farm at Øvre Ljøsne, 6 km down the Lærdal valley. The infields of Galdane lie on a very steep slope (about 55⁰) from 210 to 350 m above sea level. The cotter's farm is reached *via* an old trail along the Lærdal valley.

The bedrock consists mainly of granite. However, the soil is fairly fertile because of small outcrops of phyllite (Bryhni 1977; Lutro 1981). Soils vary greatly. Dry, rocky areas occur on the upper slopes, whereas in the valley deeper organic soils occur. Some of this humus-rich soil was probably transported by crofters to the infields to cover areas of bare rock.

The climate of Lærdal is unusual for western Norway in being dry and having warm summers but mild winters. Along the western coast of Norway annual precipitation is 2000-3000 mm. High mountains at the head of the Sognefjord create a rain-shadow, resulting in an annual precipitation of about 400 mm in Lærdal. Most summer rain is as afternoon showers (Utaaker 1978). There is little winter snow, and south-facing slopes are snow-free early in the spring. Mean summer temperatures are high (July mean 16.4⁰C) but because of the proximity of Sognefjord, winter temperatures are not low (January mean -1.8⁰C).

Cultural history

Name and history

In many places, trails through western Norwegian valleys had to cross steep rock-ledges. To safeguard such passages, half-bridges were made by laying tree-trunks parallel to the trail. There were supported on the rock in various ways, including stone-built foundations, chains and pegs hammered into the rock. Such man-made passages are called 'gald' in Norwegian. The Galdane farm got its name from such constructions in the trail near the farm.

In the late 1800's British anglers rented the famous salmon-river in Lærdal during the summer. They named Galdane as 'the Galleries', probably a misinterpretation of the local name, but still very descriptive.

The establishment of the farm at Galdane is uncertain. Its history can be traced back to 1660 when Lasse Galden lived here. Since then, the Galdane records are detailed and include such information as the names of the farmers, what they cultivated, numbers of cattle and various contracts with the main farm.

Galdane was a croft with no land of its own. The crofters rented the cultivated land and had no rights to the valuable forests nearby. However, they could cut wood from *Populus tremula* and *Alnus incana* for house repairs and fences, and could use dead trees for firewood. The crofters had to work some days at the main farm, usually a week in the spring and one in the autumn, according to an old original hire contract from 1774.

The farmer at Galdane owned his land for the period 1834 to 1856. Records from 1865 tell us that 7 people lived at the farm; the crofter had 5 cows, 14 sheep and 10 goats. He sowed about 150 kg of grain and 300 kg of potatoes. Ten years later the goats were replaced by more sheep in order to prevent the outfields from being overgrazed. The farmer could also fish for salmon and trout in the Lærdal river, and hunt reindeer in the mountains. Galdane was not one of the poorest farms in Lærdal.

Although the soil was fertile, it was hard to produce sufficient crops to support a large family. Whole families emigrated to America in the latter part of the 18th century. The long-recorded use of the cultural landscape at Galdane ended in 1947 when the farm was finally abandoned.

Paths, trails and roads near Galdane

Paths and trails have passed through the Lærdal valley and thus Galdane since prehistoric times. This was the easiest and safest route between east and west Norway. Old laws from 950 and 1274 laid down rules concerning this main road.

During the centuries many different people, from kings to outlaws, used this trail. They usually walked, rode on horses, or were carried. Increasing traffic resulted in demands for better roads where wagons could be used. Several phases of road-construction followed until the first road suitable for horse and wagon was finished in 1793 (Laberg

Fig. 2. The road at Vindhella.

1938; Skoug 1975). Later this road, called 'Kongevegen' (the King's Road) was improved with large stone terraces and bridges. The winding road-construction at Vindhella a few km east of Galdane, remains a monument to this impressive feat of road engineering (Fig. 2). The road was the main connection between Kristiania (now Oslo) and Bergen until the Bergen Railway was finished in 1909 (Skoug 1975).

Many travellers referred to Kongevegen in terms of horror and fright (e.g. Pontoppidan 1749), having to pass close to fast-flowing rivers and waterfalls, through steep mountains and narrow valleys, and over unsafe bridges. Several Norwegian painters such as Hans Dahl, Jørgen Flintoe, Thomas Fearnley and Henrik Sørensen, have also captured this wild landscape in their work (Fig. 3); some of their pictures are in the National Gallery in Oslo.

The cultural landscape at Galdane

Buildings and other constructions

On a ledge 260 m up the steep hillside there are the remains of several well-preserved farm buildings (Figs. 4, 5). The cottage is constructed of pine and sits on a solid stone foundation. Its construction is unusual for this western region. Up to 12 persons lived in this cottage of about 30 m². The cowshed, built in 1814, is the oldest building at Galdane (Lindstrøm & Lindstrøm 1981). Two barns, a

goat shed and a smithy, also survive (Fig. 5). Several of these buildings also show unusual construction methods. Most are now being
restored. Remains of an older cottage, a mill and another shed have
also been found.

Because of the low annual precipitation, the infields of Lærdal had
to be artificially irrigated. At Galdane an irrigation ditch led across
the infields from a nearby small river. A special kind of aquaduct
was necessary to lead water over screes and bare rock.

Stone walls are common in the Galdane cultural landscape. They
are here used both as fences and passages between infields and outfields. Stone-terraces support potato fields and meadows.

Cultivated fields, meadows and pastures

The relatively rich soils on the lower slopes of Galdane provided a
good substrate for growing grain and potatoes. Cultivation of grain,

Fig. 3. A sketch from Galdane
drawn by Hans Dahl in 1887.

mostly oats and barley, was common before potatoes were introduced in Lærdal at the end of the 18th century. The cultivated fields were often very small (10-50 m²) and were distributed over the area to make best use of the less steep ground with adequate moisture. In some places potato fields were supported by stone terraces to minimize soil creep and make the surface as level as possible. Extra soil was often imported to deepen the organic layer. Potato fields were commonly irrigated because of their poor water-retention.

Fig. 4. Oblique photograph of the central area of Galdane. Photo: L. Hauge.

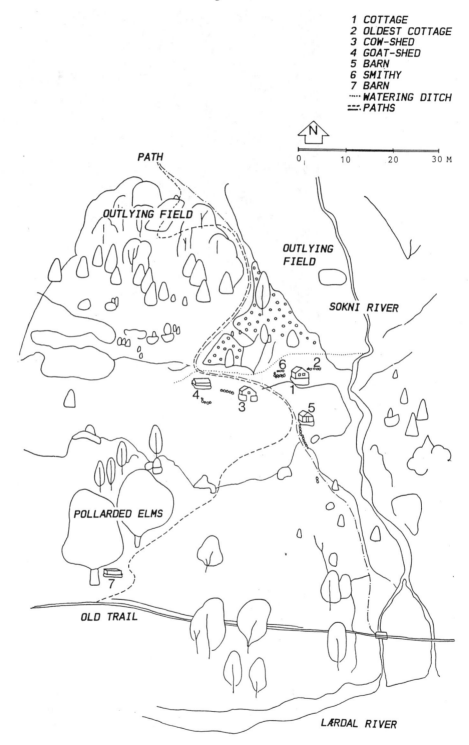

1 COTTAGE
2 OLDEST COTTAGE
3 COW-SHED
4 GOAT-SHED
5 BARN
6 SMITHY
7 BARN
····WATERING DITCH
═══ PATHS

Fig. 5. Explanation of the photograph in Fig. 4.

Fig. 6. Old pollarded elms at Galdane. Photo: L. Hauge.

The farmers used most of the open areas for haymaking. The biggest meadows were close to the buildings. Grass was cut and later dried on the ground during sunny periods, or on special racks of thin sticks ('hesjar'). Many smaller outfields were harvested and the hay was stored in small barns or under large rocks, etc. The farmers cut the meadows in July or August to obtain the maximum quantity of hay. This often resulted in poor-quality hay with stiff grasses. The late cutting would have allowed many herbs to seed, probably resulting in rich meadows.

The structure and composition of the vegetation in the meadows and outfields were related to the grazing pressure. This would have been greatest in the spring and in the autumn. In the summer most cattle were grazed in the mountains (Indrelid this volume; Kvamme this volume).

Grazing of sheep, goats, cattle and horses resulted in different vegetation in the pastures. Goats, in particular, would have formed and maintained grass-dominated areas with light-demanding species in the field layer.

Pollarding

Leaves from different deciduous trees represented an important supplement to hay during the winter. Most commonly, leaf harvesting involved cutting young branches with leaves from the main tree trunk, or harvesting young saplings (Austad this volume).

Branches were collected into bunches of twigs, and dried and stored in different ways until required as fodder. At Galdane *Betula*, *Ulmus* and *Alnus* were the main trees used for pollarding (lauving, Austad this volume). Lauving was an important part of harvesting and the Galdane crofters spent many weeks in August and September collecting fodder this way. The pollarded trees, or 'styvingstre', can still be seen along field boundaries and rivers or on poor agricultural land. At Galdane there are some especially large *Ulmus* styvingstre that form a characteristic component of the landscape (Fig. 6).

Farmers collected leaves from the ground in the autumn, even though these leaves did not have the high nutrient value of fresh or dried leaves. Trees were also harvested for twigs and bark in the spring (rising). A mixture of fresh bark and water was fed to cows to encourage production of more and better milk (Austad this volume).

Flora and vegetation

Plant geography

As a result of its unusual climate, the Galdane area supports different floristic elements. Warm summers favour species of a south-western distribution, including *Lychnis viscaria*, *Dianthus deltoides*, *Erodium cicutarium*, *Satureja acinos*, *Origanum vulgare*, *Plantago media*, *Verbascum nigrum*, *Potentilla argentea*, *Vicia tetrasperma*, *Carex pairaei* and *Cerastium semidecandrum*. Species such as *Adoxa moschatellina*, *Aconitum septentrionale* and *Pyrola chlorantha* represent a more eastern element.

Several alpines occur at Galdane, including *Oxyria digyna*, *Alchemilla alpina* and *Viola biflora*. These are probably dispersed by streams descending from the mountains. Finally, several anthropochorous species occur including *Plantago major*, *Poa annua*, *Carduus crispus*, *Polygonum aviculare* and *Arctium minus*.

Mapping of the present vegetation

The structure of the old cultural landscape at Galdane has slowly changed since the farm was abandoned in 1947. The vegetation is affected by soil conditions, nutrient influx, moisture conditions and exposure. Different combinations of these ecological factors result in distinct vegetation units (Fig. 7).

Seventeen 5 x 5 m² permanent plots have been established to map the present vegetation and document future changes in different vegetation-types. The plots will be analysed every second year. Soil chemistry has also been analyzed for each plot.

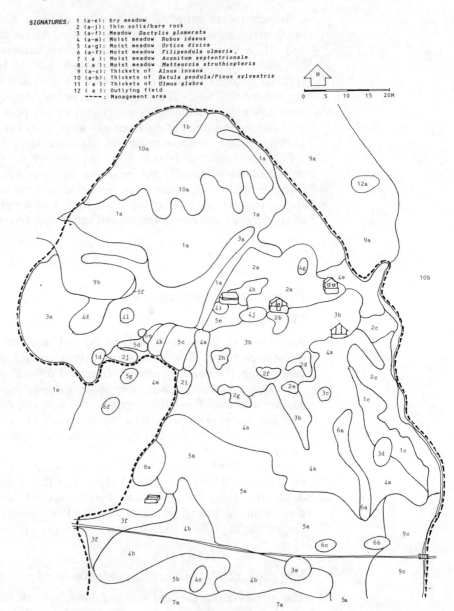

SIGNATURES:
1 (a-e): Dry meadow
2 (a-j): Thin soils/bare rock
3 (a-f): Meadow *Dactylis glomerata*
4 (a-m): Moist meadow *Rubus idaeus*
5 (a-g): Moist meadow *Urtica dioica*
6 (a-f): Moist meadow *Filipendula ulmaria*.
7 (a): Moist meadow *Aconitum septentrionale*
8 (a): Moist meadow *Matteuccia struthiopteris*
9 (a-c): Thickets of *Alnus incana*
10 (a-b): Thickets of *Betula pendula/Pinus sylvestris*
11 (a): Thickets of *Ulmus glabra*
12 (a): Outlying field
----- : Management area

Fig. 7. Vegetation types at
Galdane.

Meadows

These can be divided into three main groups on a moisture gradient;
dry, normal and moist meadows (Austad & Hauge 1988).

Dry meadows (Fig. 7) are mainly confined to the upper parts of
the infields, and in many places occur only as small fragments. The

soil, usually minerogenic, is normally very permeable. Rainwater provides most of the moisture, but also leaches nutrients downslope. Snow
cover is usually thin and the vegetation is not protected from frost.

The field-layer comprises small, creeping herbs and grasses. *Viola
tricolor*, *Erodium cicutarium* and *Scleranthus annuus* are annual and
vary in cover from year to year. Other common associates include
Dianthus deltoides, *Lychnis viscaria*, *Origanum vulgare*, *Fragaria vesca*,
Galium verum and *Pimpinella saxifraga*. Most flower in early summer,
giving the meadows a colourful appearance. The ground-layer is often
well developed, and dominated by *Abietinella abietina*, *Pleurozium
schreberi* and *Polytrichum juniperinum*.

A variant on thin soils and almost bare rock is related to the dry
meadows. Common species are *Hieracium pilosella*, *Woodsia ilvensis*,
Sedum spp., *Rumex acetosella* and *Festuca ovina*. These are often
damaged by trampling close to paths.

Grass-rich meadows (Fig. 7) are found on richer organic soils. They
are normally situated on level areas with a stable supply of water and
nutrients.

Broad-leaved mesic grasses and sedges such as *Dactylis glomerata*,
Festuca pratensis, *Anthoxanthum odoratum*, *Arrhenatherum pubescens*
and *Carex pairaei* occur. The thick grass sward partially prevents the
occurrence of many herbs, but *Pimpinella saxifraga*, *Ranunculus acris*,
Silene vulgaris and *Galium verum* occur. The ground-layer is poorly
developed.

Moist meadows are most common on the lower slopes at Galdane.
Here the organic layer is thick and regularly receives nutrients washed
from the infields above. The plant cover is well developed and reaches
1-1.5 m high. A few nitrophilous species comprise the field layer,
mainly *Urtica dioica*, *Rubus idaeus*, *Aconitum septentrionale*, *Stachys
sylvatica*, *Filipendula ulmaria* and *Galium aparine*. Small, light-
demanding plants such as *Adoxa moschatellina*, *Equisetum pratense* and
young *Urtica dioica* are only found in the spring when there is sufficient light.

Forests

Stands of different woodlands surround the meadows, the composition
of each being a result of local environmental factors. The units are
difficult to classify because of their early successional status. They are
thus divided into deciduous, alder, birch, and pine forests. The juniper
fields play a special role in forest succession.

Deciduous forest is present only as small fragments, mostly on
moist and nutrient-rich soil. *Ulmus glabra* is dominant, often growing
with *Alnus incana* and *Populus tremula*. Many younger elms grow in
the moist meadows and tree colonization is expanding rapidly now
that there is reduced grazing pressure. Three old pollarded elms with

large trunks and younger branches give the area a characteristic appearance (see Austad 1985b). The field-layer is dominated by nitrophilous herbs, mainly *Urtica dioica, Rubus idaeus, Stachys sylvatica, Filipendula ulmaria, Aconitum septentrionale, Impatiens noli-tangere* and *Matteuccia struthiopteris* (Austad *et al.* 1985; Austad & Hauge 1988).

Small examples of *Alnus incana*-dominated stands occur in several places at Galdane, mainly on humus-rich moist soil. *A. incana* is dominant along river banks, where it forms dense thickets. It is an effective coloniser of meadows, gradually appearing on the fringes of abandoned land. *Prunus padus, Sorbus aucuparia* and *Ulmus glabra* can co-exist with *A. incana* in some places. The field-layer is usually dominated by *Urtica dioica, Rubus idaeus, Circaea alpina* and *Aconitum septentrionale*.

Stands of *Betula pendula* are most common on minerogenic substrates, ranging from sand and screes to bedrock. Old trees often show signs of past pollarding and form open groves. *Betula* is slowly colonizing dry meadows and other poor sites. The field-layer is often open, sometimes with no vegetation at all. Common species include *Festuca ovina, Poa nemoralis, Hieracium pilosella, Campanula rotundifolia, Lotus corniculatus, Linaria vulgaris* and *Juniperus communis*.

Pinus sylvestris forests cover large areas near Galdane. Today young trees are colonizing abandoned dry meadows. The field-layer does not differ much from that of the birch stands, being dominated by *Festuca ovina, Deschampsia caespitosa, Potentilla erecta, Hieracium pilosella, Campanula rotundifolia* and *Juniperus communis*.

Juniper fields occur on poor agricultural land, usually in sunny, open areas (Austad 1985a). *Juniperus* is common early in the succession from dry meadows to woodland. This is partly because young individuals are seldom grazed by sheep and cattle. Associates in these fields are similar to those in the dry meadows.

Phytosociological units

The vegetation of the Galdane area reflects an intensive and longstanding utilization of natural resources. Natural and undisturbed vegetation units are seldom found and a precise phytosociological classification is thus difficult.

The rich deciduous-forest types on the lower slopes composed mainly of *Ulmus glabra* and *Alnus incana* can be grouped in Alno-Padion. This alliance is common on moist and nutrient-rich soil in western Norway. Some of the deciduous forest with *Matteuccia struthiopteris* can be classified as Alno-Ulmetum. The communities of *Betula pendula* and *Pinus sylvestris* can be placed in the association Melico-Piceetum.

The different meadow-types are difficult to classify, mainly because of their mixed composition. Dry meadows can be classified close to the order Violion caninae, richer meadows consisting of broad-leaved grasses to alliance Arrhenatherion elatoris, and the moist meadows with many nitrophilous species are a northern variant of Filipendulion.

Conclusions

Because the cultural landscape, together with the newly restored farm buildings are important features at Galdane, a management plan for the area is being produced for the long-term conservation of the farm (Austad & Hauge 1988). Restoration of the landscape must be based on a detailed knowledge of the past agricultural use and present-day vegetation structure. The cultural landscape is classified into 9 areas (Fig. 8), each with an individual management type.

Tree invasion must be controlled and overgrown woodland will be thinned to achieve wooded pastures. Old pollarded *Ulmus* must be heavily cut to restore them and lengthen their life-span. Young *Ulmus* will be topped so they may gradually replace the old trees. All *Ulmus* will be pollarded at intervals of 4-5 years. It is also planned to pollard some young *Betula* in one area.

Restoration and management must include agricultural uses that were common in former days such as raking and haymaking. Restoration of constructions such as stone-walls, terraces, roads, trails and irrigation ditches is also necessary. A small potato field will probably also be cultivated.

It is recommended that the Inspector of Nature Conservation and the Sogn Museum should take responsibility for the area and perhaps turn it into an Økomuseum (ecological museum), thereby protecting and conserving the remarkable cultural landscape and buildings at Galdane.

Summary

(1) Galdane, a crofter's farm in Lærdal, lies on a very steep slope in a landscape of flowing rivers, waterfalls, steep mountains and narrow valleys. The bedrock is mainly granite and soils vary greatly. The climate is dry with warm summers but mild winters.

(2) Detailed records for Galdane extend back to 1660 A.D. Different generations of trails have passed by the area, mainly winding trails with stone terraces and bridges. Remains of several well-preserved farm buildings can still be found. Most are now being restored. There are also remains of other constructions, such as an irrigation ditch, stone walls and stone terraces. Small potato and grain fields, meadows and outlying fields were located on the hillside. Leaves and fresh branches represented an important supplement to hay during the winter.

(3) Galdane supports several floristic elements. Different combinations of ecological factors result in distinct vegetation types. Permanent

Leif Hauge   44

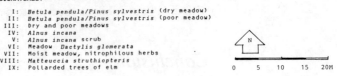

SIGNATURES:

I: *Betula pendula/Pinus sylvestris* (dry meadow)
II: *Betula pendula/Pinus sylvestris* (poor meadow)
III: Dry and poor meadows
IV: *Alnus incana*
V: *Alnus incana* scrub
VI: Meadow *Dactylis glomerata*
VII: Moist meadow, nitrophilous herbs
VIII: *Matteuccia struthiopteris*
IX: Pollarded trees of elm

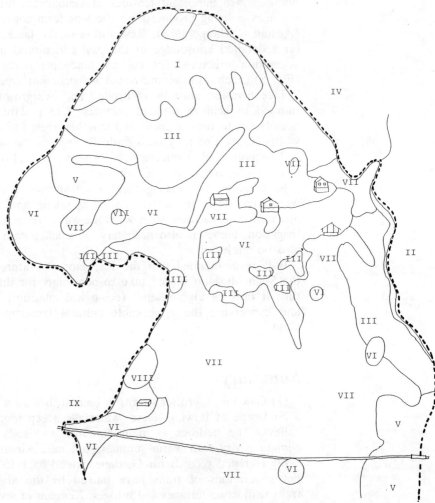

Fig. 8. Management areas at
Galdane.

plots have been established to map present vegetation and document future changes. The meadows fall into three main groups on a moisture gradient: dry, normal and moist meadows. The woodland types are dominated by *Ulmus*, *Alnus*, *Betula* or *Pinus*. The vegetation is classified into phytosociological units.

(4) A management and restoration plan for the cultural landscape is being produced for the long-term conservation of the farm, based on agricultural practices that were formerly common.

Acknowledgements

I thank Ingvild Austad for stimulating work with the original management plan for Galdane and for valuable comments on this manuscript. I also thank Gary Fry and John Birks for revising the English.

Nomenclature

Plant nomenclature follows J. Lid (1985) *Norsk-svensk-finsk Flora*, Det norske Samlaget, Oslo (vascular plants) and K. A. Lye (1974) *Moseflora*, Universitetsforlaget, Oslo (mosses).

The Farming System and its History in the Flåm Valley, Western Norway

Svein Indrelid

Introduction

The spectacular Flåm valley is 13 km long, and runs north-south between steep mountains, rising to 1400 and 1700 m above sea level (a.s.l.) in the west, and rather steep hillsides to the east (Fig. 1). Today the 350 inhabitants make their living from agriculture, tourism, transport and communications, handicrafts, and trade and community services. One hundred years ago, the Flåm Valley was remote and isolated, with little regular connection with other areas. However, there were more people living there then than today. What did people do for a living? What resources were exploited, and how far back in time can the main features of the economic system of the mid-19th century be traced?

The farms

From historical sources we know that during the last 300-400 years, agriculture has been the main occupation for people in the Flåm Valley. When looking at the steep mountains, the rocky hillsides and the comparatively modest areas of cultivated land in the valley bottom and at certain sites up in the hills, we may ask how that was possible. Additional resources, such as reindeer-hunting and river- and fjord-fishing, have been of *some* importance to *some* of the farms, but the subsistence basis for all the farms has been cattle raising and grain production.

There are 13 old, named farms (*navnegårder*) in the Flåm Valley (Fig. 1), situated in 3 distinct topographical settings (Fig. 2).

(1) Three farms (Fretheim, Brekke, Flåm) occupy the flat valley floor below 100 m. Their infields (*innmark*) are the largest and easiest to run, and have the longest growing-season, but the sun sets early because of the mountains.

(2) Five farms (Dalsbotn, Tunshelle, Berekvam, Melhus, Kårdal) are situated on the valley floor above the Holocene marine limit, between 150 and 550 m a.s.l. Their fields are smaller and steeper.

(3) Five farms (Indrelid, Ryum, Holum, Geisme, Vidme) occupy hillside shoulders at 400-450 m a.s.l. The fields are steep and difficult to run, but the greater amount of sun gives them the best conditions for cereal cultivation. They also have the shortest distance to the mountain plateau.

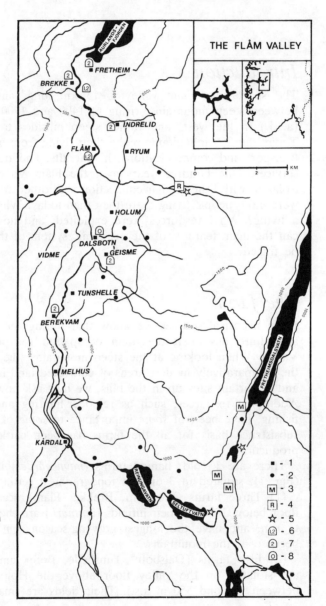

Fig. 1. The Flåm Valley.
1 = named farm (*navnegård*);
2 = summer farm (*seter*);
3 = house grounds from the Medieval Period (1050-1536 A.D.);
4 = house grounds from the Roman Period (1-400 A.D.);
5 = iron extraction site from the Roman Period (1-400 A.D.);
6 = graves from the Migration Period /Younger Iron Age (400-1050 A.D.);
7 = graves from the younger Iron Age (600-1050 A.D.);
8 = graves from the Iron Age (subperiod not specified).

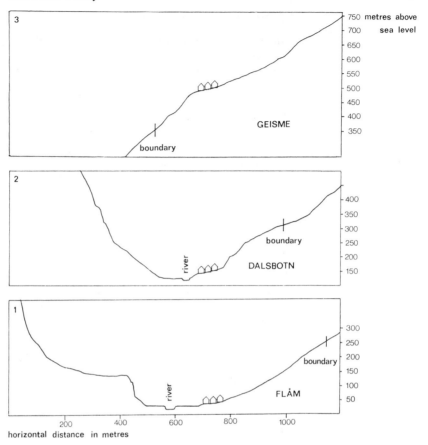

Fig. 2. The 3 types of topography for farms in the Flåm valley.

Settlement history

The flat valley bottom is old marine deposits. Holocene land uplift in the area is *ca.* 130 m. This left little suitable ground for Stone Age people. Stone Age finds are few and scattered and give no indications of settlement, nor have any traces of Bronze Age settlement been found. Late Iron Age graves (400-1050 A.D.) (Fig. 1) and other finds, particularly from the Viking Period (800-1050 A.D.) show that the main farms were occupied by then.

There is documentary evidence for 9 farms in the early 14th century. After the Black Death (1349-50 A.D.) several farms were deserted, and only 5 are mentioned in the taxation list of 1522. Recolonization subsequently occurred, and all 13 farms are mentioned in the taxation lists of around 1600.

The farming system

Most of the farms were divided into holdings (*bruk*), some as early as Medieval times, each of which provided subsistence for one family. The houses and buildings for the new family were added to the farm-yard, so that some had 50-60 buildings by the mid-19th century.

The infields for all the families of a farm surrounded the farm-yard. Some were cultivated for grain (and potatoes since the mid-18th century). The others were natural meadows producing hay.

The outlying fields (*utmark*) were on the forested valley sides, extending to the edge of the mountain plateau about 800-900 m a.s.l. There were a few scattered cultivated patches, mainly for potatoes, but the main products of the outlying fields were hay and leaves for fodder, and wood for firewood and building.

The *mountain areas* surrounding the Flåm Valley and belonging to the farms cover about 200 km. Most of the area is higher than 900 m a.s.l. Parts of it have extremely good pastures for cattle. Each farm had part of the mountain plateau, which was shared between all the holdings. There were traditional and often disputed boundaries between the farms.

Each farm had 2 or 3 summer farms (*setre*) in the mountains, where each family had a house (*sel*) (Kvamme this volume). The first summer farm was transitional, situated just above or just below the edge of the mountain plateau. It was used for 2-3 weeks in June and July, and 1-2 weeks in August/September. The walking time from the farm was 1-3 hours.

The real summer farms, as much as 8-10 hours walking time from the farm, were occupied in July and August. The cattle grazed the fertile pastures and butter was produced. The summer farms were necessary for two reasons; partly to increase butter and cheese production, and partly to keep cattle away from the infields during the growing season.

The functional farm

The combined farming system of infields, outlying fields and summer farms enabled not only an increase in production, but also the population to be kept at a maximum. I now consider how the system functioned at one of the largest farms, *Flåm*, during the second half of the 19th century.

At that time the Flåm farm had 500 decars (125 acres) of infields. The official registration lists prepared in connection with the preliminary plans for a new land register (Beskrivelse over de matrikulerede Eiendomme og Forslag til ny Skatteskyld for Herredet, iflg. Lov af 6. juni 1863. No. 258 Aurland. (Vedtatt av Herredskommissionen den 8/1 1869). Statsarkivet, Bergen) contain detailed information of the kinds and amounts of production for each of the holdings in the valley. The 1875 census tells us that 119 people in 19 families lived on the farm. Nine families owned land, between 35 and 85 decars of infields. Ten families were not land-owners, but lived as cotters (*husmenn*) on the holdings of the 9 land-owners. The 119 people obtained almost all their food supply from the Flåm farm. How was that possible?

One hundred and fifty decars were cultivated, and produced 200 barrels of barley and 305 barrels of potatoes (1 barrel = 139 litres). Three hundred and fifty decars were natural hay meadows, which produced 50 000 kg hay. This was winter fodder for 12 horses, 109 cattle and 330 sheep and goats.

This amount of hay was not sufficient for the winter. An additional 50 000 kg was collected by cutting grass in the outlying fields, including the steepest mountain sides. The bundles of grass were thrown down the mountain and collected at the bottom. Leaves were also gathered, mostly *Betula*, but also *Ulmus* and *Fraxinus* (Austad this volume). On the Flåm farm about 11500 bundles of leafy branches (*lauvkjerver*) were collected yearly in the late 1860's. This winter fodder from the outlying fields was essential for the survival of the 119 people.

It was also essential that the animals were not allowed to graze the infields or the outfields before the winter fodder had been harvested. The summer farms were therefore an absolutely essential element in the agricultural system of the Flåm Valley.

History of summer farming

House foundations and two iron extraction sites on the mountain plateau (Fig. 1) are of Roman age (*ca.* 1-400 A.D.). Pollen-analytical data indicate that cattle were also pastured nearby, suggesting that summer farming was well established and implying that the Flåm Valley was permanently settled by then. It is not unexpected that the poorly equipped graves of Roman times have not been discovered in the valley, as there has been little archaeological activity there.

If the interpretation of the Roman age house foundations as summer farms proves to be correct, one of the most characteristic elements of the traditional agricultural system can be traced back to the time of the first permanent settlement. This demonstrates quite clearly the importance of outlying fields and mountain pastures for the settlement of the Flåm Valley.

Summary

(1) During historic times the main occupation of people in the Flåm Valley has been farming, even though the areas of cultivated land are small. A substantial part of the winter fodder was leaves and hay

collected in outlying fields.

(2) Detailed fodder- and production-information, based on official registration lists from the 1860's, is given for one farm.

(3) Pastures in the mountains surrounding the valley are large, and summer farming has played an important part in the farming economy of the Flåm Valley, where permanent settlement based on farming can be traced back to the Roman Period, *ca.* 1-400 A.D.

Trees and Woodland in a Crowded Landscape - The Cultural Landscape of the British Isles

Oliver Rackham

Introduction

Historians often write of the countryside of Great Britain and Ireland as if they were merely extensions of Continental Europe. In theory this ought to be so. All the common plants and animals are found also on the Continent, and all the peoples responsible for the human component in the making of the landscape have come from thence. Processes and events which have shaped the landscape of Europe, such as the elm-decline of the 5th millennium B.C. or the introduction of open-field cultivation from the 8th century A.D. onwards, have affected the British Isles also. Why, then, should the landscapes of the nations of Great Britain and Ireland be so very different from other European landscapes and from one other?

Elsewhere I have written in detail of the history of trees and woodland, and of the cultural landscape as a whole, in the British Isles (Rackham 1975, 1976, 1980, 1986a, 1986c). In this chapter I seek to identify those differences that make up the identity of national landscapes; I am less concerned with regional differences on a finer scale, for which the reader is referred to these books.

The nations of Great Britain and Ireland

We are concerned with at least 5 nations with different cultural histories (Fig. 1). *England* in the Iron Age (750 B.C. onwards) became part of the Celtic world. It was conquered by the Romans in 40 A.D. and became a province of the Roman Empire. After 410 A.D. there comes an ill-documented period (the Dark Ages) during which the Celto-Roman culture was replaced (at least as regards language and place-names) by an Anglo-Saxon culture derived from the Germanic world. In the 8th and 9th centuries England was brought into the

Boundary of nations: ——·——·—
Division in England & Wales
between Highland zone (to N.&W.)
and Lowland Zone : ———

HIGHLANDS
OF
SCOTLAND

Glasgow

Lake
District

IRELAND

Midlands

Anglesey

Lincoln-
shire

Wayland Wood
Ditchingham
Breck-
land   Bradfield Woods
Cam-
bridge-   Staverton
shire
Hatfield Forest
Waterford

WALES

Abergavenny

Essex   Dengie
Hockley Woods

Burnham
Beeches

Somerset
Levels   New
Forest
Dart-
moor
CORNWALL

0   Km   200

Fig. 1. Great Britain and Ireland, showing the nations and places mentioned in the text.

Scandinavian world by Viking settlers and conquerors. In 1066 it was conquered by the Normans, beginning a period of involvement with France which was to continue throughout the Middle Ages.

Scotland, though a separate kingdom, was subject to much the same influences as England, except that the Roman period was brief and left little mark. The *Scottish Highlands*, however, count as virtually a separate nation, differing from the rest of the country as much as England from France. This was outside the Roman and Anglo-Saxon influence, and remained Celtic and Scandinavian - a surviving Iron Age culture - until the 18th century.

Wales is a strongly Celtic nation, with much Roman influence; it was penetrated only late and incompletely by the Anglo-Saxons and Normans.

Cornwall, although it has not been a separate kingdom for over a thousand years, is still sharply differentiated as a nation from the rest of England, both in its Celtic culture and its winter-warm climate.

Ireland escaped the Roman and Anglo-Saxon conquests, and is a Celtic nation with Scandinavian influence; English, Scots and Norman cultures reached it during and after the Middle Ages.

This chapter, unavoidably, says more about England than the other nations, mainly because England has much more documentation and has been more extensively studied.

The physical environment

Compared to the rest of north-west Europe, Britain and, even more so, Ireland, have warm winters, cool summers and strong winds. Wind and cold summers limit tree growth, especially in the north. Warm winters, especially in the west, make it possible to leave cattle and sheep out of doors, though not to dispense with domestic heating.

Great Britain is, for the most part, a very fertile island with an unusually large proportion of cultivable land. Cultivation has, at times, extended well beyond its modern limits. In contrast, the Scottish Highlands and other mountain areas have soils very lacking in mineral nutrients. The commonest alternatives to cultivation are not woodland but semi-natural grassland, heath (vegetation of Ericaceae or other undershrubs on mineral soils) or moorland (ericaceous vegetation on peaty soils). Ireland is even more fertile, and nearly the whole island has been cultivated except for the bogs.

Wildwood

In Great Britain, in the early Holocene, the prehistoric forest (*wildwood*) came to cover almost the entire country: the only places where there has never, owing to climate, been tree growth are the highest mountains and the extreme north-west (e.g. Birks & Madsen 1979). There was much regional variation with climate - every British tree species has its northern limit somewhere within the island. Compared to the rest of north-west Europe, British woodland prehistory has the following characteristics (e.g. Huntley & Birks 1983).

(1) Scarcity of conifers. Most of the Atlantic ecotypes of European conifers seem not to have survived the glaciations. In this interglacial only pine (*Pinus sylvestris*), yew (*Taxus baccata*) and juniper (*Juniperus communis*) returned to the British Isles. Pine disappeared earlier and more completely than from most of Continental Europe; by the Roman period it had died out, except from the Scottish Highlands where an endemic subspecies still forms extensive woodland.

(2) Abundance of hazel. Many British wildwood pollen deposits contain as much pollen of *Corylus* as of all other trees put together. When it is an understorey shrub hazel produces little, if any, pollen. There were, therefore, large areas of woodland in which hazel was a dominant, canopy-forming tree. This abundance of hazel has shaped British cultures from the Mesolithic, when the nuts were an important foodstuff, to the Middle Ages and since, when building and fencing

practices were adapted to using hazel wattle.

(3) Abundance of ash. *Fraxinus excelsior* has been a more abundant woodland-forming tree in the British Isles than elsewhere in Europe

(4) Scarcity of beech (*Fagus sylvatica*). Historic beechwoods are confined to the south-eastern third of England and Wales and have never been common. (The tree is now widespread throughout the British Isles, but only through 18th- and 19th-century planting.)

(5) Abundance of holly. *Ilex aquifolium* has a history of abundance in Britain. Whole woods are composed of it - an extension of temperate evergreen broad-leaved woodland to a high latitude, doubtless due to the warm winters.

Just before the onslaught of civilization in the Neolithic, the natural wildwood appears to have been a mosaic of many tree communities, which can be grouped into 5 provinces (Fig. 2):

(1) the Lime Province, covering Lowland England, predominantly limewood (*Tilia cordata*) and hazel-wood, with areas and patches of oakwood (*Quercus petraea, Q. robur*), ashwood, elmwood (*Ulmus glabra, U. minor*), etc. (Rackham 1986b);

(2) the Oak-Hazel Province, a mosaic of oakwoods, hazel-woods, and smaller areas of other types, in Highland England, Wales, and south Scotland;

(3) the Hazel-Elm Province of west Wales and Cornwall;

(4) the Pine Province - the pinewoods, with patches of birch (*Betula pubescens, B. pendula*), remaining in north-east Scotland;

(5) the Birch Province of north-west Scotland, extending (with patches of pine) to the northern limit of continuous woodland.

Ireland has a smaller range of woodland types, for several trees, such as *Tilia* and *Fagus*, never reached it. The place of *Tilia* was, to some extent, taken by a great abundance of *Ulmus*. *Corylus* and *Ilex* are even more important in Irish than in British woodland history. In the peculiar Irish climate *Ilex* can (in Co. Waterford) reach the altitudinal limit of trees, and there is an outlier of another evergreen, *Arbutus unedo*, a frost-sensitive Mediterranean tree.

Changes in the wooded area

As in the rest of Europe, the deliberate destruction of wildwood in order to make farmland began in the Neolithic, about 4500 B.C. In pollen diagrams the elm-decline identifies almost the beginning of this process.

Compared with most of Europe, wildwood disappeared remarkably early and very completely. There is virtually no evidence as to how this was accomplished; we have no written record nor any tradition or folklore of our pioneering days. In England, the latest period at

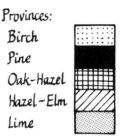

Provinces:
Birch
Pine
Oak-Hazel
Hazel-Elm
Lime

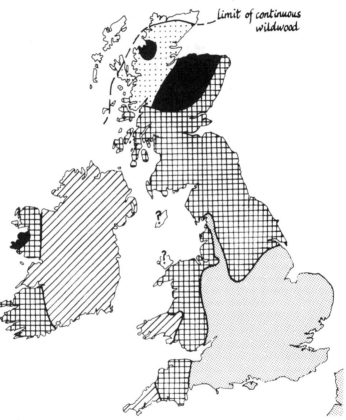

Limit of continuous wildwood

Fig. 2. Wildwood provinces in 4500 B.C., omitting small outliers (from Rackham 1986a after Birks, Deacon & Peglar 1975).

which 50% of the land could still have been woodland is the Iron Age. The earliest documents, Anglo-Saxon and Welsh charters of the 6th to 10th centuries A.D., describe a land in which woodland was already of limited extent and was managed and used. Domesday Book, which gives a systematic record of woodland in England in 1086, shows that it covered only some 15% of the land - a smaller proportion than France has now (Fig. 3).

As recent archaeological surveys show, England has a history of dense settlement and cultivation from the Bronze Age onwards (e.g. Dymond 1985). At an early date it became a crowded land. Although

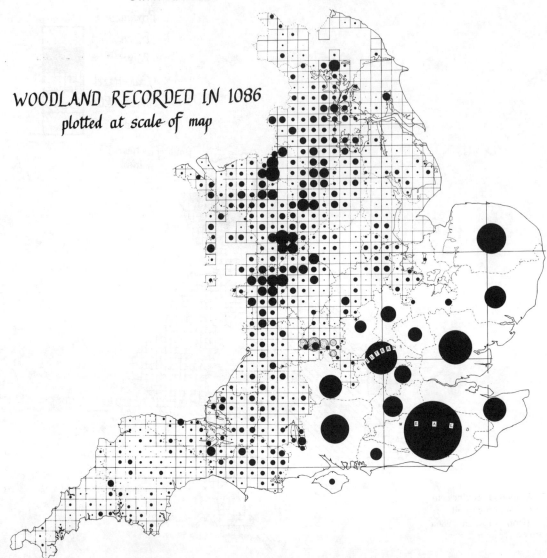

WOODLAND RECORDED IN 1086
plotted at scale of map

the pressure of population was perhaps not specially great (there were not, as in Norway, heroic attempts to create cultivated land on impossible sites), woodland had few opportunities to remain except on steep or very infertile land. Much of the woodland of 1086 was to disappear in a subsequent period of increasing population that ended with the great plague of 1349.

When land has gone out of cultivation, as has happened for various reasons down the millennia, it has often reverted to *secondary woodland*, but usually only on a local scale. Large tracts of ex-cultivated land have generally continued to be used as pastures in the form of grassland, heath or moorland. Although secondary woodland can often be formed within 20 years, its growth is less rapid than in Scandinavia or the eastern United States, and is more easily prevented by grazing. This is particularly true of moorland, whose peaty soils are now often so poor in nutrients that trees invade only slowly if at all, even where there is no grazing. Britain does not have the vast areas of secondary woodland which there are in those other countries.

For Wales and Scotland early documents are few, and the archaeological evidence is less detailed than for England, but there can be little doubt that these countries, too, lost most of their woodland in prehistory. Much of it has been replaced by moorland, especially on acid soils at high altitudes. Some moorland, as in England, arose after a period of cultivation. In north-west Scotland much of the moorland appears to result from the limit of trees moving southward through leaching of soils and the spread of blanket peat, due mainly to natural forces rather than human activity (Rackham 1986a).

In Ireland the archaeological evidence indicates a very dense population by the Iron Age. Wildwood probably disappeared as early as in England and even more completely. Moorland, as in England, grew over old fields and buried them under blanket peat (Caulfield 1978). A survey in 1654-6 A.D. records only 2.1% of Ireland as woodland, less than a third as much as England then had in relation to its area (Rackham 1986a).

The destruction of what woodland remained in the Middle Ages has been resumed in the last 2 centuries. In England there was a further onslaught between 1945 and 1975, as woods were converted either to more farmland or to conifer plantations (Kirby this volume). The ancient woodland now remaining amounts to less than 1.5% of the land area, a large minority of which consists of old secondary woodland. Wales and Scotland are similar, except that much of the woodland has been destroyed by allowing grazing animals to get into it. In Ireland, what little woodland remained in the 1650's was eaten up in the 2 centuries of extreme overpopulation that followed, and the remnants have been coniferized even more thoroughly than in England. The ancient woodland now left amounts to well under 0.1% of Ireland.

◄

Fig. 3. Area of woodland in 1086, as recorded by Domesday Book (from Rackham 1980). Where possible, woodland is mapped by 10 km squares: each black circle represents, at the scale of the map, the total woodland area possessed by places within the square. For eastern and south-eastern counties the information is less precise and has been mapped by counties, with additional black circles to represent the woodland areas in the Weald and Chilterns districts. In Oxfordshire, stippled circles represent Forests, of which an unknown proportion was woodland. Information is missing for the north and parts of the west. For details of calculation see Rackham (1980).

Management of trees and woods

In England and Wales, and to a lesser extent in Scotland and Ireland, there are 3 traditional methods of managing rural trees and woods.

(1) *Woodland*: areas of natural vegetation set aside and managed as a self-renewing resource, yielding successive crops of *timber* (trunks of trees suitable for making beams and planks) and *underwood* (poles and rods of smaller diameter than timber). Most woods are *coppices*, felled every 5 to 25 years (Fig. 4) and allowed to grow again from the stumps or *stools* to produce crops of underwood. Scattered among

Fig. 4. Coppicing in the Bradfield Woods, Suffolk. Underwood in the foreground has just been felled, revealing the great woodbank, with its external ditch, that defines the edge of the wood. To the left is an area felled one year ago and now growing again; note the scattered timber trees. In the background is underwood of 15-20 years' growth. (Felsham Hall Wood, April 1973)

▶

Fig. 5. Woods and wood-pastures as they survived into the 18th century in south-west Essex. Woodland (tree symbols) is carefully differentiated from wood-pasture (stippled) even where the two adjoin. Most of the wood-pasture was in the two big Forests of "Waltham" (now Epping Forest) and "Henhault" (Hainault Forest, mostly destroyed in 1851). Several smaller, private wood-pasture parks are shown. The small dark patches are woodland. The rest of the area mapped was farmland. Surveyed by J. Chapman and P. André, 1772-4.

the stools are *timber trees* allowed to grow for several rotations of the underwood before being felled.

(2) *Wood-pasture* (Fig. 5): areas of vegetation in which trees are combined with livestock. The earliest recorded wood-pastures, in Anglo-Saxon England, were *wooded commons*, tracts of land on which local inhabitants (often living in houses spaced round the edge of the common) had rights to graze cattle or sheep, and sometimes to cut wood. The Medieval development of the practice is bound up with the history of deer (see below). Woodland is not good pasture - woodland herbs being either distasteful or insubstantial - and most wood-pastures had areas of grassland or heath interspersed among the trees. To prevent grazing animals from getting at the regrowth of the trees, the latter were managed either by *pollarding* (cutting the trees at 2.5-4 m above ground) (Fig. 6) or by *compartmentation* (having a number of defined coppice-woods within the wood-pasture, each of which was fenced temporarily to keep out livestock from the early stages of regrowth).

(3) *Non-woodland trees*: trees and hedges between fields (Fig. 7), and trees standing within fields. These may include coppice stools, pollards or timber trees. Characteristic of most of England and Wales is the hedge consisting of a row of underwood, sometimes on an earthwork or wall, with pollards or timber trees set in it at intervals.

These 3 traditions are already mentioned, as separate practices, in English documents of the Anglo-Saxon period. The practice of making

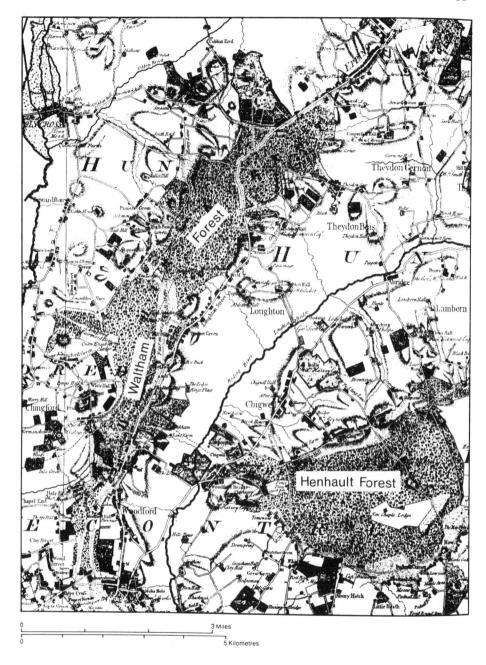

plantations of trees, which is the basis of modern forestry as the British and Irish understand it, is (with insignificant exceptions) no older than 1600 A.D.

Distribution of woodland and trees

An English characteristic is that woodland is very unevenly distributed: some areas have much more than average, others have none. This was more marked in the Domesday Book record (Fig. 3) than it is today. Earlier still, the Anglo-Saxon charters describe a state of affairs in which not every community possessed woodland, and some communities had woods located 10 km or more distant across other people's territory. These inconvenient arrangements can hardly have been intended by the Anglo-Saxons, and presumably were devised to cope with a distribuion of woodland left to them by an earlier and differently organized society. Hedges and non-woodland trees, contrary to what might have been expected, were less abundant in areas lacking woodland and cannot have made up for the deficiency.

England is a nation of shopkeepers rather than of peasants - a

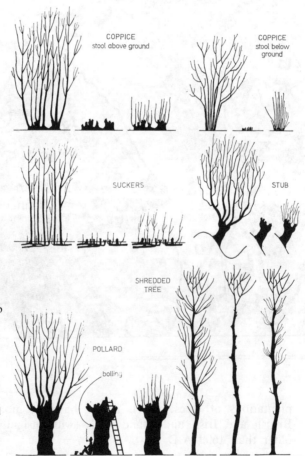

Fig. 6. Ways of managing wood-producing trees. For each method the tree, or group of trees, is shown just before cutting, just after cutting, and one year after cutting. All drawn to the same scale. A stub is a special kind of low pollard sometimes found on wood boundaries. Shredding is a well-documented historical practice, but survivals of it in the British Isles are rare (from Rackham 1976).

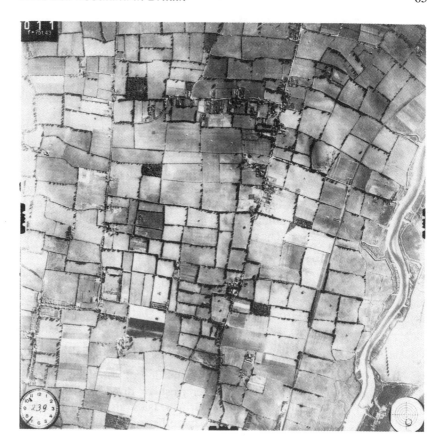

Fig. 7. Hedges and hedgerow trees. The Dengie Peninsula, Essex is covered with the remains of an immense formal landscape of not-quite-rectangular fields, attributed to the Iron Age or Roman Period and still in use. The area shown is about 4 km square. (German air photograph of World War II, September 1940. U.S. National Archives, Group no 373, GX 10019 SD 112)

characteristic which can be traced back at least as far as the 13th century (Macfarlane 1978). Instead of the local self-sufficiency of many European countries, there is a tradition of trade, specialization and different regions complementing one another. Already in the Iron Age, ordinary folk depended on things which they could not make for themselves, especially products of trees such as leather, wheels, turned vessels and boats. Trade and transport are therefore important factors in woodland history. Timber and even underwood have been moved up to 100 km or more, and wooded areas have attracted specialized industries. Because these activities involved payment, they gave rise to an earlier and more detailed record of woods than in countries where people had more direct access to woodland.

Wales, Scotland and Ireland appear to have stronger peasant traditions, and less distinction between wooded and woodless regions. Not that woodland was everywhere: there is little or no evidence of woods in historic times in the Welsh county of Anglesey (Carr 1982) or in much of the Irish Midlands or south Scotland.

Products of trees and woods

Wood has normally been the main and regular product; timber was usually less important, though most woods produced some. Wood consists of coppice poles, pollard poles and the boughs of trees felled for timber. Wood has been used as stakes and wattlework in fencing and building, and for specialized purposes such as hurdles and tools; but the bulk use of wood has been for domestic and industrial fuel. (The warm climate, reducing the need for heating fuel, has probably been offset by the failure of the British Isles to develop a wood-stove tradition).

The British Isles are very well provided with other fuels, chiefly peat and coal. Coal was widely used in the Roman Period, and was a major fuel by the 16th century. In areas with abundant peat or coal, woodland tended not to survive into the historic period, unless there was a large industry specifically using wood (e.g. the charcoal ironworks of the Lake District).

Leaf-fodder is a major product of trees and woods in many European and Asian countries; sometimes (as in parts of Norway, Austad this volume; Hauge this volume) it is the main object of management. Leafy shoots are pollarded or shredded (Fig. 6) from trees, and are fed to animals, either fresh or dried. It has often been supposed that this is true of the British Isles also; the specific pollarding of elm is a widely-accepted hypothesis to account for the prehistoric elm-decline (Troels-Smith 1960). As far as I am aware, the positive evidence for trees as fodder in the British Isles amounts to the following:

(1) The Neolithic hurdle-trackways of the Somerset Levels, dating from about 3000 B.C., are made of rods produced by coppicing hazel and ash. These rods often have an incomplete first or last annual ring, showing that they were cut during the growing season, unlike the usual historic practice. Many rods had their tops cut off at 4-6 years' growth, and were allowed to grow on for another 2-4 years before being felled and made into hurdles. These curious practices suggest that the rods were a by-product of a coppice managed mainly to produce leaves (Coles & Orme 1977; Rackham 1977).

(2) Medieval English account-rolls occasionally mention leaves along with bark, branches, *loppium et chippium* and other by-products arising when timber trees were felled (usually in early summer). Only trifling amounts were paid for them.

(3) Thomas Tusser, the 16th-century agricultural writer, advised:

> If frost doo continue, this lesson doth well,
> For comfort[1] of cattel the fewell to fell:
> From euerie tree the superfluous bows[2]
> now prune for thy neat[3] therevpon to go brows.
> In pruning and trimming all manner of trees
> reserue to ech cattel their properly fees.[4]
> If snow doo continue, sheepe hardly that fare
> Crave Mistle[5] and Iuie[6] for them to spare.

(Tusser 1573)

[1]strengthening [2]boughs [3]cattle [4]dues [5]mistletoe, *Viscum album* [6]ivy, *Hedera helix*

The context makes it clear that hedgerow or field trees are referred to. Tusser appears to follow many of the agricultural practices in the Essex of his time, and I see no reason to doubt that livestock were allowed to gnaw the bark and twigs of firewood in a hungry winter. But I know no other mention of a practice which, had it been more than a by-product, ought to have given rise to payment or litigation and thus to have been put on record.

(4) In the 16th century there are references to wood being cut for deer in parks, presumably to gnaw the bark. For example in Havering Park (Essex) in 1565 *Carpinus betulus, Crataegus* spp. and *Acer campestre* were customarily "lopped & shredd for browze", and the park-keeper had the wood when the deer had finished with it.*

(5) There is a tradition, recorded from the Middle Ages onwards, of pollarding holly as winter fodder, particularly for sheep in a hard winter. Most of the evidence comes from the highlands of England, but pollard hollies themselves are more widespread; they have been cut in recent years to feed deer in the New Forest (Hampshire) (Radley 1961; Peterken & Lloyd 1967).

This meagre tale of information, compared to what exists on other aspects of tree management, indicates that leaf-fodder has never been important in the historic period, but may have been significant earlier. In the Neolithic, when trees were abundant and grassland uncommon, the British Isles may well have shared the leaf-fodder practices of Europe. (Pollarding for leaf-fodder cannot have been more than a contributory cause of the elm-decline. The numbers of elms were too vast for any appreciable percentage of them to have been thus prevented from producing pollen. The decline consistently affected elm, and not other suitable trees such as ash and hazel, even in areas where elm

---

*Public Record Office: LRRO 5/39.

was rare. Disease is the only known influence powerful and specific enough to have been the main cause (Rackham 1980).)

When grassland came into existence, and especially after the invention of the iron scythe made it possible to cut meadows for hay, grass was preferred to leaves, perhaps because leaf-fodder demands much more labour in cutting and storing. The advantage of grass is greater in the British climate than elsewhere because it grows better and livestock can be left to graze for a longer season. By the time of written records, people rarely went to the trouble of cutting trees for fodder. They did so chiefly for starvation rations when the grass failed, and so could not store leaves in advance: tree-fodder mostly took the form of browsewood or of the evergreen *Ilex, Hedera* and *Viscum*.

The various minor uses of trees were not normally objectives of management. People continued to gather and eat hazel-nuts (*Corylus*) in the Middle Ages and down to this century, when the spread of the American grey squirrel (*Sciurus carolinensis* Gmelin) made them scarce. The fattening of pigs on acorns (*Quercus*) is known from England and Wales, though it was never as important as on the Continent because mast years are infrequent. The practice is often mentioned in Anglo-Saxon documents, and down the centuries (occasionally even to the present) advantage has been taken of the acorn crop, including that of hedgerow and wood-pasture oaks, when it happened.

The woodland tradition

The individuality of woods

This chapter is about 'woods', not 'forests'. The woodland history of England in the last 1500 years is the history of tens of thousands of woodlots, each with its own identity as a place, its name (as if it were a village or a mountain), and its boundaries and mosaic of plant communities, different from every other woodlot. Anglo-Saxon and Medieval documents show that the names of woods are of roughly the same antiquity as those of settlements, and give similar scope to philology (study of place names). Examples are Lynderswood (Old English *linde* + *hris* 'Tilia underwood'), Hearse Wood (*hyrst* 'grove on a hill'), Alsa Wood ('Ælfsige's *haga* or enclosed wood') and Alice Tails (the person Ælfsige + Norman-French *tailz* 'coppice-wood'). Here is an extract from a Medieval survey:

> Item, underwood of Tindhawe[1]. 4 score acres[2], of which 5 acres can be sold per annum for £4. 10*s*.[3]

> Item, underwood of Chertheage. 12½ acres 1 ro[o]d[4], that is ½ mark[5] per annum.

Item, underwood of Schyrifseles. 1 acre 3½ ro[o]ds[6], that is 6*d*[3]. per annum.

Item, underwood of sextingge, 4 score and 6½ acres, that is £5 per annum.

Item, Aldercarr ... 14½ acres 1 ro[o]ds, that is 8*s* per annum.

Ditchingham, Norfolk, 1270∗

[1]Missing name supplied from other contemporary documents.
[2]1 score = 20; 1 acre = 0.4 ha.
[3]£1 = 20*s*. = 240*d*.
[4]4 roods = 1 acre.
[5]3 marks = £2.
[6]0.76 ha.

In Ditchingham there were 5 separate woods with definite areas and owners, regular management (cut on average every 16 years), and an expected and sustained yield of underwood. In 1270 as today, even a wood of less than one ha had its own name. At least 3 of the wood-names are Anglo-Saxon in form and show that the arrangement was already centuries old. Woods were permanent; Tindhawe and Sextingge are recorded down the centuries and still exist, as Tindall Wood and Sexton Wood, with a reasonable approximation to their areas in 1270 (Rackham 1986b). It is not necessarily the bigger woods that survive: the *Ely Coucher Book* of 1251 recorded a "grove which is called Tykele" (again an Anglo-Saxon name) which is still there under the name Titley, and now as then is of only 2 ha (Rackham 1976).

Woodland history and ecology are very largely concerned with boundaries. An ancient wood can often be recognized on the map by its sinuous or zigzag outline. On the ground this is defined by a *woodbank*, often a great earthwork some 10 m in total width, with the bank on the wood side and a ditch on the non-woodland side (Fig. 4). Woodbanks defined the ownership boundary and made it easier to fence the wood to keep out grazing animals. Boundaries may be further defined by pollard trees at intervals. A large wood may contain many km of banks and ditches, which are a record of additions and subtractions to the wood's area and of divisions of ownership within it (Fig. 8).

Woodbanks are often mentioned in documents. Most of the big ones already existed by the 13th century, and some may be much older: the Anglo-Saxon word *wyrtruma* may mean a woodbank. It is possible to construct a chronology for them (Rackham 1986c).

∗Public Record Office: C132/38(17).

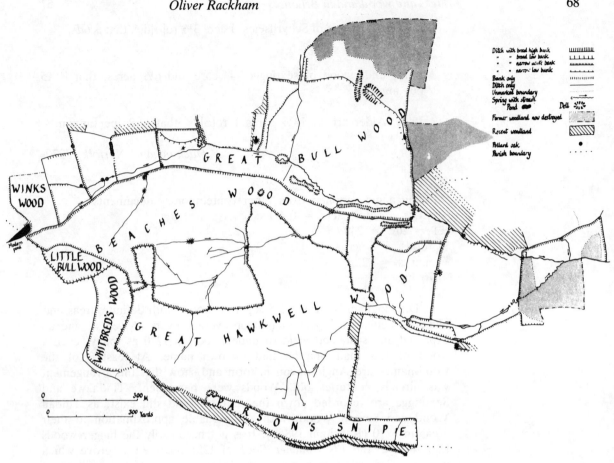

Fig. 8. The complex woodbanks that surround and subdivide Hockley Woods, Essex. They are mostly of Anglo-Saxon and Medieval date, but some may be earlier (Rackham 1986c). Each subdivision of the woods is treated as a separate wood with its own name (from Rackham 1986a).

Although best preserved in England, the tradition of permanent woodlots, each with its own individuality, is not confined to Anglo-Saxon culture. It is equally present in areas with strong Scandinavian influence, such as Lincolnshire. Many of the wood-names themselves are Norse (e.g. Wayland Wood, Norfolk, and other '-land' and 'lound' wood-names from Old Norse *lúndr* 'grove'): the Vikings in England evidently took up the tradition of embanking and naming woodlots. I have found many woodbanks in Normandy.

By the 17th century, although existing woods were still kept up and new plantations were even made in imitation of woods, the woodmanship tradition was no longer strong enough to be taken up overseas. Coppicing was introduced on a large scale to the American colonies, but not the other characteristics of woods. American woodlots do not usually have the individuality and permanence of English ones. Woodbanks are rare (although I have found examples from Massachusetts to Georgia).

Many (though not all) ancient Welsh woods are embanked coppices, like those in England, with names in Welsh. The early Welsh charters show that the practice of naming woods is at least as old as in England. In Scotland the woodmanship tradition is weaker - the native pine-woods, for example, seem not to have defined edges - but there are woods with Gaelic names, and near Glasgow I have found woodbanks and boundary pollards. In Ireland ancient woods rarely survive, and the meagre documentation for woodmanship comes mainly from the archives of the later English landowners (e.g. Jones 1986); yet a few Irish-Gaelic names for woods are known, and in Co. Waterford I have found the remains of several coppices with woodbanks round them.

The evidence of vegetation

Coppicing (and pollarding) prolong the life of a tree. The continuity and individual character of an ancient wood reside chiefly in the underwood stools (Fig. 4). Stools grow bigger with each cycle of felling and regrowth. The rate of expansion can be estimated in trees such as *Fraxinus*, which form a base above ground containing annual rings. An ash stool 2 m in diameter is likely to be over 500 years old. Many woods have *Fraxinus* stools of this size, sometimes much bigger; similar *giant stools* are formed by *Acer, Tilia, Ulmus* (*U. glabra* and other non-suckering species), *Quercus*, etc. (Fig. 9). Giant stools not only identify ancient woods in England and Wales but establish that Scotland and Ireland had coppicing traditions despite the sparse written record. They are to be found all over Europe, from ancient *Quercus* stools in Norway to ancient *Castanea sativa* stools in Crete.

Ancient woods are of many kinds. In Great Britain nearly 70 types, defined chiefly in terms of their tree communities, have so far been recognized (Rackham 1986a, pp. 104-5). This is comparable with what is known from other countries, such as Belgium and Switzerland, where sufficiently detailed surveys have been made. Many of these woodland types have their own documentary histories. Although they cannot represent a random sample of the prehistoric wildwood, they reflect much of its regional variation. For example, woods in which *Tilia* is dominant, though they are now comparatively rare, coincide exactly in geographical range with the Lime Province; *Quercus* woods are the predominant type all over the Oak-Hazel Province; *Pinus* woods are mainly in the area of the Pine Province. Woodland types are seldom the intentional result of management practices, although some of them have changed in indirect response to human activities. The long-term increase in suckering elms (*Ulmus minor, U. procera, U. stricta*) in woods results, at least in part, from the practice of planting elm hedges, from which elms have spread into adjacent woods. The increase in *Fraxinus* over the last 100 years in many

Fig. 9. A giant stool of *Castanea*, the result of centuries of successive fellings and regrowth. (West Wood, Thundersley (Essex), April 1986)

ancient woods cannot so easily be attributed to a fashion in human affairs (Rackham 1980).

It is characteristic of ancient woods that they are not, in general, composed of just one species or combination of species, but are a mosaic of tree communities. A wood of a few ha can easily contain 6 or more kinds of woodland. Herein is much of the individuality of woods.

Many plants are characteristic of woods having particular histories.

For instance, over much of England *Tilia cordata, Sorbus torminalis, Crataegus laevigata, Euphorbia amygdaloides* and *Paris quadrifolia* are characteristic of woods with long histories; *Anthriscus sylvestris* is typical of recent secondary woods. In a given region it is possible to make a list of indicator species, the presence or absence of which can be used to resolve questions of a wood's history (Peterken 1981; Rackham 1980, 1986c).

Some woodland plants are dependent, not (or not only) on continuity, but on the sequence of years of light and years of shade which is a feature of managed woodland (Rackham 1975, 1986c; Brown & Oosterhuis 1981). *Euphorbia amygdaloides*, though definitely a woodland plant, does not bear continuous shade: it grows and flowers after felling, and waits between coppicings as buried seed. Other well-known 'coppicing plants' are perennials, such as *Primula elatior, P. vulgaris* and *Hyacinthoides non-scripta*, which wait through the shade years in an attenuated form, and flower in great abundance after coppicing.

Evidence of archaeological wood and timber

England has tens of thousands of Medieval timber structures, either complete buildings or the roofs and upper floors of stone or brick buildings. Timber-framed buildings usually have panels of wattle-and-daub between the timbers. Such a structure contains evidence of the timber and the underwood trees from which it was made: the species, size, branching, growth-rates and at what age (and with what weapon) they were felled (Rackham 1972, 1986c; Rackham *et al.* 1978).

Timber buildings are very unevenly distributed; they bear no relation to whether there was woodland nearby and little to alternative building materials. In general they are more widespread in towns than in the country; they extend into towns in Cornwall, Scotland and (formerly) Ireland where there is no rural tradition of timber-framing. Whether to build in timber evidently depended on social and architectural considerations of which we know little.

Timber and underwood in a building can be used to reconstruct the structure and management of the woodland whence they came, and are an independent check on written records of woodland. Timber trees, as expected, turn out to be nearly all oak, except in small and humble buildings in which *Ulmus, Fraxinus* and even *Populus tremula* also occur. Most of the timber trees were of no more than 30 cm diameter and 50 years' growth when felled: Medieval woods produced large numbers of small oaks, and there was not (as there is now) any difficulty in replacing them. Underwood is of a variety of species, and, as expected from the documents, was most often felled at no more than 7 years' growth. But the method must be used with caution. Timber could come not only from the nearest wood but from wood-pastures or hedges or from a distant wood. Much of it is known from records to have come from Norway or the Baltic, which explains finds of Medieval *Pinus*.

Similar arguments can be applied to timber and wood recovered in excavations, or which has left impressions in concrete or stone structures. For example, impressions of wattle centrings on the undersides of vaults add to our information on the underwood tradition of Medieval Ireland. Excavated remains are the only evidence of prehistoric woodmanship, for instance in the Neolithic of the Somerset Levels.

The wood-pasture tradition

Wood-pasture systems of one kind or another exist all over the world: for example the *lövängar* (foliage meadows) of Sweden, the *Lärchen-wiesen* (larch meadows) of the eastern Alps, the phrygana set with ancient pollard *Quercus coccifera* in the mountains of east Crete, the 'woodland' savannahs of Africa or the Red Indian 'oak openings' of Michigan. They reach a particular degree of complexity in England.

The term 'wood-pasture' (Latin *silua pastilis*) begins in Anglo-Saxon documents. The distinction from woodland already existed; the archaeological record does not allow us to say how much older it may be. Anglo-Saxon wood-pastures were wooded commons, with definite rights to the land, pasturage and trees.

After the Norman Conquest, wood-pasture took a different direction through the Normans' interest in the husbandry of deer, whereby they created parks and Forests. A *park* is a piece of private land, on which the owner keeps deer (instead of, or as well as, cattle and sheep) confined by a deerproof fence called a *pale*. A *Forest* is a tract of land on which the king, or some other magnate, announces that he has the right to keep deer and to make special bye-laws for their protection. Forests are not fenced, and the deer stay there by force of habit. The deer of a Forest were added to, and did not replace, whatever else the land was being used for. Most Forests were centred on large commons, and involved landowners' and commoners' rights as well as the Crown's (Rackham 1978).

The first parks were made, and Forests declared, for the benefit of the native red and roe deer (*Cervus elaphus* L., *Capreolus capreolus* L.). Around 1100 the Normans introduced - possibly through their connection with the Normans of Sicily - an oriental species, the fallow deer (*Dama dama* L.). Fallow proved particularly suitable for parks and Forests, and ever since have been characteristic animals of England. Fallow deer were treated as woodland animals, and most parks were arranged so as to include at least some woodland. Forests, in Britain, have no necessary connection with woodland: the wooded ones (e.g. Hatfield Forest, Essex) tended to have fallow deer and the non-woodland ones (e.g. Dartmoor Forest) red deer (Rackham 1986a).

Fig. 10. A Medieval deer-park with its original pollard oaks, last pollarded in the 18th century. The area is still frequented by deer, hence the lack of low trees and shrubs. (Staverton Park (Suffolk), May 1982)

Parks and Forests multiplied. By 1200 there were about 150 Forests in England, and by 1300 about 3200 parks. Ostensibly they were to provide the king or the owner with something to eat other than mutton, beef or pork; but they soon developed other functions, notably that of being status symbols. A park was a symbol of minor aristocracy, and a Forest of aspirations to royalty.

Some wooded commons, parks and wooded Forests had compartmentation and hence resembled woodland, with timber trees, underwood and woodland herbs and undershrubs, except that the species most sensitive to grazing might disappear. Uncompartmented wood-pastures are better regarded as grassland or heath with trees (usually pollards or big timber trees) (Fig. 10); they lost their underwood, their specifically woodland herbs and often such sensitive trees as *Tilia* (Rackham 1980, p. 241). There is no sharp distinction to be drawn between uncompartmented wood-pastures and those commons and Forests which lacked trees altogether.

Although they survive less often than Medieval woods, wooded commons, parks and wooded Forests have left substantial remains. They are largely responsible for England's special reputation as a land of ancient trees and of the special lichens and invertebrates that go with places where there is a continuity of ancient trees (e.g. Rose & James 1974). Well-known examples are Burnham Beeches (Buckinghamshire, a wooded common), Staverton Park (Suffolk) (Fig. 10) and the New Forest (Hampshire), all of which have ancient pollards.

Wales, Scotland and Ireland have different wood-pasture traditions, with a less sharp distinction between wood-pasture and woodland. Parks and Forests came to Wales and Scotland soon after England. The social history of those countries being what it is, Forests were many and parks few. The connection between Forests and woodland was weak in Wales and absent from Scotland. Wood-pasture has not left much mark on the landscape, although in south-east Wales there are the remarkable montane pollard beeches of Blaenavon and the Punchbowl near Abergavenny. In Scotland the Forest tradition long outlived its decline in England and still goes on, in a moorland form, in the red-deer Forests of the Highlands.

The tradition of non-woodland trees

Woodbanks are one aspect of the preoccupation with boundaries which is a characteristic of the English landscape. Much of the time which the English have saved through not gathering leaves has been spent digging ditches. The network of hedges and roads which subdivides most of England is of a very wide range of dates, and much of it has recently been recognized as prehistoric (Bowen & Fowler 1978; Williamson 1986) (Fig. 7). For example, the present streets of the western suburbs of Cambridge are strongly influenced by a grid of fields planned in the Iron Age (Rackham 1986a, p. 176).

The existence of hedges is well attested in Anglo-Saxon documents, and there is a small but increasing body of archaeological evidence as early as the Iron Age. Many, especially in recent centuries, were planted as hedges, but this is not the only way in which they can have arisen. As can be seen on a vast scale in the United States, any field boundary not too far from an existing wood turns into a strip of woodland unless carefully prevented. It is probable that many hedges result from natural colonization of trees along what were originally fences, ditches or walls. This may explain why older hedges tend to be a mixture of different species.

Numbers of hedges increased down the centuries to reach a maximum at the end of the Enclosure-Act period (mid-19th century). Hedgerow trees were probably most numerous *ca.* 1700. Scotland, Ireland and parts of Wales have weaker and probably later hedging traditions than England.

Many hedges serve to confine grazing animals or to define ownership boundaries. Hedges are commonly managed by coppicing and pollarding, and yield much the same products as woodland. A well-known practice is *plashing* (called 'laying' by academic writers) - cutting part-way through the living stems and interweaving them to make a stock-proof boundary.

Hedges, like wood boundaries, can contain ancient coppice stools and pollards. The plant communities of old hedges are usually very different from those of the nearby ancient woods (except in some instances where a wood has been grubbed out, leaving its perimeter to form a 'ghost' hedge).

It was once a common practice to have trees standing in what were otherwise ordinary fields. Such 'fields with many trees' seldom survive, but there are some, full of old pollard oaks, round the edges of the (traditionally woodless) Breckland in Suffolk. At the junction between the Breckland and the great delta of the Fens a characteristic field and meadow tree is *Populus nigra*, which is abundantly recorded in Medieval documents but is unknown in woodland.

Conclusions

I have tried to identify and explain the differentiating features of British and Irish landscapes. They come from a wide variety of causes.

Some characteristics are probably due to the climate. The difference between the steep moorland mountainsides of the Scottish Highland sea-lochs and the steep wooded mountainsides of the Norwegian fjords probably comes from the fact that sheep can be left out of doors in a west Scottish but not a Norwegian winter. The British lack of interest in leaf-fodder may also be a consequence of the climate (see Emanuelsson this volume).

Much of the peculiar ecology of woodland in the British Isles can be ascribed to the small proportion of uncultivable land, and the peculiar situations in which woods tend to survive - for instance the clayey, waterlogged hilltops of eastern England.

Other peculiarities are inherited from wildwood. A simple example is the distribution of *Tilia*. A more subtle influence derives from the fact that the British Isles have no readily combustible tree except *Pinus*. Other inflammable plants are all non-woodland. British fire ecology, unlike that of most countries, is concerned with grassland, heath and moorland, but rarely with woodland.

Among human influences on the landscape, some are directly connected with environmental factors such as the availability of peat for fuel. Others result from historical accident. Chestnut (*Castanea sativa*), which is the characteristic woodland tree of south-east England, can be traced back to an introduction by the Romans far outside its native range. If the Normans had lost the Battle of Hastings, or had not had access to oriental beasts *via* the Sicilian connection, there might never have been a vogue for deer husbandry, with all its repercussions down to the present. The Normans' enterprise in seeking alternatives to conventional agriculture, and their love of status symbols, carried the practice of parks and Forests beyond England into Wales and Scotland.

But many aspects of the cultural landscape, especially those relating to woods and hedges, appear to 'run with the land'; they cannot so easily be traced to their origins, nor attributed to any one of the successive cultures that have come to the British Isles. The earliest documentary references show them as inherited from an earlier age still. Differences between regions (e.g. Suffolk *versus* Cambridgeshire) often increase as we go back in time, but not necessarily differences between the nations. The English attitude to woods that I have called woodmanship - the names, boundaries, management and individuality of woods - goes back at least to Anglo-Saxon times. When the Vikings came to England they adopted this attitude, as did the Normans after them (except insofar as some woods were made to subserve the function of deerkeeping). But there are sufficient remains of woodmanship in the Celtic countries to indicate that it once existed there too, and is unlikely to have been derived from England.

The beginnings of coppicing go back to the Neolithic, and probably began with the growing of leaves and of particular sizes and qualities of wood; conservation can hardly have been a motive at so early a date. Woodmanship as a whole may be of later prehistoric origin: the means whereby people came to terms with the remaining woodland in a crowded landscape. Attitudes to woods, and perhaps hedges, were taken over successively by Anglo-Saxons, Vikings and Normans; the English reached America too late for the complete set of practices to be taken there. England thus preserves what appears originally to have been a Celtic attitude to woods; this survives also to some extent in Wales, but only by careful search can the remaining traces of Celtic woodmanship still be found in Scotland and Ireland.

Summary

(1) The distinctiveness and individuality of the cultural landscape of Great Britain and Ireland compared with mainland Europe are considered. The contrasting cultural histories of the 5 nations (England, Scottish Highlands, Wales, Cornwall, Ireland) are outlined.

(2) Prehistoric forest cover differed from the rest of north-west Europe in its scarcity of conifers and beech and its abundance of hazel, elm and ash. Five wildwood provinces can be recognized.

(3) Wildwood had disappeared very early and very extensively by the Iron Age.

(4) Three major traditional approaches to tree and wood management

are woodland, wood-pasture and non-woodland trees in hedges and fields.

(5) Leaf-fodder practices have probably never been important during historical times in the British Isles, in contrast to other European countries.

(6) The history and importance of the woodland tradition are reviewed, with numerous individually named woodlots, characteristic management, and distinctive flora and vegetation.

(7) The combination of wood-pasture with deer parks and Forests after the Norman Conquest continued the wood-pasture tradition to the present day.

(8) The use of trees in hedges and woodbanks in woods are examples of the English preoccupation with boundaries.

(9) The individuality of the British and Irish cultural landscape is attributed to climatic, historical and edaphic factors, and can be traced to the Celtic attitude to woods and woodmanship.

Nomenclature

Plant nomenclature follows *Flora Europaea*, except for *Ulmus* which follows O. Rackham (1986) *The History of the Countryside*, Dent, London.

Conservation in British Woodland - Adapting Traditional Management to Modern Needs

K. J. Kirby

Introduction

Woods are important and attractive. They have a long history and they are an important habitat for native plants and animals. The Nature Conservancy Council (NCC) is primarily concerned with the conservation of British woodland as a wildlife resource, although often this coincides with other cultural reasons for protecting them.

Britain contains no really natural woodlands: perhaps the closest are scrubby fragments on cliffs and in gorges. The rest have been altered in structure, if not composition, by centuries of woodland management, grazing by livestock and burning. Important woodland mammals such as the boar (*Sus scrofa* L.) have been exterminated while others have been introduced, e.g. muntjac deer (*Muntiacus muntjak* Zimmermann) and grey squirrel (*Neosciurus carolinensis* Gmelin), with damaging consequences (Rackham this volume).

Nevertheless there are many sites which, from documentary and ecological evidence, appear to have been continuously wooded throughout the last 1 000 years, and at least some of these may be on primary woodland sites (Rackham 1976, 1980; Peterken 1981). The difficulties in tracing the history of a particular site increase the earlier the period considered, so that it has been found useful to distinguish 'ancient' sites which have been wooded since the Medieval Period (1600 A.D. is taken as an arbitrary cut-off point) from 'recent' woods which have been wooded for less than about 400 years.

The forms of woodland management, such as coppicing, that were practised during the Medieval Period and in some cases through to the 20th century, tended to retain the existing tree and shrub communities, such that at least the broad composition of a wood remained stable. Woods are defined as 'semi-natural' where the tree and shrub layer is composed of species native to the site, perpetuated by coppicing, pollarding or natural regeneration such that each tree generation is directly derived from the previous one. This contrasts with modern plantation forestry where one tree crop may be replaced by a totally different species and where neither may bear any resemblance to the former natural forest cover of the area.

These distinctions, between ancient and recent woods, plantations and semi-natural stands are illustrated in Figure 1.

In Britain, the development of a state forest service, the Forestry Commission, occurred this century, and its initial emphasis was to create a strategic timber reserve. The majority of British forests are now composed of introduced tree species, and are managed as plant-

Fig. 1. A diagrammatic history of British woodland (from Kirby *et al*. 1984).

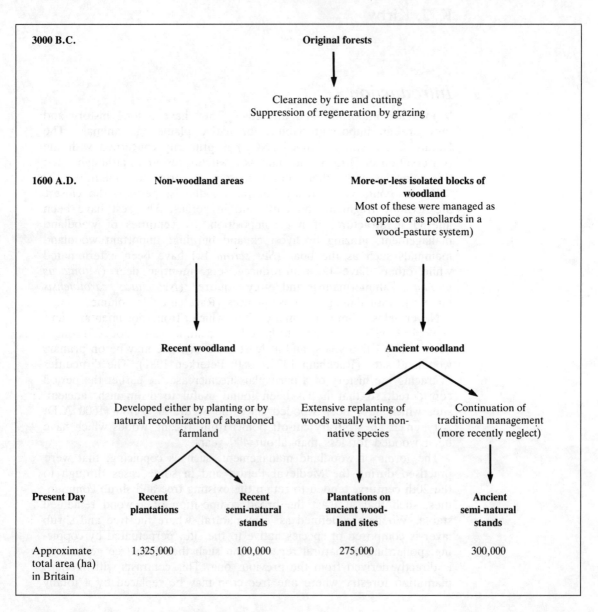

| 3000 B.C. | | **Original forests** | | |

Clearance by fire and cutting
Suppression of regeneration by grazing

| 1600 A.D. | **Non-woodland areas** | **More-or-less isolated blocks of woodland** |

Most of these were managed as coppice or as pollards in a wood-pasture system)

Recent woodland   **Ancient woodland**

Developed either by planting or by natural recolonization of abandoned farmland

Extensive replanting of woods usually with non-native species

Continuation of traditional management (more recently neglect)

Present Day	**Recent plantations**	**Recent semi-natural stands**	**Plantations on ancient wood-land sites**	**Ancient semi-natural stands**
Approximate total area (ha) in Britain	1,325,000	100,000	275,000	300,000

Table 1. A comparison of adjacent ancient and recent woods in Essex (S.D. Webster & K.J. Kirby unpublished data).
Plegdon Wood: former coppice-with-standards woodland of pre-1600 origin.
Lady Wood: oak-hazel established as coppice-with-standards in the 19th century on old agricultural land.
The adjacent woods are both now less than 5 ha. Twelve points were recorded systematically through each wood. Vascular plants in a 10 x 10 m plot were noted and basal area of trees present was assessed using the Relascope method at each point.

Relascope Results (all 12 plots combined)	Species	
	Plegdon (Ancient)	Lady (Recent)
Total (all species)	103	125
Quercus robur	47	87
Fraxinus excelsior	12	23
Betula spp.	3	15
Carpinus betulus	18	-
Ulmus glabra	6	-
Acer campestre	12	-
Salix caprea	5	-
10 x 10 m plot results		
No of species (all 12 plots combined)	51	33
Mean number of species per plot	17.3±1.8	11.6±1.0
Species *markedly* more frequent in one or other wood		
Ranunculus ficaria	11	1
Anemone nemorosa	9	-
Filipendula ulmaria	7	-
Veronica chamaedrys	4	-
Dryopteris filix-mas	4	-
Arum maculatum	9	5
Carex sylvatica	7	3
Brachypodium sylvaticum	1	8

ations on sites which were open ground 100 years ago. During this forestry expansion older, traditional forms of woodland management were frequently ignored and conifer plantations were established even in woods which had been managed as broadleaf coppice for centuries.

Various studies (e.g. Peterken 1983) have established that ancient and particularly ancient semi-natural woods tend to be more varied and richer than most woodland which has developed or been planted on open ground in the last few hundred years (Peterken & Game 1984) (Table 1). Moreover, they contain patterns of variation in plant and animal communities that reflect soil and environmental variations at least as much as any direct human influence. Nevertheless, survival of these woods and the species within them frequently depends on their continued management by traditional means.

Ancient semi-natural woods form about 15% of the present forest cover and 1-2% of total land surface. This area continues to decline because of clearance for agriculture, urban development, quarrying and

Fig. 2. Examples of woodland change in Buckinghamshire (from Kirby *et al.* 1984). The current state of the woods is indicated. In the 19th century there was limited clearance for agriculture (1). Some coppices were partially replanted with oak in the early 19th century but understorey and other broadleaf species re-grew with the oak. Small areas were replanted with conifers/conifer-broadleaf mixtures post-1885. Over the last 50 years one large block has been completely cleared for mineral extraction (2) and substantial sections have been replanted with mixtures and with pure conifer blocks (3).

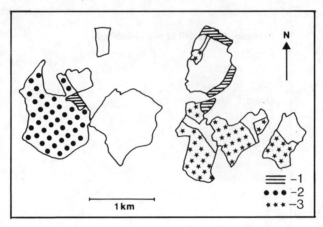

conversion of broadleaf stands to conifer plantations (Fig. 2). Therefore NCC has developed a woodland conservation programme based on the following principles: (1) recognition of ancient natural and semi-natural woodland as special categories for forestry policy purposes; (2) identification of such sites on a county-by-county basis; (3) identification and promotion of traditional management practices in these woods when this is practical and best meets wildlife conservation objectives; and (4) development of alternative approaches to woodland management that can bring similar wildlife benefits.

Identification and description of ancient woods

Rackham (1976, 1980, this volume) and Peterken (1981) describe how ancient woods may be identified from documentary and ecological evidence. Maps from *ca.* 1830, 1930 and 1970 are compared to eliminate sites that have recently become wooded. Aerial photographs, forestry stock maps and ecological surveys separate those woods that are still semi-natural from those that have been replanted. The results of these surveys are lists of probable ancient semi-natural woods, maps showing their location, and an assessment of their national distribution and extent (Fig. 3) (Kirby *et al.* 1984; Walker & Kirby 1987; Kirby *et al.* 1988).

NCC also carries out detailed surveys of ancient woods and helps with surveys by, for example, local naturalists' organizations. This information (Table 2) is used to assess the importance of particular woods for nature conservation at local and national levels (Kirby 1986). Those judged of greatest importance (*ca.* 6% of *all* woods, 21% of the ancient semi-natural areas) become National Nature Reserves or Sites of Special Scientific Interest.

Detailed surveys also provide information with which to assess the past treatment of a wood and to decide what management may best maintain its value for nature conservation in future. In a few sites it may be appropriate to do nothing at all; in some, resumption (continuation) of traditional practices will be best; in others it may be better to devise new systems that are easier to operate in modern times, but that provide similar wildlife benefits to the traditional approach (Steele & Peterken 1982).

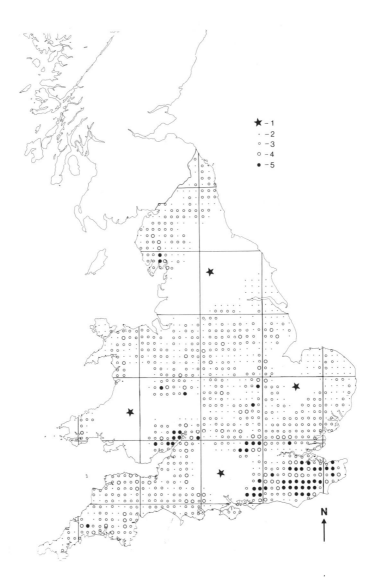

Fig. 3. The distribution of ancient woodland in England and Wales on a 10 km square basis.
1 = no information is yet available;
2 = 1-100 ha of ancient woodland per 10 km square;
3 = 101-500 ha ancient woodland per 10 km square;
4 = 501-1000 ha ancient woodland per 10 km square;
5 = 1001+ ha ancient woodland per 10 km square.

Extent of woodland
Vascular plant list
Vegetation types within the wood
Presence of rare species
Woodland structure
Present and past management
Historical information on the wood
Use of adjacent land and the transition from woodland to non-woodland habitat
Records for groups of species other than vascular plants where these are readily available.

Table 2. Types of information commonly collected in woodland surveys for nature conservation purposes in Britain and used in site evaluation (Kirby 1986).

Ways of promoting nature conservation in woods

On nature reserves, conservation organizations can directly manage the sites to benefit wildlife. 'Sites of Special Scientific Interest' are usually privately owned, but their management is the subject of consultation between NCC and the owner through the provisions of the 1981 Wildlife and Countryside Act. This Act allows for compensation to be paid to an owner if the management that NCC recommends would lead to a profit loss compared to the owner's original plans. Nature reserves and Sites of Special Scientific Interest, however, are only about 21% of ancient semi-natural woods.

In the majority of ancient woods NCC has no statutory right to determine what happens: it can only offer advice as to what might be desirable management. Therefore if nature conservation measures are to be incorporated in the treatment of these woods this must be done in other ways, for example through influence on national forestry policies.

The results from ancient-woodland surveys provided the background for NCC's submission to a recent review of policies for broadleaved woods (Forestry Commission 1984). An important outcome of that review was a recommendation that ancient semi-natural woodland merits special treatment and the Forestry Commission should be less prepared than earlier to allow such sites to be cleared for agriculture or replanted with non-native conifers (Forestry Commission 1985a, 1985b). In addition NCC tries to show that traditional management or its modern derivatives is still relevant, even though social and economic conditions have greatly changed.

Traditional management in a modern world

Traditional management may be justified on some sites purely for historical and cultural reasons, for example at Bradfield Woods (Suffolk) or Hatfield Forest (Essex), that are an exceptionally well documented working coppice and a complete Royal Forest, respectively (see Rackham this volume). I am not convinced, however, that this is a sufficient reason by itself for advocating traditional management above all else on every site.

Locally there may be strong landscape or amenity reasons for using traditional management techniques rather than adopting plantation methods. Dense coppice regrowth may be less susceptible to vandalism than young planted trees, and is an unobtrusive way of discouraging people from moving into the centre of woodland blocks that should be kept undisturbed. Pollarding may allow retention of big mature trunks with an open understorey (attractive park-like appearance) while reducing the risk that dead branches or over-mature trees will fall and cause injury.

The nature-conservation benefits from traditional management are well-described (Peterken 1981). In brief, there is continuity of (1) the tree and shrub composition; (2) the rich ground flora and invertebrate populations associated with newly cut coppice areas; and (3) the lichens and invertebrates of dead wood found in sites containing old pollards.

Landscape, nature conservation, historical and cultural benefits seldom yield much revenue. Therefore either these benefits must be sufficiently important for the owner or a public body to pay any costs involved (NCC, for example, might pay for the work to be done), the costs of the work must be reduced, which may mean departing from some aspects of traditional practice, or there must be a saleable product, even if only as an incidental part of the system. These latter two give the best hopes for widespread restoration (continuation) of traditional practices.

Much traditional woodland management in Britain in the last 10 years has been carried out using subsidized labour. This may be provided by groups such as the British Trust for Conservation Volunteers, or by people employed under government-sponsored job-creation programmes. Traditional woodland management, particularly coppicing, is labour-intensive and can be done with simple hand tools (saws and bill-hooks) that do not require as much training or safety equipment as, for example, chain-saws. It is thus better suited to this type of employment programme than plantation forestry. In counties such as Sussex (Saunders 1984), many woods that would otherwise have been left unmanaged have now been re-coppiced. Once the coppice cycle is re-started, it may generate sufficient interest, markets and income to become self-sustaining.

There is limited scope for reducing costs of coppicing or pollarding through mechanization, although chain-saws speed the rate of cutting and there is little difference in stump regrowth compared to the use of hand tools (Phillips 1971). Small portable saws may be used to convert large timber in the wood and so reduce costs of transporting timber, but much of the product from coppices is not of planking size or quality. Tractor-mounted log-chippers provide another outlet for coppice material, but may require that a larger area of woodland be

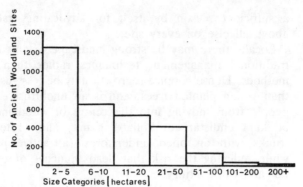

Fig. 4. Size distribution of
ancient woods in Essex,
Humberside, Northampton-
shire, Northumberland,
Gwynedd, Somerset and
Surrey (Kirby *et al*. 1984).

cut at one time to be economic, so reducing some of the nature conservation benefits from the coppice system.

The markets for the products from traditionally managed woods are partly the same as in the past; fencing, thatching spars, turnery poles, tool handles; partly recent marked revivals as with firewood or charcoal; partly new ones such as pulp or brushwood jumps for horses (Garthwaite 1977; Driver 1985; Crowther & Patch 1980). Often they are quite localized. As most ancient woods are also small and scattered (Fig. 4) and have poor access, it may be difficult to bring market and crop together. Nevertheless the potential for creating or retaining employment in rural areas in lowland Britain may be as great as from broad-scale forestry enterprises, major sawmills, pulp mills, etc.

Modifications to traditional practices and high forest systems

Prior to the 19th century, except for the Scottish pinewoods, there was little tradition in Britain of treating woodland of native species as high forest. In areas such as the Chilterns an almost complete change to high forest management occurred in the 19th century and with changes in the pattern of demand for wood and wood products many owners now prefer to have at least a proportion of their timber growing on long rotations. From a purely historical viewpoint this change from mainly coppice to high forest may always be undesirable, but it is not necessarily so in nature conservation terms: indeed there are potential wildlife benefits.

One risk, when a former coppice wood is allowed to develop towards high forest, either through neglect or deliberate singling of stems is that species that flourished in open conditions following a coppice cut are lost. Plants can survive as buried seed (Brown & Oosterhuis 1981), but some invertebrates, particularly butterflies, may become extinct (Stubbs 1982). This can be avoided if substitute open areas are created in the form of wide grassy tracks with adjacent narrow strips of coppiced woodland. Such tracks may also be good for pheasant rearing and help provide an additional source of revenue from shooting.

A second potential problem with converting coppice woods to high forest is that selection for just one or two components in the tree and shrub community may reduce variation within and between woods. (The extreme is where a site is completely replanted with a single species). This effect can be reduced by encouraging natural regeneration rather than planting over much of the wood. If planting does take place then it should be at wide spacing (6 m or more between trees) and with trees of local provenance. In general the aim should be to create a fine-scale mosaic of age classes, although few British native trees grow well in an intimate mixture of ages as might occur in a selection forest. Limited planting of oak standards occurred in many woods in the past so that enrichment by planting is not a new process.

Old parkland and wood-pasture areas (Rackham this volume) where the trees were formerly pollarded may also be allowed or encouraged to change to high forest. Any grassland or heathland around the trees will be destroyed and this itself may be undesirable for conservation. If the pollards, such as at Boconnoc Park (Cornwall), are rich in lichens, then increased shading of the lower trunks and branches by the surrounding young growth may also lead to species losses. On the other hand, there may be more dead and dying wood suitable for beetles and flies that occupy such habitats in humid conditions. Therefore some woodland development around old pollards can be desirable provided they are not overshaded or killed.

The history of woods in north and west Britain is less well documented than for those in the south and east (Rackham this volume). Coppicing and pollarding were practised, but in the acid oakwoods which often predominate, there appears to be less benefit from the restoration of such treatments except in a small number of sites for historical or demonstration purposes, as, for example, next to an iron furnace that consumed local charcoal.

After the decline of the markets for tan bark and charcoal in the 19th century a new tradition arose of using the woods as sheltered grazing for stock, particularly sheep. Limited grazing may help maintain glades, increase the abundance of bryophytes and create patches of bare ground that serve as establishment sites for tree seedlings (Miles & Kinnaird 1981). Long-sustained heavy grazing prevents tree and shrub regeneration and greatly increases the predominance of grasses compared to broad-leaved herbs such as *Endymion non-scriptus, Anemone nemorosa* or, on acid soils, *Vaccinium myrtillus* (Pigott 1983). Use of the woodland by stock is often an important part of the farming system, but how much should be tolerated and how it can be controlled, if the woods and their communities are not to be permanently impoverished, need further research.

Pressure from animals - mainly red deer (*Cervus elaphus* L.) - is also a problem in managing some Scottish pinewoods. Very high numbers are encouraged on the open hills. These tend to concentrate in the woodland in bad weather and thereby restrict pine regeneration. 'Traditional' management of pinewoods varies on different estates, but it has not created a distinctive set of habitats comparable to coppice woods that it is desirable to maintain. Despite much modification these pinewoods may come as close to a natural forest structure as any sites in Britain with patches of over-mature trees, middle-aged stands, open glades and regeneration areas. Therefore apart from the need to control grazing levels, emphasis is on defining relatively large minimal intervention zones; areas where extraction, but little other forestry work, takes place; and on trying to use planting as a means of extending the overall area of pine woodland rather than restocking within existing sites (Forster & Morris 1977).

Conclusion

Ancient semi-natural woods are widespread in Britain but very limited in total extent. They are attractive and important historical features and very rich in wildlife. Hence NCC has made great efforts to identify, describe and protect them.

The survival and current state of a wood are related to its past use and treatment. Restoration or continuation of traditional management is one way to ensure that the wood and its wildlife continue to survive. However, social and economic conditions are not the same as when traditional management practices developed. Therefore we must consider whether we wish to maintain traditional practices purely for historical reasons or because of certain conditions associated with or created by those practices. If the latter, then it is reasonable to look at alternative treatments that may meet nature conservation objectives more easily, more cheaply, or better than traditional practices.

Summary

(1) Although Britain has no completely natural woodland, *ca.* 1-2% of the surface area is covered by ancient semi-natural woodland; that is woods composed mainly of native species that have not been planted, and which have been woodland since at least 1600 A.D. Some of these woodland sites may be primary.

(2) Ancient semi-natural woods are the most important for nature conservation. The woods and species within them have survived under traditional management, particularly coppicing. Modern forestry has concentrated on new plantations mainly on open ground and using non-native species. Consequently the wildlife value of ancient woods

has declined through neglect, clearance or conversion to plantations.

(3) Maps, historical sources, aerial photographs and field surveys are used to identify ancient semi-natural woods. Changes in forestry policy have been encouraged to protect sites not within nature reserves.

(4) Ways in which coppice management is being promoted are described, and nature conservation problems (and benefits) associated with the conversion of coppice woods to high forest are considered. In north and west Britain woodland conservation is complicated by the recent use of many woods for stock shelter and grazing, which limits their regeneration.

Nomenclature

Plant nomenclature follows A. R. Clapham, T. G. Tutin & E. F. Warburg (1962) *Flora of the British Isles* (2nd edition), Cambridge University Press, Cambridge.

Upland Hay Meadows in Britain - Their Vegetation, Management and Future

Jo Hughes and Brian Huntley

Introduction

Very little grassland in Britain can be regarded as natural, since most, if unmanaged, soon becomes scrub and often ultimately woodland. Most semi-natural grassland is maintained by agriculture, either for grazing of livestock (pastures) or for producing grass fodder fed to stock during the winter (meadows).

Northern and western Britain is composed primarily of hard igneous and metamorphic rocks. The high altitude and latitude of much of this area results in the uplands (Fig. 1) experiencing high precipitation, low temperatures and long periods of frosts. Except in sheltered valleys, arable agriculture is rarely possible. Most upland farms have a mixed stock of sheep and cattle. With increasing altitude, dairy cattle give way to beef cattle and on the highest farms only sheep are kept. Farming in the uplands has been affected less by modern innovations than lowland agriculture; there is less room for experimentation and change within the hostile environment of the hills. In most areas, the farm traditionally encompasses land at a range of altitudes and consists of three areas of exploitation: (1) enclosed fields near to the farm buildings in the valley; (2) unenclosed or partially enclosed rough grassland on the lower slopes, grading upwards into (3) rough moorland on the upper slopes. In summer, sheep graze on the moors but during the winter and certainly during the lambing period, the animals are brought down to the enclosed valley fields. In the severest periods, cattle especially are housed indoors. Even when the grass is free from snow, winter temperatures are too low to allow grass growth and the ground is sometimes too wet even for sheep to be allowed on to the fields. Thus, supplementary feeding, whether indoors or on the fields, may be required for up to 6 months of the year. Production of adequate winter fodder is thus probably the most important single concern of the upland farmer: the condition of the stock and thus the health of the farm are primarily controlled by winter feeding.

Fig. 1. Map of Britain, showing the extent of upland areas and the geographical localities mentioned in the text.

History of meadow management

From the earliest domestication of grazing animals in northern Europe, winter fodder has been required. Neolithic people (4000-2300 B.C.) are believed to have collected tree branches and leaves for their stock (a tradition now completely unknown in Britain; Rackham this volume). Only during the Iron Age and Roman Periods (750 B.C.-450 A.D.) did grass cutting begin. The evolution of metal axes to clear woodland and scythes for mowing allowed the replacement of forest by meadows and pastures.

During Medieval times, large hay fields with common or shared ownership were widespread. Many traditions and laws applied to these vitally important meadows, few of which survive in Britain today. A well-known example is Port Meadow near Oxford. Since as early as the 12th century, common or shared ownership of agricultural land progressively gave way to individual ownership of discrete holdings.

Later many Enclosure Acts were enacted with a peak in the late 18th and, especially in upland Britain, the early 19th centuries. The present system of small farm units, with at least some enclosed valley land, evolved through this progressive enclosure. Moorland grazing is, however, often still held in common ownership.

In the uplands, much inbye land (enclosed land close to the farm buildings) is cultivated for grass fodder. Traditionally, field-cured hay was the grass-fodder crop. Sheep and lambs grazed the fresh, spring grass-growth in the hay meadows until mid-May, when they were moved to higher grasslands and moorlands. In August, the hay crop was cut, dried upon the field and removed to a store. After a few weeks' growth, the late season's grass provided valuable grazing for sheep and cattle until the onset of winter.

Arable agriculture occasionally expanded, either during times favourable for producing more cereals and/or root crops, or when some emergency required the production of more crops, for example during the Napoleonic Wars of the early 19th century. This resulted in cultivating some meadows for a short period. When these fields were subsequently put back to grass, barn-sweepings were used to re-establish the sward, thus helping to reconstruct the original flora or at least the species diversity of the field.

These meadows, the lynch-pins of the farm, were created and maintained by a system of farming in long-term balance with the environment. Nutrients removed in the hay crop were returned by applying farmyard manure during winter and early spring (Emanuelsson this volume; Olsson this volume). Given their vital importance to the farming system, meadows were treated with respect and care. Over the last 40 years, however, agriculture has become a 'science' and an industry. This chapter does not discuss the complex economic, historic and social factors and events that have driven these changes. Their consequence, however, is that in the 1980's, winter fodder production has become part of a high-input:high-output cropping system. Large doses of inorganic fertilizers are used to stimulate grass growth. Since few plants of traditionally-managed meadows respond well to high fertilization levels, many meadows have been ploughed and winter grass-fodder is now produced from agricultural grasslands - fields of 2 or 3 highly N-responsive grass species. These swards are soon 'exhausted' (they are inherently unstable ecosystems, prone to invasion by ruderals and competitive, if less N-responsive, perennials); hence the fields are ploughed and reseeded about every 5 years. The rapid encroachment of 'weeds' into an open agricultural grass-sward leads to a need for regular herbicide treatment. Drainage, too, is part of meadow management today, and is particularly important for maintenance of the artificial swards. Since ploughing destroys the fibrous mat of decaying leaves and surface roots, and also leads to loss of the bryophyte cover which many traditionally-managed swards support, artificial swards are much less able to sustain grazing pressure under wet soil conditions. Many remaining traditionally-managed meadows support some fen species, favoured by seasonally-waterlogged soils: drainage threatens their survival in meadows.

The lush, tall- and fast-growing grasses of agricultural mixes are well suited to silage production, when grass is fermented in order to preserve it. No longer does the farmer anxiously await the rare 3-5 days of dry weather required to allow cutting, drying and collection of high quality hay. Silage cutting tends to occur much earlier in the season than hay cuts, when the grass is less dry and more digestible. In some, more favourable areas, multiple silage cuts can even be achieved.

Some farmers are unwilling or unable to plough and reseed their fields but few can afford to miss the opportunity of increasing winter grass-fodder production. Consequently almost all have adopted one or more of the management practices mentioned above. Thus, a spectrum of 'meadow' types can be seen, ranging from unimproved semi-natural grasslands through to artificial agricultural grasslands. Most farmers use inorganic fertilizers on their meadows, albeit often at low dosage (125 kg ha^{-1} of 20:10:10 N:P:K). Very few fields remain undrained. Silage production is an attractive proposition where persistent summer rain can ruin a hay crop.

The botanical and zoological interest of 'meadows' declines rapidly along this gradient of increasingly intensive agricultural management. Species most responsive to exogenous nutrient application, particularly N (mainly grasses) grow vigorously and exclude traditional hay-meadow species, many of which have declined as a result of changing management. Traditionally-managed meadows are thus a resource of increasing conservation value. Early cutting of fertilized meadows prevents the set and dispersal of mature seed by many species, further reducing species-diversity within meadows. It is thought that many seeds are shed whilst the hay crop lies drying on the field. The more-or-less immediate removal of cut grass for silage-making further reduces regeneration by seed of meadow species.

Vegetational composition of meadow communities

In the lowlands of Britain, fertile riverine areas are exploited as so-called flood meadows, a practice also common in much of continental Europe. Studies of the flora and vegetation of such meadows in Britain include Fream (1888) and Baker (1937). At Durham, as befits its location within Britain, our work concentrates on upland meadows. Our studies are still in progress and so the descriptions which follow should be regarded as preliminary.

Meadows in the whole of upland Britain are being surveyed; Figure 2a shows the distribution of the 152 fields examined to date. Data have been collected from the whole range of fields that are exploited to provide hay or silage crops. Vegetation data have been collected from 2 m x 2 m quadrats within each field using phytosociological techniques. All vascular plant, bryophyte and macrolichen species present are recorded and their cover/abundance estimated using the 10-point Domin scale. Basic physical data - altitude, slope, aspect, soil pH, etc. - are also collected. Details of current management practices are elucidated wherever possible by interviewing farmers. Management history is also investigated, especially ploughing history. Figure 2b shows the distribution of the 72 fields examined so far which are believed not to have been ploughed within living memory.

A total of 377 relevés has been collected to date containing 228 taxa. These data have been classified using Two-Way INdicator SPecies ANalysis (Hill 1979). This classification provides the basis for distinguishing 13 noda within our data-set, as shown in Table 1; taxon constancy values are given using five constancy classes (I - 1-20% constancy; V - 81-100% constancy - 'constants'). Taxa with a maximum constancy value of I in 3 or fewer noda are omitted from the table.

The primary division is between *Festuca rubra-Rhinanthus minor-Trifolium pratense* herb-rich, species-rich grasslands (noda 1-7) and the relatively species-poor *Phleum pratense-Alopecurus pratensis* grasslands (noda 8-13):

Fig. 2a. Distribution of meadows sampled by 10 km square. Boundaries of the regional subsets defined for the canonical correspondence analysis are indicated.

Fig. 2b. Distribution of sampled meadows not ploughed within living memory.

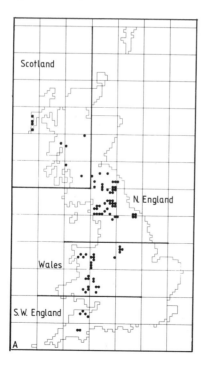

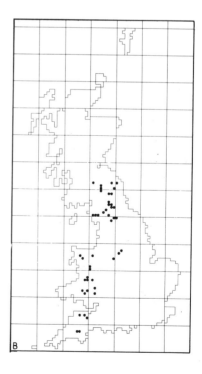

Nodum number	Festuca rubra-Rhinanthus minor-Trifolium pratense							Phleum pratense-Alopecurus pratensis					
	1	2	3	4	5	6	7	8	9	10	11	12	13
Holcus lanatus	V	V	V	V	V	V	V	III	II	.	V	V	V
Ranunculus acris	V	V	IV	IV	IV	IV	V	IV	I	III	II	IV	III
Rumex acetosa	III	II	V	IV	V	V	III	V	II	III	IV	IV	IV
Anthoxanthum odoratum	V	V	V	V	V	II	IV	IV	I	.	IV	III	III
Cynosurus cristatus	V	V	I	V	V	.	V	III	II	.	V	.	II
Trifolium repens	III	V	IV	V	V	.	V	IV	IV	II	V	V	II
Cerastium holosteoides	.	IV	IV	V	V	.	V	V	IV	I	V	V	IV
Brachythecium rutabulum	III	III	II	IV	III	II	II	II	II	.	II	.	II
Poa trivialis	.	II	II	III	IV	III	V	III	III	V	V	III	V
Lolium perenne	.	I	I	II	V	.	I	V	V	V	V	V	IV
Ranunculus repens	.	I	I	I	II	III	V	II	II	I	V	V	V
Eurhynchium praelongum	.	II	II	III	II	II	.	II	I	.	I	I	.
Rhinanthus minor	V	V	IV	V	III	.	III	I	I	I	.	.	II
Trifolium pratense	V	V	III	V	IV	.	III	II	I	.	II	.	I
Festuca rubra	V	III	V	V	IV	IV	II	II	.	.	II	.	II
Plantago lanceolata	V	III	V	V	IV	II	II	III	I	.	.	.	.
Agrostis tenuis	.	I	V	IV	III	II	I	II	.	.	IV	III	.
Equisetum arvense	III	III	.	I	I	III	II	.	.	.	.	.	.
Lathyrus pratensis	.	III	II	II	II	IV	I	I	.	I	.	.	.
Juncus effusus	.	III	.	.	I	II	I	I	.	.	.	.	.
Galium palustre	.	II	.	.	I	II	II	.	.	.	.	.	.
Arrhenatherum elatius	.	.	.	I	I	II	I	I	.	.	.	.	.
Prunella vulgaris	V	IV	I	III	I	.	.	I	.	.	.	.	.
Lotus corniculatus	V	III	IV	III	I	.	.	.	.	.	.	.	.
Pedicularis sylvatica	V	I	I	.	.	.	.	.	.	.	.	.	.
Briza media	V	.	I	I	.	.	.	.	.	.	.	.	.
Pseudoscleropodium purum	III	.	I	I	.	.	.	.	.	.	.	.	.
Ajuga reptans	III	.	I	I	I	.	.	.	.	.	.	.	.
Lophocolea bidentata	.	II	II	I	.	.	I	.	.	.	.	.	.
Chrysanthemum leucanthemum	.	I	I	III	I	.	.	I	I	.	.	.	.
Lychnis flos-cuculi	V	V	I	I	I	.	IV	I	.	.	.	.	.
Succisa pratensis	V	V	IV	I	I	II	.	.	.	.	.	.	.
Agrostis stolonifera	V	IV	.	I	I	.	IV	I	I	.	.	.	II
Luzula campestris/multiflora	V	III	IV	IV	II	.	II	I	.	.	.	.	.
Rhytidiadelphus squarrosus	V	II	IV	II	I	.	I	I	I	.	.	.	.
Hypochoeris radicata	V	I	III	IV	I	.	II	.	.	.	.	.	.
Cirsium palustre	III	I	.	.	I	II	.	.	.	.	.	.	.
Carex nigra	III	V	.	.	I	.	V	.	.	.	.	.	.
Molinia caerulea	.	V	II	.	.	III	.	.	.	.	.	.	.
Vicia cracca	.	V	I	I	I	.	III	I	.	.	.	.	.
Dactylorchis fuchsii	.	V	.	I	.	II	III	.	.	.	.	.	.
Potentilla anserina	.	V	.	I	.	.	IV	.	.	.	.	.	.
Euphrasia officinalis agg.	.	IV	II	IV	I	.	II	.	.	.	.	.	.
Equisetum fluviatile	.	IV	.	.	.	.	IV	.	.	.	.	.	.
Leontodon autumnalis	.	III	I	IV	II	.	II	I	I	.	.	.	.
Carex flacca	.	III	.	.	I	.	I	.	.	.	.	.	I
Plagiomnium undulatum	.	II	I	I	I	.	I	.	.	.	.	.	.
Calliergon cuspidatum	.	II	I	I	I	.	II	.	.	.	.	.	.

Table 1. British upland meadows - synoptic table of noda (1984-5).

Nodum number	Festuca rubra-Rhinanthus minor-Trifolium pratense							Phleum pratense-Alopecurus pratensis					
	1	2	3	4	5	6	7	8	9	10	11	12	13
Potentilla erecta	.	I	V	II	I	IV	.	.	.	.	.	.	.
Centaurea nigra	.	.	V	IV	II	IV	.	I	.	.	.	.	.
Conopodium majus	.	.	V	IV	III	III	.	III	.	.	I	.	II
Veronica chamaedrys	.	.	II	II	II	IV	.	III	I	.	I	.	.
Myosotis discolor	.	.	.	II	II	.	II	II	II	.	.	.	I
Filipendula ulmaria	.	.	III	I	I	V	.	I	.	.	.	.	I
Juncus articulatus	.	.	I	.	I	V	.	.	.	.	.	.	.
Deschampsia cespitosa	.	.	II	I	I	IV	.	.	.	.	.	.	I
Geranium sylvaticum	.	.	.	I	I	IV	.	I	.	.	.	.	.
Bromus lepidus	.	.	.	.	I	IV	.	I	.	.	I	.	.
Achillea ptarmica	.	.	II	I	I	III	.	.	.	.	.	.	I
Geum rivale	.	.	.	.	I	III	.	.	.	.	.	.	.
Crepis paludosa	.	.	.	.	I	III	.	.	.	.	.	.	.
Juncus acutiflorus	.	III	I	I	I	.	V	.	.	.	.	.	.
Polygonum amphibium	.	III	.	I	.	.	IV	.	.	.	.	.	.
Caltha palustris	.	III	.	.	I	.	IV	I	.	.	.	.	III
Cardamine pratensis	III	.	.	.	I	II	II	II	I	.	II	I	III
Taraxacum officinale agg.	III	.	.	II	III	II	.	V	IV	V	II	I	I
Bellis perennis	.	.	.	II	IV	.	II	V	IV	III	III	.	III
Dactylis glomerata	.	.	II	IV	IV	IV	.	V	IV	V	III	II	II
Poa pratensis	.	.	I	I	II	.	IV	IV	II	I	.	II	.
Ranunculus bulbosus	.	.	I	III	II	.	.	III	I	II	I	.	.
Veronica serpyllifolia	.	.	I	.	I	.	.	III	II	.	I	I	.
Bromus mollis	.	.	.	III	IV	.	.	IV	IV	V	II	III	II
Phleum pratense	.	II	.	I	II	.	I	III	V	II	I	.	I
Poa annua	.	II	I	I	I	.	II	III	V	I	II	.	II
Rumex obtusifolius	.	.	.	I	I	.	.	II	III	I	I	I	II
Alopecurus pratensis	.	.	.	I	III	III	.	II	I	V	IV	III	V
Stellaria media	.	.	.	.	I	.	.	II	III	IV	II	IV	II
Anthriscus sylvestris	.	.	.	.	I	.	.	I	.	IV	.	.	I
Cirsium arvense	.	.	.	.	I	.	.	I	I	.	I	I	.
Vicia sepium	.	.	.	I	I	.	.	I	I	.	.	I	.
Carex panicea	V	I	.	.	.	.	.	.	.	.	.	.	.
Cardamine spp.	V	.	.	I	.	.	.	.	.	.	.	.	.
Carum verticillatum	V	.	.	.	I	.	.	.	.	.	.	.	.
Nardus stricta	V	.	.	.	.	.	.	.	.	.	.	.	.
Cirsium dissectum	V	.	.	.	.	.	.	.	.	.	.	.	.
Carex echinata	III	I	.	.	.	.	.	.	.	.	.	.	.
Potentilla reptans	III	.	.	I	.	.	.	.	.	.	.	.	.
Juncus conglomeratus	III	.	.	.	.	.	.	.	.	.	.	.	.
Carex pallescens	III	.	.	.	.	.	.	.	.	.	.	.	.
Carex pulicaris	III	.	.	.	.	.	.	.	.	.	.	.	.
Epilobium spp.	III	.	.	.	.	.	.	.	.	.	.	.	.
Hydrocotyle vulgaris	.	IV	.	.	.	.	II	.	.	.	.	.	.
Senecio aquaticus	.	III	.	.	.	.	II	.	.	.	.	.	.
Plantago maritima	.	III	.	.	.	.	.	.	.	.	.	.	.
Oenanthe lachenalii	.	II	.	.	.	.	I	.	.	.	.	.	.
Ranunculus flammula	.	II	.	.	I	.	I	.	.	.	I	.	.

Table 1. (continued)

Nodum number		Festuca rubra-Rhinanthus minor-Trifolium pratense							Phleum pratense-Alopecurus pratensis					
	1	2	3	4	5	6	7	8	9	10	11	12	13
Festuca ovina	.	I	I	.	I	.	.	I	.	.	.	.	I
Hylocomium splendens	.	I	II	.	.	.	.	.	.	.	.	.	.
Achillea millefolium	.	.	III	II	I	.	.	I	I	.	.	I	.
Campanula rotundifolia	.	.	III	I	.	.	.	.	.	.	.	.	.
Alchemilla glabra	.	.	II	I	I	.	.	I	.	.	.	II	.
Heracleum sphondylium	.	.	I	III	I	.	.	I	.	I	.	.	.
Leontodon hispidus	.	.	I	II	I	II	.	I	.	.	.	.	.
Trifolium dubium	.	.	I	II	I	.	.	I	I	.	.	.	.
Endymion non-scriptus	.	.	I	II	I	.	.	I	.	.	.	.	.
Alchemilla xanthochlora	.	.	I	I	II	.	.	II	.	.	.	.	.
Galium aparine	.	.	.	.	.	IV	.	.	.	.	.	.	.
Angelica sylvestris	.	I	I	.	.	III	.	.	.	.	.	.	.
Trollius europaeus	.	.	I	.	.	III	.	.	.	.	.	.	.
Carex spp.	.	.	I	I	I	II	.	.	.	.	I	.	.
Cirsium heterophyllum	.	.	I	.	.	II	.	.	.	.	.	.	.
Lathyrus montanus	.	.	.	I	I	II	.	.	.	.	.	.	.
Galium cruciata	.	.	.	.	I	II	.	.	.	.	.	.	.
Ranunculus ficaria	.	.	.	.	I	II	.	I	.	.	.	.	.
Valeriana officinalis	.	.	.	.	.	II	.	.	.	.	.	.	.
Equisetum palustre	.	.	.	.	.	II	.	.	.	.	.	.	.
Iris pseudacorus	.	I	.	.	.	.	II	.	.	.	.	.	.
Potentilla palustris	.	I	.	.	.	.	II	.	.	.	.	.	.
Eleocharis palustris	.	I	.	.	.	.	II	.	.	.	.	.	.
Myosotis caespitosa	.	.	.	.	.	.	II	.	.	.	.	.	.
Epilobium palustre	.	.	.	.	.	.	II	.	.	.	.	.	.
Dactylorchis incarnata	.	.	.	.	.	.	II	.	.	.	.	.	.
Dactylorchis cf. purpurella	.	.	.	.	.	.	II	.	.	.	.	.	.
Veronica officinalis	.	.	.	.	I	.	.	I	III	.	.	.	.
Cardamine flexuosa	.	.	.	.	I	.	.	I	II	.	.	.	.
Cerastium glomeratum	.	.	.	.	.	.	.	I	II	.	.	.	.
Agropyron repens	.	I	.	I	I	.	I	I	I	II	.	.	.
Montia fontana	.	.	.	.	I	.	.	I	I	.	II	.	.
Festuca pratensis	.	.	.	I	I	.	.	I	.	.	.	.	II
Number of taxa (total 217)	39	75	80	120	118	47	65	79	57	24	38	27	36
Number of relevés (total 377)	2	6	11	49	173	4	8	67	30	5	9	5	8
Mean number of taxa/relevé	31	33	25	28	23	23	25	20	14	12	15	13	13

Table 1. (continued)

Festuca rubra-Rhinanthus minor-Trifolium pratense species-rich grasslands

Within this group, noda 1-4 are the most traditionally-managed meadows, with high species diversity, including *Euphrasia officinalis* agg., *Hypochoeris radicata*, *Prunella vulgaris*, *Lotus corniculatus*, high values of *Trifolium pratense*, a restricted occurrence of *Centaurea nigra*, and localized occurrences of many other 'tall-herbs'.

1. Nardus stricta-Anthoxanthum odoratum-Cirsium dissectum nodum

One mid-Wales meadow contains an unusual community of nutrient-poor, damp grassland species. A short, species-rich sward of *Nardus stricta*, *Anthoxanthum odoratum*, *Festuca rubra*, *Holcus lanatus*, *Cynosurus cristatus*, *Briza media* and *Agrostis stolonifera* harbours a range of herbs including *Carum verticillatum*, *Pedicularis sylvatica*, *Cirsium dissectum* and *Succisa pratensis*. *Luzula campestris*/*L. multiflora* and *Carex panicea* are also constant.

This nodum shows affinities with the *Succisa-Trollius europaeus* nodum of Jones (1979), which he found on less productive neutral soils. It also falls within the Centaureo-Cynosuretum of Alcock (1982) and is related to the Centaureo-Cynosuretum of the Burren, Ireland (Ivimey-Cook & Proctor 1966). The constancy of *Cirsium dissectum*, *Carex panicea*, *Pedicularis sylvatica* and *Lychnis flos-cuculi*, and the occurrence of other species of damp soils, suggest relationships with fen communities, for example the *Schoenus nigricans-Cirsium dissectum* association of Ivimey-Cook and Proctor (1966) but *Schoenus* and the characteristic fen bryophytes are absent. There are some relationships with the rich-fen meadow communities of Wheeler (1980) although again without clear affinity to any one of his units.

2. Holcus lanatus-Succisa pratensis-Carex nigra nodum

These samples came from herb-rich, damp meadows on South Uist, Outer Hebrides, with one sample from an unusually damp and rich area of a mid-Wales old meadow. *Succisa pratensis*, *Ranunculus acris*, *Trifolium repens*, *T.pratense*, *Rhinanthus minor*, *Lychnis flos-cuculi*, *Potentilla anserina* and *Dactylorchis fuchsii* are commonest amongst the *Holcus lanatus-Anthoxanthum odoratum-Cynosurus cristatus-Molinia caerulea-Carex nigra* sward. *Hydrocotyle vulgaris* and *Equisetum fluviatile* are present, attesting to the moist substrate.

MacKintosh (1984) described similar vegetation from the Shetland Islands under her Vegetation Type 6. Our nodum has affinities with some Junco (subuliflori)-Molinion noda of Wheeler (1980) but does not closely approach any one of his units.

3. Agrostis tenuis-Festuca rubra-Centaurea nigra nodum

Several sites from the North Pennines, North and mid-Wales and mid-Scotland exhibit this *Rhytidiadelphus squarrosus*-rich *Agrostis tenuis-Festuca rubra* grassland, with *Holcus lanatus, Anthoxanthum odoratum, Conopodium majus, Rumex acetosa, Centaurea nigra, Potentilla erecta, Plantago lanceolata, Succisa pratensis, Rhinanthus minor, Lotus corniculatus* and a range of other species.

From a broad perspective our nodum has affinities with *Festuca-Agrostis* grasslands described from the British uplands by many authors. It differs from most, however, in its constancy of *Centaurea nigra* and occurrence of 'tall herbs' rare in short *Festuca-Agrostis* swards and more closely associated with, for example, the tall-herb nodum of McVean and Ratcliffe (1962) and related communities of others. For example our nodum resembles Haffey's (1983) Vegetation Type III (*Geranium sylvaticum/Rumex acetosa*) of meadows with relatively high nutrient levels, in the north of the Northumberland National Park. The National Vegetation Classification (J. C. Rodwell unpublished) describes similar vegetation under its broad Centaureo-Cynosuretum cristati association. Our nodum could be considered a species-poor variant of Alcock's (1982) Centaureo-Cynosuretum, or a species-rich version of his Anthoxantho-Festucetum rubrae within the alliance Cynosurion in Yorkshire. Jones (1979) describes a similar short, calcareous grassland within his *Festuca-Centaurea nigra* nodum.

4. Cynosurus cristatus-Trifolium-Euphrasia officinalis nodum

This *Cynosurus cristatus-Anthoxanthum odoratum-Festuca rubra-Holcus lanatus-Agrostis tenuis* grassland has the following associated herbs: *Trifolium repens, T. pratense, Leontodon autumnalis, Euphrasia officinalis* agg., *Cerastium holosteoides, Plantago lanceolata, Rhinanthus minor, Ranunculus acris, Conopodium majus, Rumex aceotsa, Centaurea nigra* and *Hypochoeris radicata*. It is widespread within upland Britain, stretching from Exmoor and Dartmoor in south-west England, through Wales, Derbyshire, the Lake District, North Pennines and the Cheviots, to the Outer Hebrides.

The nodum has affinities with many published vegetation types: a *Leontodon autumnalis-Anthoxanthum odoratum* traditional meadow of fertile soils in the Yorkshire Dales (Smith 1983); Haffey's (1983) Vegetation Type II (*Euphrasia officinalis-Anthoxanthum odoratum*) in Northumberland, also related to Smith's *Trifolium dubium-Anthoxanthum odoratum* vegetation type; Vegetation Type 2 in Shetland (MacKintosh 1984); and Group 3 vegetation types on the Uists (MacKintosh & Urquhart 1984). Our nodum falls within Centaureo-Cynosuretum cristati of the National Vegetation Classification and Centaureo-Cynosuretum of Alcock (1982). Jones (1979) defines an

Agrostis-Plantago lanceolata vegetation type with a calcareous species component with similarities to our nodum.

Noda 5-7 are also traditionally-managed meadows, but they are characterized by the absence of many 'short-herbs', particularly *Prunella vulgaris* and *Lotus corniculatus*. Instead, these meadows have abundant 'tall herbs', each nodum having a characteristic set.

5. *Lolium perenne-Festuca rubra-Trifolium repens nodum*

Over 1/3 of our samples fall into this widespread and frequent type. These grasslands are reasonably species-rich. Individual species often grow commonly in more intensively managed fields, giving the nodum the closest affinity of any noda within the *Festuca rubra-Rhinanthus minor-Trifolium pratense* group with the other major group, the *Phleum pratense-Alopecurus pratensis* grasslands. Thus, *Lolium perenne, Holcus lanatus, Anthoxanthum odoratum, Cynosurus cristatus, Festuca rubra, Bromus mollis, Poa trivialis* and *Dactylis glomerata* are constant grasses with *Conopodium majus, Taraxacum officinale* agg., *Cerastium holosteoides, Trifolium repens, Rumex acetosa, Bellis perennis, Ranunculus acris, Plantago lanceolata* and *T. pratense*. Most samples come from northern England and southern Scotland, e.g. North York Moors, Lake District, North Pennines, Cheviots and Derbyshire. Some are from mid-Wales, Dartmoor and Exmoor.

This nodum is similar to the *Holcus-Plantago lanceolata* vegetation type of Jones (1979), although *Conopodium majus* and *Geranium sylvaticum* are less frequent. It falls within Lolio-Cynosuretum cristati of the National Vegetation Classification, and more precisely within Lolio-Cynosuretum sub-association Anthoxanthetosum of Alcock (1982). Its affinity to more intensively managed grasslands is reflected by its lack of clear affinities with non-meadow upland communities.

6. *Juncus articulatus-Dactylis glomerata-Filipendula ulmaria nodum*

This damp *Juncus-Filipendula* vegetation has *Holcus lanatus, Dactylis glomerata, Festuca rubra* and *Deschampsia cespitosa* growing within it. *Rumex acetosa, Galium aparine, Veronica chamaedrys, Geranium sylvaticum, Ranunculus acris, Lathyrus pratensis, Centaurea nigra* and *Potentilla erecta* are also present, suggesting an acid but reasonably nutrient-rich substrate. There are only four samples in this nodum, from Cheviots and central Scotland.

The wide range of 'tall-herbs' present, including *Geum rivale, Crepis paludosa, Angelica sylvestris, Trollius europaeus, Cirsium heterophyllum* and *Valeriana officinalis*, as well as others mentioned above, confers affinities with tall-herb vegetation of mountain ledges, for example the tall-herb nodum of McVean and Ratcliffe (1962), as well as with fen-meadows (Wheeler 1980). Jones's (1979) *Juncus acutiflorus-Filipendula ulmaria* nodum and Alcock's (1982) Centaureo-Cynosuretum typical sub-association *Filipendula ulmaria* variant have some similarities.

7. *Juncus acutiflorus-Carex nigra-Holcus lanatus* nodum

These samples from South and North Uist have *Ranunculus repens, R. acris, Trifolium repens, Caltha palustris, Lychnis flos-cuculi, Equisetum fluviatile, Potentilla anserina* and *Polygonum amphibium* within a swrd of *Juncus acutiflorus, Carex nigra, Holcus lanatus, Poa trivialis, Cynosurus cristatus, P. pratensis, Anthoxanthum odoratum* and *Agrostis stolonifera*.

Unlike the previous nodum, this nodum shows affinities with several published vegetation types, for example the Group 4 base-poor fen community on the Uists (MacKintosh & Urquhart 1984) and Vegetation Type 6 on the Shetland Islands (MacKintosh 1984). Jones (1979) defines a *Juncus acutiflorus-Filipendula ulmaria* nodum that shows affinities with this and the previous nodum. Haffey (1983) described a *Juncus effusus/Festuca rubra* vegetation type on wet, acidic soils in Northumberland, that shares many floristic characteristics with the present nodum. Alcock's (1982) work in Yorkshire described two vegetation types with which this nodum has affinities: Centaureo-Cynosuretum sub-association Juncetosum effusi and Holco-Juncetum effusi association. Neither precisely matches our Nodum 7, which has affinities to some rich-fen meadow noda of Wheeler (1980).

Phleum pratense-Alopecurus pratensis grasslands

Noda 8-13 represent more intensively-managed meadows that tend to be less species-rich, particularly with a more restricted range of herbs. As a group they are characterized by the frequent occurrence and high constancy in some noda of the grasses *Bromus mollis* and *Dactylis glomerata* and the ruderals *Stellaria media* and *Rumex obtusifolius*. These species occur in some relevés of the *Festuca rubra-Rhinanthus minor-Trifolium pratense* noda but not frequently, and the latter two ruderals are almost differential for the *Phleum pratense-Alopecurus pratensis* grasslands. Noda within this group have little affinity with any non-meadow vegetation types described from the uplands.

8. *Lolium perenne-Bromus mollis* nodum

This *Lolium-Bromus* grassland contains few other grasses, only *Dactylis glomerata, Poa pratensis* and *Anthoxanthum odoratum* being consistently found. The dicotyledons found are those of more strongly agricultural-influenced grasslands, for example *Taraxacum officinale* agg., *Bellis perennis, Cerastium holosteoides, Rumex acetosa, Ranunculus acris* and *Trifolium repens*. The vegetation type occurs in scattered meadows from south-west Scotland, mid-Wales and Dartmoor but is primarily restricted to the North York Moors, Lake District, North Pennines, Cheviots and North Wales.

Similar communities are described from the Shetlands (MacKintosh 1984, Vegetation Type 1) and northern Pennines (Jones 1979, reseeded *Holcus-Plantago lanceolata*, sub-group *Lolium-Bromus mollis, Lolium perenne* variant).

9. *Lolium perenne-Bromus mollis-Trifolium repens nodum*

The composition and distribution of this type are very similar to the previous nodum. However, *Poa annua* and *Phleum pratense* are constants, and *Rumex acetosa, Ranunculus acris* and *Anthoxanthum odoratum* are rarer than in nodum 8.

Jones's (1979) *Holcus-Plantago lanceolata* group, sub-group *Lolium-Bromus mollis, Phleum pratense* variant of reseeded, productive, mid-altitude meadows contains similar vegetation samples.

Fig. 3. Landscape of small, walled meadows and pastures, with hay barns between and within those fields traditionally exploited as hay meadows, near Keld, Upper Swaledale, North Yorkshire.

10. *Lolium perenne-Bromus mollis-Anthriscus sylvestris nodum*

The presence and abundance of *Anthriscus sylvestris* distinguish these samples from those of other *Lolium-Bromus* noda. Associates include *Poa trivialis*, *Dactylis glomerata*, *Alopecurus pratensis*, *Taraxacum officinale* and *Stellaria media*. These indicate high nutrient levels and heavy agricultural exploitation. Samples are from the Cheviots, North Pennines and Dartmoor. *Anthriscus sylvestris* is especially favoured by high levels of manuring in the absence of inorganic fertilization (Birch *et al.* 1988).

Smith's (1983) *Anthriscus sylvestris/Lolium perenne* vegetation type, which he found only in the Yorkshire Dales under systems of high manuring or high fertilization, is similar. Alcock (1982), also from the Yorkshire Dales, described an Alopecuro-Festucetum pratensae, *Alopecurus pratensis-Lolium perenne* nodum which, despite the absence of *Anthriscus sylvestris*, shows affinities to our nodum.

11. *Lolium perenne-Trifolium repens nodum*

This nodum comprises samples from the Cheviots and North Wales, with one from Exmoor. The *Lolium perenne-Holcus lanatus-Cynosurus cristatus-Anthoxanthum odoratum-Agrostis tenuis-Poa trivialis-Alopecurus pratensis* sward has few herbs; only *Ranunculus repens*, *Cerastium holosteoides*, *Trifolium repens* and *Rumex acetosa* are present with any constancy. This nodum falls within Lolio-Cynosuretum sub-association Anthoxanthetosum of Alcock (1982), although it is a depauperate relative.

12. *Lolium perenne-Holcus lanatus-Ranunculus repens nodum*

This degraded *Lolium perenne-Holcus lanatus* grassland has *Ranunculus repens*, *Cerastium holosteoides*, *Trifolium repens*, *Stellaria media*, *Ranunculus acris* and *Rumex acetosa* as frequent associates. All indicate disturbance and agricultural improvement. North Wales and central Scotland provided these samples.

13. *Alopecurus pratensis-Poa trivialis-Ranunculus repens nodum*

Apart from one sample from Dartmoor, this nodum is restricted to the Cheviots and northern Pennines. *Alopecurus pratensis*, *Poa trivialis*, *Holcus lanatus* and *Lolium perenne* form the sward, with *Ranunculus repens*, *Cerastium holosteoides* and *Rumex acetosa* the only common herbs. Poor, acid, heavily exploited ground is indicated.

This nodum is a species-poor variant of Alopecuro-Festucetum pratensae, *Alopecurus pratensis-Lolium perenne* nodum of Alcock (1982).

Relationships between management and composition

In order to assess the ways in which various management regimes influence the vegetation by altering its composition, the technique of canonical correspondence analysis (CCA) (ter Braak 1986) was used. Management variables used were: time since last ploughing, date of cutting, period of shut up (between end of grazing and cutting), inorganic fertilizer application, manuring, liming, grazing by sheep, grazing by cattle, harvesting as hay, and harvesting as silage. In addition various geographical variables (altitude, slope, grid easting, grid northing) were analysed, along with sampling date. The results of this preliminary analysis revealed strong systematic relationships between geographical location and management regime, and between location and sampling date. These relationships obscured any relationships between composition and either location or management.

In order to overcome these problems, the data-set was split geographically into four subsets (see Fig. 2a) - south-west England, Wales, northern England and Scotland. Each subset was then analysed using those management and position variables that showed significant independent variation within the samples of that subset. The results are summarized in Table 2; for each region and for the first three CCA axes, the management and geographical variables are listed according to whether their biplot scores were negative or positive, and in order of decreasing absolute scores. The significance level (on an arbitrary scale) of the regression relationship between variables and axes is also indicated. The results reveal marked, but as yet not fully understood, differences between regions in the relationships between management, location, and sward composition.

In both Wales and northern England, the two largest data-subsets, the primary management/location gradient (which is axis 2 of the northern England analysis) corresponds to a gradient from relevés of *Festuca rubra-Rhinanthus minor-Trifolium pratense* grassland to relevés of *Phleum pratense-Alopecurus pratensis* grasslands. The management variables with high scores and significant regressions are, however, different, indicating that relationships between management and composition are complex and vary between regions. This finding is potentially important in formulating successful strategies for conservation of 'traditionally-managed' meadows. Any policy that prescribes a standard conservation management for all upland meadows would seem likely, from our evidence, to be doomed to fail in many cases by being inappropriate to local conditions and traditions under which the present composition has evolved.

Region	Axis 1		Axis 2		Axis 3	
	-ve	+ve	-ve	+ve	-ve	+ve
S.W. England	Grid northing** Liming** Inorganic fertilising Slope	Period of closure* Cutting date* Soil pH Cattle grazing** Altitude** Grid easting**	Grid northing Slope Soil pH Liming Grid easting	Cutting date Altitude Cattle grazing Inorganic fertilising Period of closure	Liming Inorganic fertilising Altitude Soil pH Grid northing Cutting date Period of closure Slope	Grid easting Cattle grazing
Wales	Hay making Cutting date Slope* Manuring** Cattle grazing** Altitude	Inorganic fertilising** Silage making* Grid northing* Liming Sheep grazing	Cutting data** Slope* Grid easting** Altitude Hay making Grid northing* Liming	Manuring* Silage making Inorganic fertilising* Sheep grazing* Cattle grazing**	Slope** Sheep grazing Altitude*** Inorganic fertilising Silage making* Cutting date Liming	Grid easting** Grid northing Cattle grazing** Hay making
N. England	Sheep grazing* Cattle grazing Hay making Liming Period of closure* Grid easting	Slope** Manuring Altitude Cutting date** Age since ploughing Silage making Grid northing	Grid northing* Cutting date Manuring** Altitude** Period of closure** Silage making*	Cattle grazing** Grid easting Slope* Age since ploughing** Hay making* Liming Sheep grazing	Cutting date** Silage making Cattle grazing* Sheep grazing Period of closure Slope	Age since ploughing* Altitude** Hay making Grid northing Liming Grid easting* Manuring
Scotland	Grid northing Sheep grazing** Cutting date	Grid easting* Altitude* Manuring Slope** Period of closure	Grid easting Altitude* Slope* Manuring Period of closure	Sheep grazing* Grid northing Cutting date	Altitude* Grid easting**	Manuring* Cutting date** Period of closure Grid northing** Sheep grazing** Slope

Value and importance of meadows

In agricultural terms, herb-rich, diverse hay meadows resulting from and maintained by a traditional low-input system can be regarded as relics of an anachronistic style of farming that is economically unjustifiable. However, in recent years, even the grant-aiding structure of the European Economic Community has begun to reflect the view that it is irresponsible of society to support a high-output system of farming in an already wastefully over-producing area, particularly when this system is also environmentally damaging. In Britain, degradation and loss of traditionally-managed meadows are considered to be one of the most alarming consequences of modern agriculture.

Hay meadows are biologically important. Some of Britain's rarer species find refuge in the few remaining traditionally-managed hay fields. The value of a habitat for wildlife conservation is, in part, a function of its species-diversity and hay meadows provide some of the most diverse communities in Britain, not only in terms of plants but also of invertebrates. These, in turn, support a large number of increasingly uncommon mammals and birds. The late, tall grass provides cover for nest-sites for many birds, for example partridges and the elusive corn-crake. The wildlife value of meadows is also, in large part, a reflection of the paucity of alternative habitats, and especially the rarity of natural habitats, for the species they harbour. Many tall-herbs are intolerant of grazing and dense shade. Their natural habitat of open, upland woodlands has been almost totally destroyed by millenia of intense grazing and woodland clearance.

The aesthetic value of meadows in summer is rarely matched - the fields are a riotous mix of colours, textures and heights. The change, associated with enclosure, from shared ownership of large fields has produced an appealing landscape in many upland valleys, where small fields are walled, often with little hay barns between adjacent fields (see Fig. 3). No longer, however, are farm workers employed to maintain the walls; the additional movement towards a more intensive farming system with frequent ploughing and/or reseeding and the use of large silage machines has encouraged removal of walls, whilst old barns are redundant and left to decay. Thus, the maintenance of traditional methods of hay production is important for landscape conservation over and above its value for biological conservation. As Tittensor (1985) argues, landscape conservation is an important part of the conservation of our 'cultural heritage' as well as an essential aspect of nature conservation (see also Austad this volume).

Much of Britain's archaeological heritage lies in the sub-soil of the uplands. Many Iron Age and Early Roman sites have been destroyed by deep-ploughing. Many of the few remaining unimproved, traditionally-managed hay fields are thus also sites of archaeological importance. Medieval field systems, with ancient ploughing ridge-and-furrow, are

◄
Table 2. Canonical correspondence analysis biplot scores of management and geographical variables for four geographical sub-sets. The variables are listed in order of decreasing absolute scores. * = t-value of canonical coefficient 2.1-3.0, ** = t-value of canonical coefficient >3.0; (see ter Braak 1987 pp. 48-9.)

confined today to those few meadows and pastures whose traditional management has kept them free from ploughing in recent centuries.

The future for meadows

The value which herb-rich meadows have on many levels is based on characteristics that can probably not be recreated. Those who suggest that a valuable field ploughed today will have the same species-rich community in 100 years' time that it had yesterday have yet to be proved right or wrong - and in 100 years it will be too late to revise our attitudes and actions in the light of the results of this 'experiment', given the rate at which meadows are being ploughed. It can, of course, be argued that some meadows judged valuable today were ploughed during the Napoleonic wars, and 150 years later they include some of our richest fields. There are, however, few parallels between a brief period of management as emergency arable land, and the treatment that a meadow receives today after ploughing, drainage, reseeding with artificial seed mixes (rather than barn-sweepings), herbicide treatment and heavy dosage with inorganic fertilizers, none of which were practised in Napoleonic times! In addition, given the absence of many meadows with an unbroken history of use as hay meadows over centuries, it is impossible for us to assess species loss and community degradation resulting from even this 'gentle' break in meadow exploitation of fields. If we are to conserve the wildlife they sustain, and that part of our cultural heritage which they represent, then we must conserve and continue traditional management of our few remaining herb-rich meadows.

The delicate and essentially unstable plant communities of 'seminatural' meadows are controlled by many management and environmental factors. These communities are created and maintained by 'traditional' methods of farming - the precise details of this management vary greatly between areas and have been developed by farmers as those best suited to the conditions and requirements of an individual farm. Unfortunately, details of these carefully evolved management systems are being lost at, perhaps, an even greater rate than the meadows themselves.

We suggest that conservation of the few remaining hay meadows in upland Britain should be addressed immediately. When long-practised farming methods have been maintained to the present, they should be retained. Elsewhere, it is perhaps best to adopt a cautious approach to future management until research on meadow vegetation begins to elucidate more fully the relationships between management practices and vegetation.

Of course, we must never forget that these fields were created by farmers and that their role, as the producer of winter fodder, is of unrivalled importance on the upland farm. One option, of course, is to 'write off' these fields as far as economic fodder production is concerned, relying on alternative, less valuable fields on the farm that could be managed in a high-input:high-output fashion. This, however, further marginalizes the traditional hay field, rendering it even more likely to find a place only as a 'museum' piece. One cannot expect the increasingly financially pressed hill-farmer not to respond to the lure of potentially more than trebling his grass-fodder crop, as has been achieved in some areas following reseeding and intensively managing the resultant artificial sward.

For the future, one can hope that research will demonstrate that meadow communities can be sustained within an economically viable variant of modern agricultural practices or that agriculture within Britain will move away from such a heavy emphasis on economics towards a more environmentally-balanced agriculture.

Summary

(1) The British uplands suffer harsh winters, necessitating feeding of livestock, especially sheep, for several weeks; winter fodder production is thus vitally important.

(2) Past management for hay production provided conditions suitable for the establishment and maintenance of highly diverse plant communities in meadows.

(3) Since 1940 more intense fodder-production methods involving heavy and frequent inorganic fertilization, drainage, early and sometimes multiple cuts of the sward, increasingly widespread silage production and often the ploughing and reseeding of the sward, have produced a range of heavily anthropogenic grasslands with much reduced botanical interest.

(4) The present range of plant communities found in upland meadows is described and the relationships between management and the nature of meadow vegetation assessed.

(5) Traditionally-managed hay meadows are increasingly important refuges for animals, as well as plants, and their protection has become a priority in the British uplands.

Acknowledgements

Attendance by J. C. H. at the Sogndal Symposium was made possible by support from the Natural Environment Research Council, British and Foreign School Society, Robert Hughes and Barbara Hughes. This research was supported by the Natural Environment Research Council.

Nomenclature

Plant nomenclature follows A. R. Clapham, T. G. Tutin & E. F. Warburg (1962) *Flora of the British Isles* (2nd edition), Cambridge University Press, Cambridge (vascular plants); A. J. E. Smith (1978) *The Moss Flora of Britain and Ireland*, Cambridge University Press, Cambridge (mosses); and E. V. Watson (1981) *British mosses and liverworts* (3rd edition), Cambridge University Press, Cambridge (liverworts).

A Model for Describing the Development of the Cultural Landscape

Urban Emanuelsson

Introduction

It is now well known that the vegetation of southern Scandinavia, and other parts of northern Europe, is determined both by factors controlling the occurrence of natural biotopes and by the history of human impact. The detailed studies of these two aspects now available should be synthesized and put into a theoretical framework, so that useful conclusions can be drawn.

Various theoretical frameworks have been proposed. Welinder (1984), Harris (1975) and Rasmussen & Reenburg (1980) have proposed general theories about the interaction between people and nature, mostly from the point of view of people, and how they relate to certain land-use situations. This chapter presents a model for the interaction between humans and nature in which the consequences for nature are emphasized. The socio-economic variations between different human societies are given less importance than the ecological factors acting in the cultural landscape.

The model

Background

The population size of an organism is regulated by its predators, its competitors and the available resources. The 'available resources' in the case of animals is generally equated with the amount of food available. The term 'carrying capacity' has often been used to describe the number of individuals of a species which can survive in a certain area. The carrying capacity is related to the available resources. The population size of big organisms with a slow reproductive rate closely follows the carrying capacity. Small, fast-reproducing organisms often increase beyond the carrying capacity and then decrease drastically.

The population dynamics of primitive *Homo* species was probably similar to that of other big mammals with a diverse diet (Harris

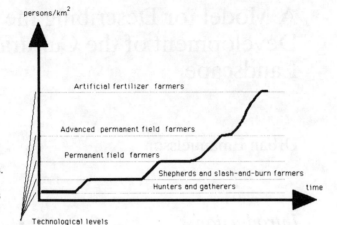

Fig. 1. Hypothetical development of the size of a human population reaching different technological levels. These levels are dependent upon the use of the nutrients in the territory of the human population.

1975). The population size remained closely related to the same carrying capacity over long periods. With increasing brain size, early humans began to use tools and to find alternative sources of food. There were probably many occasions when the carrying capacity for humans rose without any change in the natural environment, because they had discovered a new source of food.

Throughout their history, humans have made inventions which increase the carrying capacity for their populations. The new carrying capacity levels and population increases are reflected by a stepwise change in the landscape (Berglund 1969, 1985b) (Fig. 1).

The mechanisms behind the stepwise increases in human populations have been debated for nearly 200 years. Malthus argued in his classical study that 'technological' inventions lead to larger human populations. A further rise must be preceded by another new invention. Opposing this view, Boserup (1973) argued that inventions are made independently of increases in population, but were only used when the carrying capacity was exceeded. For example, in a situation where normal food resources failed, a previously known technique for increasing food production came into general use. After a while, the population recovered to its previous level. Then it continued to increase to the new carrying capacity set by the new technique. The drawback for the human individual was that, with a new technique, the time spent on food production was initially longer (Harris 1975). This was an important factor in delaying the introduction of new techniques for higher food production until the need arose. The stepwise increases in population can be seen in Broserup's terms as a long series of technological inventions.

What were the main steps on this technological ladder of inventions? Naturally, they were not always the same in different parts of the world, but the principle behind them seems to be similar over very large areas, namely the increase in the amount of inorganic nutrients in the ecosystem which can be directed into human food production. From a plant ecologist's point of view, the important technological steps were those which had a marked effect in changing the natural environment into a cultural landscape. I propose that there are 4 main steps between 5 technological levels (Figs. 1, 2).

Level 1: Hunter-gatherer society

Such a society will concentrate a small amount of nutrients from the environment into the area around their habitations. The impact on the environment will be similar to that produced by other big omnivores. The balance between negative factors such as leaching and denitrification and positive factors such as weathering and nitrogen fixation will not be essentially affected. However, there are examples of hunter-gatherer societies which have changed their environment to such a degree that they must have indirectly changed the nutrient balance (Martin & Wright 1973).

In fertile parts of Scania, southern Sweden, it can be estimated that 0.5-2 persons km^{-2} could survive at Level 1. These figures are based upon assumptions given by Larsson (1984), Welinder (1977) and Harris (1975).

Level 2: Shifting cultivation and/or pastoralism

For the same fertile area of Scania, the population density at Level 2 is about 20 persons km^{-2}. The impact on the environment is much more intense than at Level 1. This figure is based upon data from Dahl (1942) concerning relict shifting cultivation in Scania in the 18th century, but it is probably comparable to earlier periods when shifting cultivation was the dominant form of land-use.

Compared with Level 1, there is a much greater net loss of nutrients, particularly by leaching, which is not balanced by weathering and nitrogen fixation. During periods of human exploitation, nutrients in an area would be seriously depleted. However, if the intervening periods of non-exploitation were long enough, the nutrient balance could be restored (Rasmussen 1985).

In the case of pastoralism, the environmental impact can be similar to that of Level 1, but if grazing is intense, the effect can be comparable to that of shifting cultivation (Cloudsley-Thompson 1977).

Level 3a: Farming in permanent fields

The permanent fields require manure, so the territory is divided into one part for farming and one part for producing materials which can be converted into manure. There are various kinds of manure-producing areas. Fodder from hay meadows, different wooded meadows, and areas used for leaf collection is converted into manure by livestock consuming hay and leaves during the winter. Grazing areas can provide manure for the fields if the livestock are gathered together at night. Finally, plant material and soil can be collected from grazing areas and spread on arable fields (e.g. plaggen-culture in north-west Germany, Behre *et al.* 1982; Scottish Highlands, Dodgshon this volume).

In a fertile area in Scania, 50 persons km^{-2} could survive at Level 3a. This is estimated from figures for the 18th century from Campbell (1927) and Dahl (1942). The ratio between shifting agriculture and permanent manured fields of *ca.* 2.5 can also be found in other parts of the world according to the data of Harris (1975).

During the 18th century, and probably also earlier, management of the manure was not optimal. A fairly large proportion of the nutrients was lost between the source areas (meadows, grazing areas, etc.) and the sink areas (fields and some meadows), due to bad management. Even so, the use of nutrients in Level 3 was more efficient than that in Level 2, where leaching was pronounced during intensive periods of cultivation (Fig. 2).

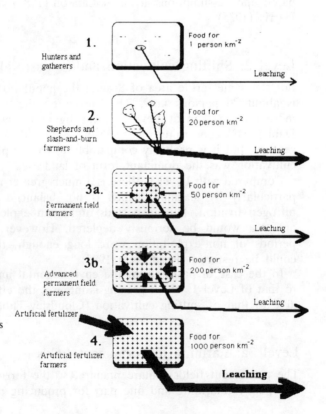

Fig. 2. The general structure of the 5 technological levels. - = areas from which nutrients are taken; + = areas which will have a positive net balance of nutrients. For further explanation see text.

The time of the introduction of Level 3a is uncertain. Widgren (1986) says that village-like structures have been found in Sweden that are about 2000 years old. Thus, Level 3a could have been in existence by 2000 B.P.

Level 3b: More efficient use of permanent fields

This level is very similar to 3a, but the transfer of nutrients from the source areas to the arable fields is much more efficient. This allows the arable fields to be expanded at the expense of the nutrient-source areas, such as meadows and pastures (Olsson this volume).

Level 3b developed in Holland in the 17th century, and then in Great Britain. In the Nordic countries, this level of technology was attained during the 19th century, and was fully developed for only a short period (Osvald 1962; Hannerberg 1972; Eskeröd 1973). This level was also more or less jumped over in other remote areas, where Level 3a was directly followed by Level 4.

Level 3b could support *ca.* 300 persons km^{-2} in Scania (Emanuelsson & Möller in preparation).

Level 4: Use of artificial fertilizers

Using artificial fertilizers, it is in theory possible to use the total land area as arable fields. Nutrient-source areas are redundant, and meadows and pastures are also artificially fertilized. Manure from the livestock largely has its origin in artificial fertilizers. Level 4 agriculture in Scania can support 1000-5000 persons km^{-2} (based on data in Jord-bruksstatistisk Årsbok 1985).

The use of nutrients in artificial fertilizers is not very efficient (Fig. 2). Much of the nutrients is lost by leaching before it can be used by plants.

Regression periods

A pollen diagram illustrates the technological levels through which a landscape has passed. The phases of retreat of forest trees and the introduction and expansion of pasture plants and cereals have been transformed by, for example, Berglund (1969, 1985b, this volume) into diagrams showing the hypothetical changes in human populations (see Birks *et al.* this volume). These show expansion stages associated with new technology, and also a number of stages where it seems that the human population has decreased, and the environment has re-covered by natural successions.

What are the reasons for these declines in human populations? Over-exploitation of the soil is one obvious explanation, and catastrophic events such as wars and diseases can also be invoked. Whatever the reasons, Kjekshus (1977) has pointed out that the course of events in the cultural landscape is similar. His main concept can be summed up in the term 'ecological control'. This is the management carried out by the farmer or shepherd to maintain the landscape in its current condition, such as pasture, meadow or arable field. Continuous hunting of livestock predators is a kind of ecological control (Szabo 1970). The critical feature of ecological control is that it is much easier to maintain it, than to regain it should it be lost. For example, it is much more work to make a wooded meadow out of a wood than it is to keep it as a wooded meadow. Rough estimates show that between 3-10 times more work is required to bring a piece of land under ecological control than to maintain it as such (Emanuelsson in preparation). The variation depends upon the period of neglect or the type of land-use.

When ecological control is lost, for example during a lengthy war, the ecosystems are not managed, and will therefore not produce food. Even if no-one is killed in the war, there will be too few people to restore the food-producing ecosystems to their previous level. Famine results, and people die, thus reducing still further the ability to restore food production. However, some areas will still be productive, and the people with food will then use their surplus manpower to restore more areas, and so on. This pattern of events is illustrated in Figure 3.

The population may decrease as a result of over-exploitation of the natural resources. Without going into a detailed discussion, I propose that over-exploitation will often trigger a situation where ecological control will be lost. Loss of control will then make the food shortage more severe, and famine will be considerable. When the surviving population restores ecological control, the landscape will have meanwhile recovered some of its fertility.

Geographical modifications of the model and examples of the model

Within all 5 levels of landscape exploitation, large differences occur between different geographical regions in the ways in which the environment is utilized. Such geographical differences within one level often have a climatic basis, which is not always immediately obvious. In northern Europe, the climatic gradient from oceanic to continental is very important. In many parts of the world, precipitation gradients have controlled the use of the landscape. In some cases, extreme climate has made it impossible to change from one level to another. In most cases, climatic gradients have the strongest influence at Levels 3a and 3b. Within a climatic region, geological differences can control different kinds of land-use within one level.

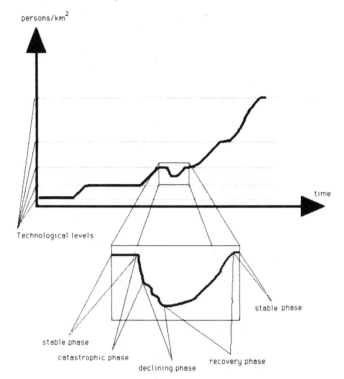

Fig. 3. A regression in the population size in an area can be triggered by many different events (catastrophic phase). In the declining phase the decline in the number of people will continue as a result of lost ecological control over the landscape. In the recovery phase the man-power surplus will be used to gradually bring back more and more parts of the landscape into the food-producing system. In the stable phase the carrying capacity level is reached.

I will use data from Scania to illustrate the 5 levels of land-use in the model.

The use of wooded meadows and coppice-woods in north-western Europe will serve as an example of a geographical modification of land-use within one level, Level 3a.

Five levels of land-use in Malmöhus län, Scania (Fig. 4)

Up to around 6000 B.P. Scania was inhabited by hunters and gatherers (Level 1) (Burenhult 1982), and no major transformation of the ecosystem is assumed to have occurred. The size of the population of Malmöhus län at this time is extremely hard to estimate. Using data from Welinder (1977), Burenhult (1982) and Larsson (1984), one can guess at a population of between 500 and 3000 persons. This gives a population density of 0.1-0.6 persons km^{-2} (Emanuelsson in preparation). The density was probably much higher in favourable areas, such as at the coast and beside shallow lakes. The fertility of the soil was unimportant compared with later levels.

During the Younger Stone Age, the Bronze Age and the Early Iron Age, people in Malmöhus län survived by pastoralism and slash-and-burn agriculture (Level 2) (Burenhult 1982). Large inland areas

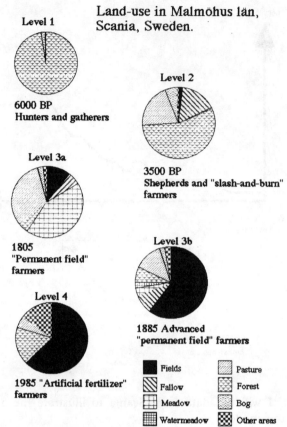

Land-use in Malmöhus län, Scania, Sweden.

Level 1

6000 BP
Hunters and gatherers

Level 2

3500 BP
Shepherds and "slash-and-burn" farmers

Level 3a

1805
"Permanent field" farmers

Level 3b

1885 Advanced "permanent field" farmers

Level 4

1985 "Artificial fertilizer" farmers

Fields   Pasture
Fallow   Forest
Meadow   Bog
Watermeadow   Other areas

Fig. 4. Land-use in the province Malmöhus län at the 5 technological levels shown in Fig. 2. The land-use in Levels 1 and 2 are very rough estimates made on the basis of a number of pollen diagrams and archaeological sources (including Berglund 1969; Burenhult 1982). The data for the other levels are based upon official statistics used by Emanuelsson & Müller (in preparation).

were still not populated at all, and used very little for food production. The estimated Level 2 density of 20 persons km^{-2} gives a reasonable estimate for the populated areas of Malmöhus län in 2000 B.P. For the total area, 5 persons km^{-2} can be estimated for favourable periods. In Level 2, note the large area of fallow in comparison to the field area. The forested area was probably, to some extent, later successional stages of slash-and-burn activity.

1805 is the first time we have reliable figures for the land-use of the whole of Malmöhus län. At this time, the transition from Level 3a to Level 3b was beginning, so earlier figures would have been preferable. However, in 1805, the landscape of Malmöhus län was dominated by potential manure-producing areas, such as meadows and pastures. The area of arable land is much smaller, and areas of forest and 'other use' are insignificant (Fig. 4). Using the official 1805 statistics, the population density of Malmöhus län was 30 persons km^{-2}.

During the 19th century large areas of meadows and pastures were brought under the plough (Bringéus 1964), with a consequent loss of manure-producing land. To compensate, the manure was managed more efficiently, nitrogen-fixing plants such as legumes were used in the crop rotation, marling was used for a short period, and large areas of highly productive water-meadows were constructed (Zachrison 1922; Bengtsson *et al.* 1973). In 1885 this efficient agriculture without artificial fertilizers had reached its peak. Thereafter, artificial fertilizers began to be used (Olsson this volume).

The population density at this time was 64 persons km^{-2}. However, much food was being exported, so the area could potentially support a larger population. Much of the income from food export was invested in industrial enterprises which started at this time.

In 1985, Malmöhus län is a good example of an area with a large food production almost entirely supported by artificial fertilizers. Unfertilized areas for manure production are extremely small, but the area of forest has increased (Fig. 4). The population density of 152 persons km^{-2} is unrelated to the food production, which could support 2000 persons km^{-2} (Jordbruksstatistisk Årsbok 1985).

The gradient of wooded meadows used for fodder or wood production

In north-west Europe there is great variation in the use of coppice woods and wooded meadows. In a transect from central Scandinavia, with long severe winters, to England and north-west France with shorter, mild winters, a general pattern can be seen (Bergendorff & Emanuelsson 1982) (Fig. 5).

In the northern part of the transect, the greatest problem for the farmers was to get enough winter fodder for the animals when they were stalled. There was no shortage of fuel, so trees were not managed in any way. Often whole trees were felled to collect the leaves for winter fodder, and in wet areas trees were removed to provide good wet meadows (e.g. Hicks this volume). There was no particular relationship between collection of winter fodder and fuel.

Further south, where the winters were less severe and less winter fodder was required the trees were more valuable. Multiple-use wooded meadows were developed which provided hay, and leaves from trees which were pollarded or shredded (Hæggström 1983; Austad this volume; Hauge this volume), and also wood for fuel and special purposes.

In most of southern Sweden and some parts of Denmark and Germany, fuel production from wooded meadows was more important than winter fodder (Bergendorff & Emanuelsson 1982). These meadows resembled coppice woods of Britain and France, but they were still used for some fodder production.

At the southern end of the transect in Britain, fuel production dominates, and some grazing takes place in the woods (Rackham 1980, this volume).

Conclusions

The model presented here consists of three parts: (1) a description of 5 levels of human use of the landscape, with emphasis on the use of soil nutrients; (2) the use of the concept of 'ecological control' to explain, to some extent, regressions in human populations; and (3) climate as a differentiating factor explaining spatial modifications of land-use within the 5 levels.

This model does not attempt a full explanation of the interaction between human use and the ecology of the landscape. However, it can help to describe different stages in population expansions and declines in an area. In historical studies it is essential to realise the degree of dependence of humans on the environment. The concept of soil nutrients as determining factors can be very useful. Also, when classifying the landscape, for example for purposes of nature conservation, it can be valuable to distinguish different types of cultural landscapes as remnants from different levels in the model, e.g. a wooded meadow is a remnant from Level 3a, and a water-meadow is a remnant from Level 3b, etc.

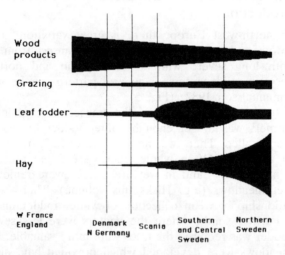

Fig. 5. Example of a climatic gradient (long, severe winters - short, mild winters) which will modify a certain part of the land-use system (in this case the alternative area for livestock fodder or fuel) at a certain technological level (in this case Level 3a) (Partly after Bergendorff & Emanuelsson 1982). Similar but shorter gradients are to be found in the Alps and the Pyrenées.

The model as presented here is preliminary. Refinements have to be made concerning the geographical modifications and the theory for human population regressions. For the latter particularly, there is a need for more basic data.

Using the overview provided by the model, it is of interest to make a chronology for the different technological levels. For example, when was slash-and-burn agriculture introduced into an area? It is often possible to use pollen-analytical data to answer this question (e.g. Hicks this volume; Edwards this volume). Much less attention is, however, paid to the question of the change from Level 2 to Level 3a, the introduction of permanent manured fields (e.g. Gaillard & Berglund this volume). To answer when this very significant change took place in different areas must be one of the most important tasks for historical ecology in the future.

Summary

(1) A three-part model is described which relates human exploitation of the landscape to population size and soil nutrient utilization at five technological levels.

(2) The five levels are (1) hunter-gatherer economy, (2) slash-and-burn agriculture and pastoralism, (3a) the use of permanent manured fields, (3b) a more efficient use of permanent manured fields, and (4) dependence on artificial fertilizers.

(3) The concept of 'ecological control' can be partly used to explain population regressions.

(4) Climate is important in controlling geographic variations of land-use within one technological level.

(5) The first part of the model is applied to the historical ecology of Malmöhus län, southern Sweden.

(6) The third part of the model is applied to the geographical variation in the use of wooded meadows along a transect from central Scandinavia to southern Britain.

(7) The usefulness of the model in historical ecology and in landscape conservation is discussed.

Acknowledgements

I am grateful to H. J. B. Birks, H. H. Birks, C. Bergendorff, B. E. Berglund and J. Möller for helpful comments concerning both the content and the language of this chapter.

Nutrient Use and Productivity for Different Cropping Systems in South Sweden during the 18th Century

Gunilla Olsson

Introduction

Up to the late 19th century Scandinavian agriculture was based upon animal husbandry for producing meat and dairy products. Since winter grazing is possible only in limited areas in the extreme west and south, the focus on animal husbandry resulted in a great need for fodder-producing areas, both for summer grazing and hay-collecting. The village organization of land-use in Scandinavia, following the principle of a main division into infields and outfields, dates back to the Viking Age (Uhlig 1961; Christiansen 1978). Generally, the infields were manured and used for arable crops and hay-harvests. They were protected from grazing until after harvest. Implicit in this system is a continuous transfer of plant nutrients from the outfields, the common grazings, as well as from the meadows, to the arable land (Fig. 1). During the pre-industrial period commercial fertilizers were absent and plant nutrients had to be restored by addition of manure or other organic materials.

In some areas, however, there was an early shift towards cereal production at the expense of animal husbandry. This occurred in the fertile plains of the province of Scania in southern Sweden where most of the land was owned by a few manorial landowners. These areas were involved in a north European trade market and cereals were exported. Taxes to the state and rents to the manors were also paid in grain. This emphasis on cereals as the main agricultural product was accompanied by the introduction of an intensive cropping system, the three-course rotation system (Dahl 1942; Smith 1967). Large arable fields with minor fallowed areas and small meadows and pastures were characteristic features of this system. The commons used for grazing were largely converted to arable land.

Within 15 km of such areas in southern Scania there were 'traditional' animal husbandry areas using less intensive crop rotations. Here, fairly

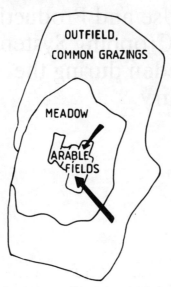

OUTFIELD,
COMMON GRAZINGS

MEADOW

ARABLE
FIELDS

Fig. 1. General relationships
between land-use types and
nutrient flow (arrows) within
the infield-outfield system.

small arable fields prevailed, but the farmsteads had large meadows
and extended common grazings for livestock maintenance (cf. Dahl
1942).

Since 1982, an interdisciplinary research project at the University
of Lund, has been studying the development of the South Scandinavian
rural landscape during the past 6000 years (Berglund this volume).
The following questions have been identified as vitally important:

(1) To what extent can the decline and expansion of human settlement and farming be related to the use of environmental resources:
viz, are periods of decline in intensity of land use or reorganization
periods a result of exceeding the carrying capacity of that area
(Emanuelsson this volume)?

(2) Are differences in settlement patterns and farming practices for
villages related to differential availabilities of ecological resources?

This study focuses on the ecological implications of the use of two
different cropping systems for two villages in Scania. The following
questions are considered:

(1) How was the productivity of the arable fields maintained in
areas where the stock size was kept low due to greatly reduced fodder
areas?

(2) Was the focus on animal husbandry, which produced relatively
large amounts of manure, reflected by a greater productivity of the
arable land compared to the mainly cereal-producing villages?

To address these questions, nutrient-use efficiency (NUE) for two
principal plant nutrients, nitrogen and phosphorus, was estimated for
two villages with different physical geographical and ecological charact-

eristics. The time period studied was mid-18th century. The term nutrient-use efficiency relates the quantity of added manure to the response in yield of cereals (cf. Vitousek 1982).

Site description

The villages Bjäresjö (13°45'N, 55°27'E) and Beden (13°37'N, 55°32'E, Fig. 2) are situated in the province of Scania, southernmost Sweden. In Bjäresjö the bedrock is Tertiary limestone overlain by thick (>50 m) deposits of clay till and clayey-sandy till with a high lime content (*ca.* 20% in an unweathered profile; Daniel 1986). Soil pH ranges between 6.5-7.0 (Kindstrand & Jungbeck 1984). The area is characterized by gently rolling topography with hillocks 10-20 m high surrounded by small depressions with lacustrine sediments or peat.

Beden, on the south-eastern slope of the PreCambrian hill Romeleåsen, is on till formed by siliceous rocks and shales. The clay content of the soils is generally low. Sandy till with a relatively high boulder content occurs in the northern parts (Daniel unpublished). Soil pH ranges between 5.5-6.5 (Kindstrand & Jungbeck 1984).

The Bjäresjö area was settled during prehistoric times. It contains remains of Mesolithic settlements and numerous artefacts from the Neolithic period (Larsson 1985). Indications exist for the predominance of an intensive cropping system with cereals as the main product as early as the 13th century (Campbell 1928; Skansjö 1983). Large manors whose owners possessed most of the land and used the labour force of the tenants in the villages were a characteristic feature of

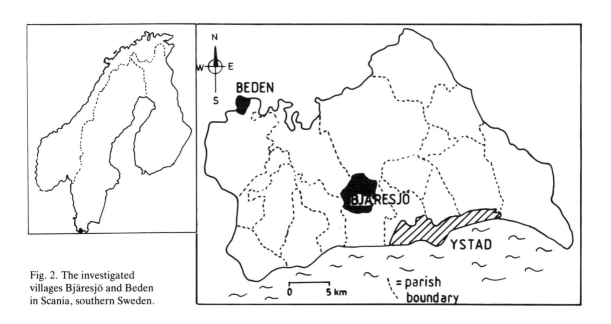

Fig. 2. The investigated villages Bjäresjö and Beden in Scania, southern Sweden.

YEARLY DISTRIBUTION OF LANDUSE; INFIELDS

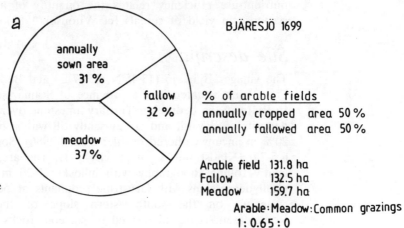

Fig. 3. Distribution of land-use
areas within infields in
(a) Bjäresjö 1699, and
(b) Beden 1750. ▶

this area from the 16th century onwards (Skansjö 1985). Villages were large, consisting of 20 farms or more.

Beden, in contrast, was probably not settled until late Medieval time, 1450-1550 A.D. There are no prehistoric finds and the name Beden is first documented from 1547 (Hallberg 1975). The settlement pattern is characterized by isolated farmsteads or small villages.

Different cropping systems were used in the two villages. Bjäresjö was characterized by three-course rotation (Figs. 3a, 4), the general principle of which was a division of the arable fields into three main parts. Two parts were used annually for cereal crops while one part was laid fallow (Fig. 4). Over the 3-year rotation period every field was fallowed once. Cultivated crops were barley and winter rye. The fields were manured in autumn every third year before the fallow was ploughed. The cultivation of oats followed a different rotation with only one year of cultivation in the 3-year cycle. Oat fields were never manured and comprised half of the total field area (Fig. 4). The meadows in Bjäresjö were non-forested and equalled the area of the annually cultivated fields (Fig. 5a). The villagers had no pastures or common grazings and were obliged to graze their cattle on the fallows, the stubble and the meadows after harvest. The ratio of arable: meadow:commons was 1:0.65:0 (Fig. 3a).

Beden used a less intensive cropping system; quadrennial two-course rotation (cf. Dahl 1978), in which 2 years of cultivation were followed by 2 years of fallow. However, various exceptions to this general rule occurred, and 2 or 3 years of cultivation followed by as much as 4, 6, 7 or 8 years of fallow was also practised (Table 1). The cropping system was very intricate and included different fallow-regimes for the different soil types within the village. Only 28% of the total field area was cultivated annually; most of the fields lay fallow (Fig. 3b). The same

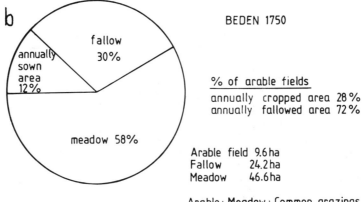

BEDEN 1750

b

fallow 30%

annually sown area 12%

meadow 58%

% of arable fields
annually cropped area 28%
annually fallowed area 72%

Arable field 9.6 ha
Fallow 24.2 ha
Meadow 46.6 ha

Arable : Meadow : Common grazings
1 : 1.4 : 1.5

crops were grown in Beden as in Bjäresjö but with the addition of buckwheat (*Fagopyrum esculentum* Moench.). The fields were manured just before the fallow was ploughed. Oats and buckwheat were probably never manured. Of the meadows, half were coppice areas where hay was collected beneath the managed trees and shrubs (Fig. 5b). The Beden farmers had access to a large common grazing area shared with several other villages (cf. Persson 1986a) (Fig. 5b). The ratio of arable:meadow:common grazing was 1:1.4:1.5 (Fig. 3b).

Methods

Historical sources

Information on acreage, cropping systems used, land-use and distribution of land-use categories was obtained from land-survey maps for Bjäresjö (1699) and Beden (1750) (Swedish land survey board, Archive,

Table 1. Cropping systems in Beden 1750. * = percentage of total arable field area.

| Crop | %* | Number of years | |
		cultivation	fallow
Rye	32	2	1
and		3	6
Barley	37	3	7
Oats	19	3	6
		2	6
		2	8
Buckwheat	2	2	8

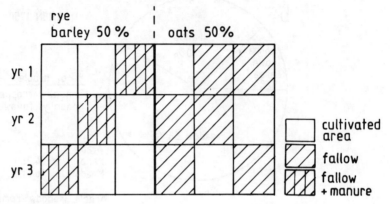

Fig. 4. Cropping system and manure regime in Bjäresjö 1699. The figure indicates the relative magnitude of the areas for the different crops.

Gävle: Bjäresjö 3[1], Villie 4[1]). The numbers of inhabitants and cattle in the villages were obtained from inventories of the goods and cattle of the deceased and from tables of land surveys and taxes (A. Persson unpublished) (Tables 2, 3). Estimates, based on tithe tables, of seed: yield ratios for rye, barley and oats cultivated in the two villages (Table 4) were obtained by Persson (1986b and unpublished).

Calculations of harvest and available manure

Seed:yield ratios and information on the cropping system used allowed the annual harvests of different crops to be calculated.

Data on livestock numbers in Beden were available for 1671 (both farms) and 1768 (one farm). Means for these two times were used in the calculations. For Bjäresjö no figures were available for the whole village. Instead data for two neighbouring villages, Varmlösa (1671) and Gårdlösa (1671) (Skansjö 1981), were used and adjusted according to the different sizes of the villages (Table 2). Data on cattle size appropriate for the 18th century were obtained by combining cattle figures from the following sources: Noring (1936, 1942), Åkerblom (1891), Nannesson (1914), Husdjursskötsel (1919), Seebohm (1927) and Slicher van Bath (1963). The mean weight for cows (south Scandinavia, England, north France, north Germany) was estimated at 246 kg live weight, amounting to 50% of the values for Swedish cows in the 1950's (Johansson 1953). In calculating the amount of manure produced, 18th century cattle were assumed to produce the same amount of manure per unit time and weight as modern cattle. Furthermore, it was assumed that the nutrient concentrations of the manure were the same. Another assumption was that at Beden, only 1/2 to 2/3 of the cattle were kept over winter and thus contributed to the amount of farmyard manure available for arable land (cf. Linné 1749; Dahl 1942). The amounts of human manure were estimated from the number of inhabitants in the two villages (Table 3).

Table 2. Number of livestock per farm in the villages Bjäresjö and Beden. For Bjäresjö the figures are mean values of weighted means per farm 1671 based on data from two neighbouring villages Varmlösa (A. Persson un-published) and Gårdlösa (Skansjö 1981). For Beden the figures are means for two counts; 1671 and 1768 (A. Persson unpublished).

Livestock	Bjäresjö	Beden
Horse	2.5-3.3	9.4
Cow	2.1-4.5	16.5
Pig	1.3-4.3	11.5
Sheep	3.4-5.9	13.8
Hen	1.8-3.3	3.5

Table 3. Number of farms and number of inhabitants in the villages Bjäresjö (1699) and Beden (1768).

	Bjäresjö	Beden
Number of farms	20	2
Number of inhabitants	90	15

Table 4. Estimated seed:yield ratios for the villages Bjäresjö (1699) and Beden (1750) respectively (Persson 1986)

	seed:yield	
Crop	Bjäresjö	Beden
Barley	1:4	1:4
	1:5	
Rye	1:4	1:2.5
	1:5	
Oats	1:2.8	1:4
	1:3.5	

Table 5. Annual seed input and content of N and P (in seed; N = 1.7% P = 0.25%). Annual production of cereals (kg grain ha^{-1}) and total amount of N and P in harvested above ground biomass (in grain + straw: N = 1.05% P = 0.25%) and in the annual manure input. Losses of N and P in manure due to volatilization and leaching during storage were 50% and 20% respectively. I = per annually cultivated area. II = per total field area. Ranges for yield and manure input figures refer to minimum and maximum figures on estimates of seed:yield ratios and number of cattle kept in the villages.

Annual seed input	Bjäresjö		Beden	
	I	II	I	II
Grain, kg ha^{-1}	239	-	235	-
N-tot kg ha^{-1}	4.1	2.0	4.0	1.1
P-tot kg ha^{-1}	0.6	0.3	0.6	0.2
Annual cereal production	I	II	I	II
Grain, kg ha^{-1}	858-1072	428-534	842	239
N-tot kg ha^{-1}	18.0-22.5	9.0-11.2	17.7	5.0
P-tot kg ha^{-1}	4.3-5.4	2.1-2.7	4.2	1.2
Annual manure input	I	II	I	II
N-tot kg ha^{-1}	8.6-14.4	4.3-7.2	34.9-45.1	9.5-12.3
P-tot kg ha^{-1}	2.6-4.4	1.3-2.2	10.9-13.8	3.0-3.8

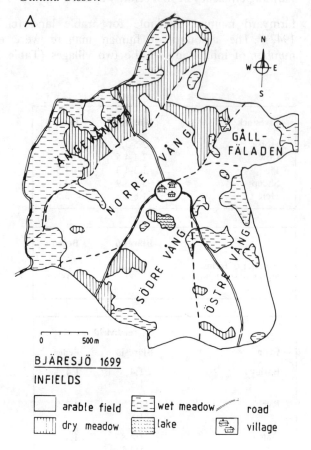

Fig. 5. Land-use map for
(a) Bjäresjö 1699 and
(b) for Beden 1750. ▶

Calculations of nutrient-use efficiency

Nitrogen- and phosphorus-use efficiency for the two villages were calculated as quotients between the total N- and P-content in annually harvested crops and added N and P amounts in manure. The nutrients in the seed-input were subtracted from the nutrient content in the crops (= grain + straw). Thus, estimates of leaching and runoff losses for the different soil types could be avoided. Losses (volatilization and leaching) due to inefficient storage and distribution of manure were estimated to be 20% for P and 50% for N (cf. Bondorff 1939; Schrøder 1985). Nitrogen and P contents in harvested biomass were obtained from Mengel & Kirkby (1981) and Fitter & Hay (1983). Nutrient contents in manure were obtained from Jansson (1956), Bengtsson & Kristiansson (1955), Grahn & Hansson (1959) and Steineck (1984). Both the total area of the arable fields, and the annually cultivated area were used as a basis for the calculations (Table 5).

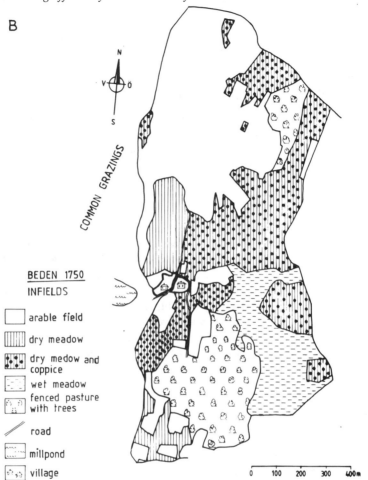

B

COMMON GRAZINGS

BEDEN 1750
INFIELDS

| | arable field |
| dry meadow |
| dry medow and coppice |
| wet meadow |
| fenced pasture with trees |
| road |
| millpond |
| village |

0 100 200 300 400m

Results

The distribution of crops in percentage weight of annual production differed in the two villages. For Bjäresjö rye and barley dominated, each contributing 37% of the yearly total (Table 6). In Beden, barley was the dominant crop followed by oats and buckwheat while rye was a minor constituent (Table 6). This is an effect of the different rotation systems used but is also linked to the different yields in the two villages (Table 4). The total seed input for the different crops was the same in both villages. There was, however, a difference in seed input between crops and since the relative distribution of crops differed, the mean weight of seed input (Table 5) is not exactly the same for Bjäresjö and Beden.

Table 6. The fraction
(% weight) of different crops
of the total annual production
in Bjäresjö and Beden.

	Bjäresjö	Beden
Rye	37	19
Barley	37	45
Oats	26	36
Buckwheat	0	

Table 7. Nutrient-use
efficiency (NUE) for N and P
in the infields of Bjäresjö and
Beden.

	Bjäresjö	Beden
N-NUE	0.97-2.14	0.32-0.41
P-NUE	0.82-1.85	0.26-0.33

The annual yield of grain per unit of the annually cultivated areas was about the same in the two villages, but when expressed per unit area of the total arable land the yield was twice as large in Bjäresjö as in Beden (Table 5). The amount of added manure was, however, almost 4 times as high (calculated per unit of cultivated area) in Beden as in Bjäresjö (Table 5).

The NUE was three- and four-fold higher for P and N, respectively in Bjäresjö than in Beden (Table 7). This implies that the production of an equivalent amount of cereal in the two villages demanded 4 times greater input of manure and a double field area in Beden than in Bjäresjö.

Discussion

Soil factors

One explanation for the great differences in NUE between the two villages may be the different soil fertility (sandy soil *vs* lime-rich, clayey till). The sandy soil in Beden is much more susceptible to leaching than the clayey till soils of Bjäresjö. From modern studies, a doubling of leaching rate for N can be expected for a sandy soil (Gustafson 1982; Brink 1983).

The lower yield figures for Beden can also be related to a greater susceptibility to harvest variations due to short-term climate fluctuations. The easily-drained, sandy soils and the greater slope of the fields probably rendered them more susceptible to drought during dry summers. The clayey till soils in Bjäresjö have a much better storage capacity for water and nutrients, hence reducing variation in yields between years.

Another feature of the sandy soil in Beden might be a smaller content of earthworms. Generally, earthworms have a positive effect on the soil by increasing porosity, aeration and microbial activity. Evidence of the beneficial effect of earthworms on crops is presented by Graff & Makeschin (1980). Earthworms, however, do not thrive in sandy soils with low (<5) pH (Lofs-Holmin 1983).

Phosphorus is less mobile in the soil than N but there is a risk of soil erosion, especially from clay soils with substantial slopes (Ulén 1985). The slopes of Bjäresjö are, however, only moderate and the small (1-1.7 kg ha^{-1}, cf. Table 5) deficit in P can be expected to be covered by release from the soil pool. The high P-efficiency in Bjäresjö can thus be referred to a combination of three factors: (1) clayey soil, (2) fields with only moderate slopes, and (3) short fallow periods.

Cropping system

The extensive cropping system used in Beden had long fallow periods of 6-8 years. Manure application in the last fallow year was followed by 2 or 3 years of cultivation. This probably resulted in inefficient use of the added nutrients since manure has a relatively slow decomposition rate (Widdowson *et al.* 1982). During the first year after application, only 50% of the manure is decomposed; even after 8 years 20% still remains undecomposed (Hansen & Kyllingsbaek 1983), suggesting that mineralization of N and P from the manure probably continued during the following fallow period. Consequently, much of the nutrients in the manure was not directly available for the crop. Some was leached during the first year of its application on to the fallow, and still more of the mineralized N was utilized by weeds in the following fallow cycle.

The use of a more intensive cropping system would probably have resulted in a higher NUE, but would have yielded only a small increase in production per cultivated area. Considering the greater demand for labour necessary to practise an intensive cropping system, and the small increase in yield, the 2 families living in Beden probably optimized their available resources.

Additional nitrogen sources

The high NUE for Bjäresjö cannot be explained solely by its better soils and different cropping system. Although clayey soils preserve nutrients better, there must have been substantial annual leaching of the manure-nitrogen. Modern figures for equivalent soil types and climatic regions are *ca.* 15 kg ha^{-1} yr^{-1} (Brink 1983). Obviously, therefore, there must have been additional N sources. In pre-industrial times, the contribution of N from wet and dry deposition would have been of minor importance although only scattered data exist: Rothamsted, UK, 1877-1919, 4.5 kg N ha^{-1}; Jönköping, Sweden, *ca.* 1900, 5.4 kg N ha^{-1}; near Copenhagen, Denmark, 1880 *ca.* 1925, 8-14 kg N ha^{-1} (Bondorff 1939). Considering the uncertainties in the analytical methods used and the lack of direct evidence from the 18th

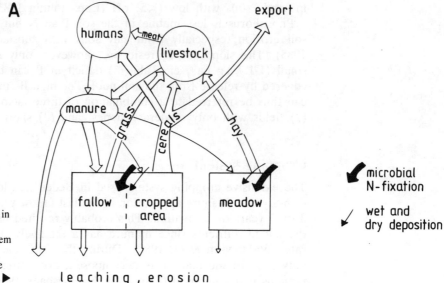

Fig. 6. Nutrient flow in
(a) the cereal-based system in
Bjäresjö and
(b) the livestock-based system
in Beden. The size of the
arrows indicates the relative
magnitude of nutrient flow. ▶

century, it seems reasonable to believe that the contribution from
air-deposited N could have been, at most, 4 kg ha^{-1} yr^{-1}.

Finally, biological N-fixation must be considered. This includes
symbiotic root-nodule bacteria of legumes and free-living soil bacteria
and blue-green algae. Pure stands of white clover (*Trifolium repens*
L.) can fix 150 kg N ha^{-1} yr^{-1} (Gray & Williams 1971). There is no
evidence of regular cultivation of legumes in the two villages during
the time period studied, although Dahl (1942) mentions the infrequent
cultivation of pulses in the area. Linné (1749), while travelling in this
province, described the fallows as overgrown with leguminous weeds
such as *Ononis* spp. Gaillard & Berglund (this volume) found a high
frequency of both *Trifolium-* and *Ononis*-type pollen in Bjäresjö lake
for the period 1200-1800 A.D. These pollen values can be directly
related to the infields of Bjäresjö village since the catchment area of
the lake corresponds to the infield area. The symbiotic N-fixing bacteria
have a pH optimum around 6.5-7.5 (Etherington 1975). On the sandy
and more acid soils at Beden it seems less probable that symbiotic
N-fixation was important.

Recent investigations of the importance of free-living soil bacteria
report greater N-fixation in clay soils (3-9 kg N ha^{-1} yr^{-1}) than in sandy
soils (1-3 kg N ha^{-1}) (in Hansen & Kyllingsbaek 1983). The activity
of these bacteria is negatively affected by low soil moisture (*op. cit.*)
again putting the sandy soils with low water-holding capacity at Beden
in an unfavourable position.

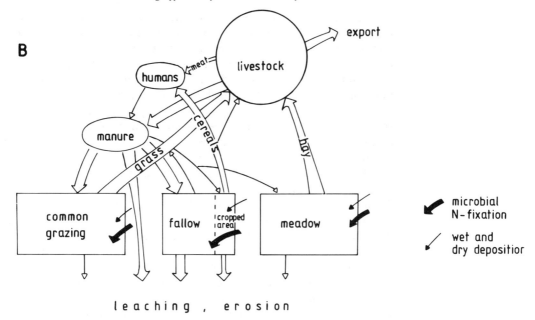

I suggest that the annual N-deficit in the Bjäresjö arable system, calculated to be *ca.* 10 kg ha-1 (Table 5) was compensated by (1) wet and dry deposition (*ca.* 3 kg ha-1); (2) biological N-fixation by free-living bacteria (*ca.* 3 kg ha-1); and (3) biological N-fixation by leguminous weeds and infrequently cultivated pulses (>4 kg ha-1).

Chorley (1981) proposed that the major N-source for crops before the introduction of legume cultivation was non-symbiotic fixation, mainly by blue-green algae. I have found little support for this, but it seems evident that within temperate, pre-industrial agro-ecosystems with low productivity and a low amount of N being recycled, the contribution from biological N-fixation was crucial for maintaining productivity over several centuries.

Conclusions

Comparing the entire system in the two villages (Fig. 6), it seems that there were substantial differences in environmental resources. Bjäresjö, settled in prehistoric time, had a much more fertile soil, and hence more favourable conditions for insurance against long-term nutrient exhaustion. Five hundred years of an intensive cropping system (albeit with a low yield) had been practised. Nutrient depletion from the arable fields was probably balanced largely by biological N-fixation. The poorer soils in Beden would not have permitted the same practice over such a long time period. By the beginning of the 18th century, it seems that the Bjäresjö system had depleted its resources. This was due to the conversion of common grazings into arable land. Therefore in summer the animals grazed the fallows and the stubble. This limited the number of animals and thus the amount of manure-nutrients available (Fig. 6a). The possibilities of increasing the yield in the absence of artificial fertilizers within this system were thus very small. Introduction of leguminous ley cultivation would have been one possibility, but it took another 80-100 years before it was commonly practised in this area (cf. Osvald 1962).

For the Beden system, the greater availability of manure was not reflected in larger yields, probably due to unfavourable soil conditions. There were probably some possibilities for increasing the yield by adopting a more intensive cropping system and at the same time devoting a greater proportion of the annual production to winter crops, which could act as catch crops for the mobile N in the manure (cf. Gustafson 1985). The environmental resources in the form of meadows and common grazing might have permitted a certain increase in livestock, at least in combination with grazing the fallows (Fig. 6b). On the other hand, the poor soils at Beden with their low clay content and low pH had small reserves of organic N and P and a low capacity to release those nutrients by mineralization. A shortage of P might perhaps have most limited the crop production.

Summary

(1) The ecological implications of using two different cropping systems, the three-course rotation which utilizes the main part of the arable land for annual cultivation, and the two-course rotation system where major parts are in fallow, were investigated for two villages in Scania, southern Sweden. The time-period chosen is the mid-18th century.

(2) For the two villages, based on grain production or animal husbandry, respectively, the following questions are considered: (i) Are differences in settlement patterns and farming practices for villages related to different availabilities of ecological resources? (ii) How was the productivity of the arable fields maintained in areas where the stock size was kept low due to heavily reduced fodder areas? (iii) Was the focus on animal husbandry which produced relatively large amounts of manure, reflected by greater productivity of the arable land compared to the mainly grain-producing village?

(3) Nutrient budgets for nitrogen and phosphorus were constructed for the two villages. Nutrient-use efficiency (NUE) was calculated (nutrients in harvested crops:added nutrients in manure).

(4) For the grain-producing village, N-NUE was 0.97-2.14 and P-NUE

was 0.82-1.85. For the livestock-economy village, N-NUE was 0.32-0.41 and P-NUE was 0.26-0.33.

(5) The annual cereal yield per annually cultivated area was about the same in the two villages (*ca.* 850 kg ha^{-1}).

(6) The yield per total arable area was twice as large in the three-course rotation village as in the two-course rotation village. Almost four times as much manure was added per unit of arable land in the latter village. Hence, to provide the same amount of cereals about four times as much manure and twice as much land was needed in the two-course rotation village with livestock economy as in the three-course rotation village with cereal economy.

(7) The cereal economy resulted in a net N loss of about 10 kg ha^{-1}. This was probably replaced by biological N-fixing sources, particularly legumes.

(8) When the environmental conditions of the two villages are taken into account, the results support the hypothesis that environmental resources are principal constraints in the development of pre-industrial agrarian activity.

Acknowledgements

The study is part of the Ystad project, and was funded by the Bank of Sweden Tercentenary Foundation, I am indebted to A. Persson for generously providing unpublished data on numbers of cattle and inhabitants in the villages and seed/yield data for one village. S. Karlsson, A. Persson and H. Staaf made valuable comments on the manuscript. M. Varga drew the figures.

The Ecological Basis of Highland Peasant Farming, 1500-1800 A.D.

Robert A. Dodgshon

The case for Highland land pressure

Descriptions of the Scottish Highlands and Islands during the late 18th century depict the region as suffering from acute land pressure. Yet with limited pockets of settlement surrounded by vast areas of uninhabited waste, the problem can easily be overlooked. What mattered was society's struggle against the insufficiency of land suitable for cultivation rather than any gross shortage. By the 1790's, not only were there many estates on which the amount of arable per family was barely adequate for their subsistence in a good year, but also a high proportion comprised land that - by any definition - was marginal, suffering from a combination of high relief, broken, rock-strewn surfaces, thin, waterlogged and acidic soils, and wet, windy climate. Despite the region's greater suitability for livestock production, local farm economies revolved around the cultivation of such land, employing an almost garden-like technique in return for poor, uncertain yields. If their farm strategy could be summarized in a sentence, then it would be that they sustained their limited and precious reserves of arable by drawing on what they had in abundance, namely labour and the resources of their non-arable sector.

Environmental limitations

Land pressure is a two-sided relationship. On the one hand, there are the physical limitations of the region and the severe restrictions that they impose on cultivation. Although we can speak of soils that are relatively fertile, like those developed on the Old Red Sandstone of Orkney or the basalts of Skye, and of soils that are inherently infertile, like those developed on Lewisian gneiss, the question of whether land was cultivated prior to 1800 depended more on general site conditions, including soil depth and drainage and steepness or elevation of land. Acting together, such conditions greatly fragmented the possibilities for cultivation. Early farmers responded by being wholly opportunistic, confining their efforts to a patchwork of sites that offered some positive advantage, so that we find them not only exploiting the better-drained, more easily-worked soils on the lower

slopes of the broader straths or raised-beach platforms, but also seeking out sites where wind-blown shell-sand had offset the acidity of
peat soils, where an alluvial fan offered a depth of soil that was
otherwise lacking or where a pocket of fertile soil remained. Around
such sites settlements developed that were either part of a narrow
string of townships or part of a cluster that fanned out from a cultivated core towards more marginal ground. If we take the assessed area
of townships - or that part measured in terms of customary land units -
as denoting their Medieval extent, then it is clear that these restricted
corridors or pockets constituted the maximum extent of cultivation at
ca. 1500 (Dodgshon 1981, 1983). But no less important in shaping the
character of resource exploitation was that beyond these strips or
pockets of easily worked soil lay ground which was capable of cultivation, provided there was sufficient investment of labour in preparing
the soil and provided there was an acceptance of significantly lower
returns. In effect, such ground acted as a high-pressure safety valve,
an area that ordinarily would not be cultivated but that could be if
communities accepted its steeply pitched opportunity costs.

Evidence for population growth

As this last point implies, no matter how much the environmental
limitations of the region are stressed, they alone do not generate land
pressure. It also requires conditions in which townships were able and
willing to meet the considerable energy costs demanded by cultivation
of marginal ground. I want to argue that such conditions developed
over the 16th, 17th and 18th centuries. The primary cause was an
increased demand for land brought about by an overall increase in
population. We cannot evidence this increase directly as census data
for the period are lacking. For the 16th and early 17th centuries, we
are forced to infer increased demand from the creation of new settlements and the expansion of arable into outfield (Dodgshon 1981,
1983). In a subsistence economy, both trends can be taken as reliable
indicators of a growing population. Indeed, an early 17th-century reference to Shetland depicts it as already suffering from acute overpopulation (Peterkin 1820, no. 111). During the late 17th and early
18th centuries, lists of inhabitants and tenants provide a more direct
if localized measure of increasing numbers. For instance, lists of male
inhabitants in Rannoch (Perthshire) show an increase from 104 to 163
between 1695-1745 (SRO GD150/156). The first census of sorts was
compiled in 1755 by Dr Webster (Kyd 1950). When compared with
the first official census of 1801, it reveals further increases over the
second half of the 18th century. The population of Ardnamurchan and
Sunart, for instance, grew from 1719 to 2552 between 1755-1801. In
fact, we can see the growth of Ardnamurchan and Sunart's population
over a still longer period, for a local census in 1723 gave its population

as 1352 (Murray 1740, plate VI). Tiree's population grew from 1509 in 1755 to 1676 in 1768, and 2443 in 1792 (Kyd 1950; Argyll Estate 1964; OSA 1794, vol. X). Of course, we cannot assume that these increases were continuous or sustained, only that overall, the period 1500-1800 saw a cumulative growth in numbers (Flinn 1977). This does not provide a sufficient explanation for land pressure. We still need to explain why land and society functioned as a relatively closed system so that growth was absorbed locally despite increasing cost. In particular, we need to clarify why, when faced with exploitation of increasingly marginal land, outward migration did not operate to regulate numbers at a more optimum level and why, for their part, landlords allowed the increase to accumulate so freely on their properties since - in the circumstances - any constraint on the sub-division of holdings or colonization of waste would surely have caused the average age of marriage to have risen and birth rates to have fallen.

Role of the clan system

The explanation lies in understanding how the clan system affected the relationship between society and land. It has been said that the clan system was concerned as much with the cultivation of men as of land. What clan chiefs desired most was a large and thriving tenantry. This gave them the socio-political advantage of a large following and enabled them to gather in vast quantities of grain and livestock in the form of food rents and feudal dues. Little of this produce was marketed prior to the late 17th century. Instead, chiefs used it to further enhance their socio-political status. There were three ways of doing this: (1) they could use it to support a personal retinue of followers, including fighting men, armourers, harpers, pipers, poets and so on; (2) they could use it to sustain lavish displays of hospitality and feastings; and (3) they could use cattle in payment of tocher-gude or bridewealth, contracting favourable marriage alliances (Dodgshon 1987). In these different ways, clans with access to substantial areas of fertile land, e.g. Islay, were especially well placed to build an elaborate hierarchy of power around themselves. Such complex chiefdoms were ephemeral and can be seen as cyclical eruptions and collapses of power (Ekholm 1977; Freidman 1982). The Lordship of the Isles, the last chiefdom in the Hebrides that can properly be described as a complex chiefdom, collapsed in the 1490's, when central government finally asserted its authority over the region. In its place, there emerged - for a time - a pattern of lesser or petty chiefdoms, each competing for enhanced status through constant feuding and feasting. Recent anthropological work has described this sort of chiefly competition as "fighting with food" (Friedman 1982), for feuding invariably led to the theft of cattle and grain and their consumption *via* roasting and fermenting. By the early 17th century though, government legislation sought to undermine even this system. Chiefs were prevented from using their food rents to maintain a retinue of fighting men, each man being required to live by farming or a trade. The much abused custom of cuid-oidche - whereby tenants had to provide so-many nights of hospitality for the chief and his retinue - was abolished (Dodgshon 1987). Yet despite these attempts to undermine it, the basic ideology of the clan system survived. The saying that a man without a clan was a broken man just as a clan without land was a broken clan was just as relevant in the 17th as it was in earlier centuries. Even after the crushing of the '45 rebellion, when the political hopes of Gaelic Scotland were finally dashed, it could be argued that it was economic rather than political reality that did more to change the ideology of the clan system. In relation to my argument, this long persistence of special bonds between a chief and his tenants provides the closed relationship between society and land, and helps explain why landlords accepted the relentless increase in numbers and why, for their part, tenants were prepared to live with the rising opportunity costs that came with the colonization of marginal land.

Evidence for increased population density

That these trends led to acute land pressure by the second half of the 18th century is borne out by a variety of evidence. The most direct consists of contemporary references to the poverty of particular communities (Highlands 1750, 1898; Annexed Estates 1973), to the frequency with which poor harvests plunged them into sharp subsistence crises (Johnson 1971) and, most telling of all, to the grain deficits which remote areas like Strathnaver and Assynt experienced every year (Highlands 1750, 1898; Burt's Letters 1754, vol. 1; Macfarlane 1905; Sutherland Estate 1972). Exposing the root of the problem are the statistical data showing that the number of tenants per township had increased (e.g. Duke of Argyll 1883) and that, despite the extension of arable, the amount of arable per family had fallen significantly over the century (Gray 1957). Even without taking into account the quality of land involved and the fact that much of it had to be delved with a spade or caschrom, the 53 acres of arable shared between 39 families (= 249 people) in Barrisdale during the 1760's must surely have left its inhabitants on the verge of survival (Statistics of Annexed Estates 1973; see also Duke of Argyll 1883). Equally indicative of the excess carried by estates was the increasing number of landless or those simply with a cottage, the so-called scallogs or cottars, a problem that was particularly acute on Skye (Gray 1957).

Methods employed to combat land pressure

The problems of increasing land pressure are also manifest through the sort of land colonized between 1500-1800, with patches of cultivation stretching up hill-sides, pushed out over water-logged ground and carefully shoe-horned into what Samuel Johnson (1971) called the "intricacies among the craggs". By the mid-18th century, there are also many instances of distant shieling grounds being cropped (Gaffney 1960). Nor was the marginality of recently-colonized ground the only indicator. Yields on old arable were reduced. Tiree, for instance, was long renowned as *terra ethica* or land of corn. Yet in 1699, Martin Martin (Martin 1716) talked of it being "less fruitful than formerly" and a survey carried out in 1737 for the Argyll estate similarly reported yields down and much of the arable infested with weed (CCR 1886, appendix A; cf. Sutherland Estate 1972; Walker 1980). The underlying cause was not lost on the estate, for it instituted plans to remove what it called the "supernumeries" though it did not begin to implement its plans until the 1780's (Argyll Estate, 1964, vol. 1). Its attempt to solve the problem by diversifying the island's economy into fishing was typical of other attempts to devise a solution over the closing decades of the century, one that recognized that subsistence farming was - by then - no longer sufficient in itself. Change of a different kind can be detected in the region's diet, with the adoption of the potato. By the time the parish reports for the Old Statistical Account were being compiled in the 1790's, potatoes accounted for over 50% of local diets in some areas (OSA 1793, vol. 8; 1799, vol. 20). As a crop that grows well in peaty, acidic soils and which, compared with oats or barley, could support a family on less land, the potato served to increase the ·resource potential of local farms (see Walker 1980). It is tempting to argue that this generated the more rapid growth rates of the late 18th century. A more likely interpretation, however, is that, as in Ireland, the rapid spread of the potato through the region was a response to, rather than a trigger for, late 18th century growth rates.

The myth of archaism

Turning to the question of how townships coped with the problem of growth during 1500-1800, we must look at broad strategies as well as specific techniques of husbandry. Here, I challenge the entrenched view that these strategies and techniques are invariably deep rooted, a measure of the archaism of the region's culture prior to 1800. This particular orthodoxy owes much to the thinking of scholars like Cyril Fox, who saw the upland areas of northern and western Britain as a refuge for early cultural traits. Fox (1932) talked of new cultural traits being absorbed alongside older traits, so that - in modern times - we find the two side by side. Subsequent scholars, however, have tended

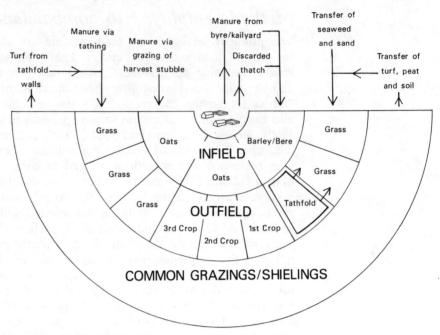

Fig. 1. Pattern of
infield-outfield cropping.

to mis-state his argument and to impose a blanket of cultural archaism over regions like the Highlands and Islands. Early travellers helped foster this view. When A. Campbell visited the region in 1802, he had hardly stepped foot inside the Highlands when he began talking about "rude implements", "dwarfish" cattle and a people "in the unpolished state of infant society" (Campbell 1802, vol. 1). The strategies and techniques of husbandry discernable between 1500-1800 have been forced into this interpretative framework. In other words, the region's response to increasing land pressure is seen as an extension of time-honoured methods. Communities are conceded no resourcefulness or ability to shape their husbandry to circumstance. There is, in fact, a case for arguing differently and for seeing key strategies and techniques of husbandry as an adaptation to growing land pressure.

The creation of outfields

The most fundamental strategy adopted was the creation of an outfield cropping sector. In all townships, arable appears divided into two sectors by the 18th century: infield and outfield (Fig. 1). We can treat their differences in terms of husbandry as an attempt to match the varying potential of cultivable land with different levels of exploitation. Infield, usually the smaller of the two, was cultivated intensively, receiving all the manure available from the byre together with the manure produced by the grazing of harvest stubble, plus a variety of fertilizers or manurial inputs that ranged from discarded roof thatch to seaweed. This preferential treatment had a purpose, for infield was maintained in permanent cultivation, yielding crop after crop of oats and bere (= barley). Early descriptions of arable as "manurit land" confirm that this emphasis is not misplaced (Monro 1961; Macfarlane 1905). Outfield was a more extensive system of cropping. Each year, part would be brought into cultivation and subjected to 3 or 4 years of cropping before being abandoned back to grass. In any one year, over a third might be in crop. The only manure that it received was provided by tathing or the nightly penning of stock on that part of outfield scheduled to be brought into cultivation, with the tath-folds - or enclosing walls constructed of turf - also being ploughed into the soil before cultivation.

The standard view of infield-outfield has long been that it combines the two most primitive types of husbandry, shifting and constant cultivation. By extension, it is argued that since outfield is the more primitive, then it must pre-date infield, the system *in toto* evolving out of an initial pattern of shifting cultivation through the creation of a more intensive sector permanent cultivation. So well entrenched is this assumption of antiquity that even prehistorians now freely talk of outfield systems. Yet against this, I argue that if we concentrate on those systems which, historically, are labelled as infield-outfield, they offer no support for such a conclusion. If we look at Scottish evidence - which has always been accepted as the most explicit - we find that infield and outfield were, first and foremost, distinctions in the basis of land tenure. Infield was that part assessed as arable during Medieval times. By this I mean it was an area of the township specifically and rigidly defined as arable and that, in return, bore a measured burden of feudal rents and dues. In the Highlands and Islands, it invariably delimits strips or pockets of good land. Outfield, meanwhile, was the difficult, marginal land that lay outside. The earliest charter references date from the 15th century, but do not become commonplace until after 1500. It represents an expansion beyond the bounds placed on arable during the Medieval Period. The breaching of these bounds, with all its significance for customary rents and dues, can be seen as a prime strategy for coping with the pressures of rising demand between 1500-1800. That it involved different concepts of tenure is underlined by the fact that whereas existing arable was held in return for food rents, new land or outfield was held for a cash augmentation of rents (Dodgshon 1981).

Explaining the character of outfield cropping

This still leaves us with the problem of explaining why it was cultivated on a shifting basis. The answer lies in establishing the resource strategy open to townships at the point when they breached their assessed land. As an area rigidly delimited as arable, we can expect assessed land, or infield, to have been intensively cultivated before tenants took in new land and burdened themselves with an extra cash rent. Logically, one approach was simply to extend the system of cropping already practised on infield. In a system crucially dependent on manure though, this would have spread the resources of the byre more thinly. An alternative - and the strategy actually adopted - was to utilize the manure produced by stock during summer when they grazed the common waste or hill pasture (Fig. 2). What we find - and the details are widely documented (e.g. SRO, GD50/135) - is that stock were folded over-night on that part of outfield selected for cultivation. Once in cultivation though, such tathed land was removed from further manuring during the very time that stock were available on the common grazings. In other words, the waste of a potential resource for arable re-appeared unless the tath fold was moved to a new section. The consequence was that outfield cultivation became a shifting system, continually accumulating then exhausting the manure provided by tathing (Fig. 1).

Resource exploitation of non-arable sector

A further strategy was the funnelling of resources from the extensive non-arable sector of the township on to the limited arable sector (Fig. 1). Logically, the manurial resources transferred by stock was an important factor. There were, however, other more visible transfers. Some coastal townships drew on local reserves of machair, a calcium-rich shell-sand that provided an ideal additive for peaty soils (Darling 1955). Many also used seaweed (Brand 1701; Pococke 1886; Skene 1876; Fenton 1986) whilst a 16th-century account talks of fish being used (Skene 1876). Inland, other exploitable manures, such as turf, peat and soil, were harvested in abundance from rough pasture ground. Turf or feal was cut extensively for building, with domestic dwellings inter-leaving turf and stone in the construction of their walls and using turf as roofing material (e.g. Sutherland Estate 1972). An English officer, visiting Sutherland in the far north during the early 18th century, remarked that back home they built their houses of stone and pastured their cattle on turf but that here, in Sutherland, they built their houses of turf and pastured their cattle on stone (OSA 1793, vol. 8). Like the turf used in tath folds, turf used for walls or roofing was eventually re-used as manure (OSA 1793, vol. 8). However, townships also cut turf for more immediate use as a manure, so much so that not only do we find estates expressing concern over the way it seriously degraded pasture (e.g. SRO GD50/136/1), but there was even an Act of Parliament of 1685 expressly outlawing the practice. The Act named the eastern Highlands as its main area of practice (APS, vol. 8), but the large area of 'skinned' land or gearraidh

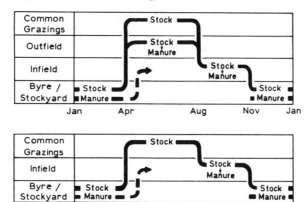

Fig. 2. Diagrammatic representation of the movement of livestock and the use made of their manure within a system consisting of infield-outfield plus common grazings (upper diagram); and within a system consisting of infield plus common grazings (lower diagram).

in the Western Isles confirms that there too, the cutting of turf was widespread (Darling 1955). Communities also dressed their arable with peat and soil quarried from their rough pasture, usually mixing it in with dung before applying it (Skene 1876, vol. 3). On the Breadalbane estate, tenants were actually required by the local barony court to carry "thrie score leadis of earth" to "thair middingis for guiding of ilk merkland" through an enactment passed in 1627 (Black Book 1855; see also Sutherland Estate 1972).

The increasing primacy of arable

A less obvious strategy concerned the balance struck between arable and livestock. Initially, or *ca.* 1500, the balance between them would have attuned, on the one hand, to the basic subsistence requirements of the peasant economy and, on the other hand, to the higher need of chiefs. Subsequently, this pattern of need changed. First, the cumulative increase in population between 1500-1800 caused the peasant economy to become concerned with arable as a first priority. It was arable that could best sustain any increase: though pincered tightly between limited supply and heavy demand, it could only do so if extended or intensified. By comparison, the abundance of hill pasture meant that its supply could be taken for granted. In short, the one was carefully husbanded, the other carelessly exploited. Second, as Highland chiefs were forced to disband their retinues and moderate their lavish displays of feasting, it created the problem of how the large quantities of grain and stock collected as rent could now be disposed. There are signs that some chiefs tried to market such produce as early as the mid-17th century (e.g. SRO, GD201/1/54). However, by the early 18th century, the main course of action was conversion of rents-in-kind into cash rents, thus transferring the burden of marketing produce on to tenants (e.g. SRO GD1332/541, GD128/

11/1, GD64/1/84). The latter's response was to raise the cash now needed by marketing a few cattle each year (Gray 1955). This had the effect of shifting the rent burden from grain and stock on to stock alone, enabling tenants to secure a greater proportion of their arable output for their own subsistence. But by dichotomizing their field economy in this way, townships demonstrated that the Higlands could be more profitable if filled with stock rather than people.

Intensive use of labour: hand-tool cultivation and lazy beds

The most salient feature of husbandry techniques in the Highlands and Islands between 1500-1800 is their labour intensiveness, with communities "Buying their comforts at a dear rate" as one commentator put it (Campbell 1802). The labour required by the transfer of additives and manures like sand, seaweed, peat, soil and turf from the non-arable to the arable sector of townships is indicative of this. No less revealing is the importance of hand-tool cultivation, with the ordinary spade, the caschrom (= spade with a long 'L'shaped or curved handle, a foot-rest on one side and a heavy but narrow blade) and the mattock being commonplace. As early as 1549, Donald Monro wrote of Toronsay in Argyll that "all the tilth is delved with spades, except so much as a horse plough will till" (Monro 1961; Martin 1716). "Delvit" land, or land worked with a spade was, in fact, present in many parts of the Western Isles at this date. An exact measure of its importance during the early 19th century is provided by figures for Ardnamurchan, where 2071 acres were reported as ploughed and 2063 acres as delved (NSA Argyll 1845). Though it gave no figures, a mid-18th century survey of Lewis reported an even greater preponderance of delved land (Walker 1980). Bound up with the use of spades and caschroms was the construction of lazy beds - narrow ridges of soil formed by the heaping of turf into narrow ridges. The technique was ideally suited to areas of shallow or water-logged soil for it created a deeper, better-drained soil. Their construction provided an opportunity for mixing in dung from the byre, sand, seaweed or peat with the soil. Given that the combined use of spades and lazy beds enabled cultivation to be extended out over difficult ground, picking out small patches of soil between rock outcrops or boggy land, there is a case for arguing that their use was a response solely to the cultivation of such poor ground. This, however, states only part of the problem. The social ecology of their use is equally relevant. The use of intensive, garden-like techniques presupposes that there was not only a pressing need to use them in the physical circumstances in which they were used, but also, a sufficient supply of labour to offset their cost in energy terms, a supply ensured by an increasing population and a general reduction in average holding size per family (cf. Walker 1980). The effort involved in the use of spades, caschroms and lazy beds certainly impressed contemporaries. A mid-18th century estimate suggested that, even allowing for the fact that the relatively light plough of the region made fairly heavy use of labour, cultivating land with the caschrom needed three times as much effort but increased output by only one third (Walker 1980). Clearly, labour was not treated as a scarce commodity. Labour-intensive techniques were not only more appropriate to the difficult ground now being colonized and to the smaller holdings now emerging, they enabled output to be maximized, if only in a marginal sense. In effect, increased population became its own resource, enabling the increased demand for food to be met - in part - by an increased investment of labour. As a 16th century report on Shetland put it, "the industrie of the poor labourers doeth exceed the fertilitie of the ground" (Peterkin 1820, no. 111), with the abundance of the one being used to offset some of the deficiencies of the other. Logically, when tenant numbers were eventually reduced and farm size increased over the 19th century, there occurred a reciprocal shift from spade to plough husbandry (e.g. Duke of Argyll 1883).

Intensive use of labour: harvesting techniques

Comparable points can be made about harvesting techniques. The sickle - an implement ideally suited to small plots of corn - was used throughout the region, but in some localities we also find grain being harvested by hand, an indication perhaps of poverty rather than primitiveness, with remote and crowded island communities having freer access to labour than to iron. Once harvested, such 'pulled' crops were burnt so as to separate the grain from the straw and to help dry it, a technique called graddaning (Martin 1716; Macculloch 1819, vol. 1; NSA Inverness 1845; Annexed Estates 1973), and one that clearly set a higher priority on obtaining and drying the grain than preserving the straw (see also Walker 1980). Parts of the Highlands used a small horizontal water mill (Fenton 1976) and querns were widely used for grinding corn, especially by what one observer called the "poorer" sections of society (cited in Fenton 1976). The use of the quern is one technique whose prehistoric origin is not in dispute, yet to stress its primitiveness above all else misses a vital point about the context of its use. The quern - like the horizontal mill - was particularly suited to the ecology of grain production that became more established over the period 1500-1800. So much so, that whatever the extent of its use *ca.* 1500, we might expect it to have become more widely used afterwards.

Land pressure as a study in adaptation

Once we acknowledge the link between cultivation of marginal land, surplus labour and hand-intensive methods, then any assumption about the blanket archaism of Highland husbandry becomes less plausible as an interpretation of its character. Key elements can be seen as a response to the growing pressure on land and the need to cultivate smaller holdings and more difficult ground between 1500-1800. Even when dealing with simple hand-intensive methods of husbandry, we should be more cautious when deciding what was long-established and what was a response to conditions of labour abundance and the cropping of marginal land. There is, for instance, no archaeological or historical evidence for the cashrom - the hallmark of the system - until the 16th century, whilst even the archaism of the lazy-bed has yet to be authenticated. In short, we must be wary of allowing the use of such methods to run ahead or out of step with the social conditions that made them feasible and the ecological context that made them necessary. Significantly, Irish sites where ridges of a lazy-bed type have been located in a prehistoric context show them submerged beneath peat, suggesting they were a response to growing environmental degradation.

The argument that Highland society first delayed, then coped with its problems of land pressure between 1500-1800, by making significant adjustments to its strategies and techniques of husbandry hardly squares with established descriptions of the region as culturally conservative prior to 1800. However, it stands comparison with anthropological work on chiefdoms that stresses their flux rather than stagnation, with some arguing for cyclical swings between simple and complex forms (Ekholm 1977; Freidman 1982). The work of Kirch (1980) is particularly relevant, for it underlines the importance of seeing such sociopolitical structures in their ecological setting. Basing his ideas on Polynesian chiefdoms, he argues that the organizational shift from simple to complex forms involved changes in their adaptive strategy or in the way they exploited their resources. A key part of his argument is the r-K selection continuum. As population expands in a region, it does so freely, without constraint. Its organization is shaped by r-selection patterns. However, as the limits of its environment are approached, we can expect K-selection patterns to come to the fore, with conscious regulation of population, greater competition between groups resulting in a more pronounced social hierarchy, tighter control over production and more economic specialization. Between these two strategies, we find a phase during which efforts are made to increase the carrying capacity of the environment by devising new, more intensive husbandries that invariably make use of the greater pool of labour available (Kirch 1980; Emanuelsson this volume). An obvious implication is that we must expect any drift towards a state of land pressure to produce qualitative and not just quantitative changes in the relation between land and society. Such ideas clearly have a bearing on the Scottish Highlands. Both regions functioned as relatively closed systems of population. Both experienced phases of growth that involved developing new strategies and techniques of husbandry and, ultimately, diversification. Their main difference was that by the late 18th century, the socio-political structure of the Highlands and Islands had been linked to a wider political and economic system, one that cut across existing relationships and adaptive strategies. The K-selection patterns that now came into effect were not associated with a kin-based system of chiefdoms, but with a market-based society in which commercial sheep production displaced peasant subsistence farming.

Summary

(1) Between 1500-1800, population in the Scottish Highlands and Islands increased substantially. How the region absorbed the increase is as much an ecological as an economic problem.

(2) Despite severe environmental limitations, township economies depended heavily on arable. Suitable ground had largely been colonized by *ca.* 1500, so any subsequent growth in arable had to cope with land that was generally steep, stoney and waterlogged.

(3) We can explain why communities accepted the high cost of cultivating such ground by acknowledging the ability of the clan system to create a relatively closed relationship between land and society, yet one within which there was a constant pressure for growth.

(4) Combined with the overall decrease in holding size, cultivation fostered new strategies of husbandry, with the extensive resources of the non-arable sector being used to sustain arable and a greater reliance on labour-intensive techniques.

Air Photo Interpretation and Computer Cartography - Tools for Studying the Changes in the Cultural Landscape

Margareta Ihse

Introduction

In order to understand the cultural landscape of today it is necessary to know how the landscape appeared previously. During prehistoric time human impact on the landscape was slight. Since then the landscape has gradually changed. For nearly 1000 years, from the Viking Age to the beginning of the 19th century, the organization of the landscape was very stable. Small villages surrounded by meadows and pastures dominated. However, in the early 19th century, villages were split up and farms scattered over the whole landscape. During the next 100 years the impact on the landscape slightly increased through the introduction of new agricultural methods, and cultivated areas were at· a maximum.

Since the Second World War, changes in the landscape have been very rapid and drastic. During this 40-year period, which is very short when seen in a historical perspective, modern agricultural techniques have given people enormous possibilities for drastically changing the landscape. Small biotopes such as fragments of former meadows and pastures, are particularly threatened because they are considered as obstacles to cultivation. These fragments are often important biotopes and the only refuges for certain flora and fauna.

This chapter describes a method for studying small biotopes and documenting the changes in them during the last 4 decades. The biotopes are interpreted in aerial photos from 3 periods, *ca.* 1940, 1965 and 1980, and the changes are presented and analysed by computer cartography.

The full implications of these changes are not yet known, but are being studied in joint research projects with the zoological, botanical, ecological and geographical departments at the University of Lund and the Agricultural University in Ultuna. Some consequences are discussed.

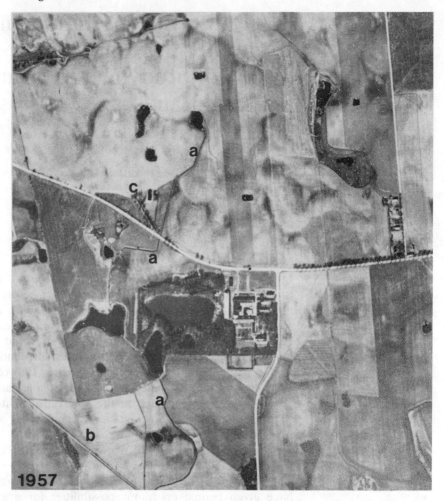

Air-photo interpretation

The memory of human beings is fantastic, but often too adaptable and therefore rather unreliable, at least for recalling unimportant things. Small biotopes have been considered unimportant and this indifference is reflected in the lack of statistics concerning them. We therefore need a tool to describe the biotopes and to document the changes in the cultural landscape. The air photo is such a tool.

Material

Since 1930 the Swedish Board of Land Survey has photographed the whole country about every 10 years. The central archive contains *ca.*

Fig. 1. Air photos from 1957 and 1978 showing many of the biotopes from the Jordberga estate. Original scale 1:30 000. Several small ponds (dark dots) have been drained and around several remaining ponds spruce has been planted. The linear elements have diminished due to subsoil drainage of the small river (a), the removal of the railway (b), and the removal of the road and house (c). The pasture around the river (a) has been converted to a cultivated field. In the lower right corner, some ponds for industry have appeared.

one million photos. The air photos are objective documents of the landscape at a particular time. For nearly every part of Sweden it is thus possible, using air photos, to study changes in the landscape at intervals of *ca*. 10 years.

The photos are principally used for photogrammetric purposes and the film is mostly panchromatic black-and-white. For different interpretation purposes infrared colour film (IR-colour film) has been frequently used in the last 15 years. The whole of Sweden will be photographed with IR-colour film during the 1980's. This film is superior for thematic interpretation, especially for vegetation mapping (Ihse 1978). Because single photos are difficult to interpret correctly, stereo pairs are used. This means that the landscape looks like a three-dimensional model when the photos are viewed in a stereoscope.

Information from air photos and classification system

The air photos are taken from an altitude of 3000-4000 m, which gives a scale of 1:30 000. The smallest object that can be detected is *ca*. 1 m. Linear elements of *ca*. 0.5 m can also be detected. It is thus not possible to detect threatened species, but it is possible to detect the biotopes or habitats associated with these species (Ihse & Nordberg 1984).

The amount of information presented in a single air photo is quite high, about 56 million pixels (picture elements). It must be sorted and interpreted according to a relevant classification system. Any interpretation must be based on previous experiences from air-photo interpretation and from ground truth as well as knowledge about the objects under study. For example, the classification system used for habitats of game consists of areal elements and linear elements, representing about 40 different units (Table 1).

Many of the detectable biotopes are shown in Figure 1. Cultivated fields, meadows and forests are all easy to detect in air photos. Cultivated fields often have straight borders and always have straight patterns due to cultivation practices such as ploughing, sowing and subsoil drainage. Grass fields are most often used as pastures. They appear bright and smooth, with an uneven structure resulting from

Areal elements

10 cultivated fields	51 'island' in field with spruce
11 meadow, dry-fresh	52 'island' in field with bushes
12 meadow, moist	53 'island' in field with trees
13 meadow, wet	61 mire, wetland
14 meadow, planted with spruce	62 pond with only grass bordering it
20 meadow, sparsely with trees	63 pond with bushes
31 bushes, dense	64 pond with coniferous trees
32 bushes, sparse	65 pond with deciduous trees
41 coniferous forest	71 house, garden
42 deciduous forest	72 park
43 clear-felled area	73 industry
	80 water
	99 not mapped

Linear elements

101 open ditches with grass	205 stonewall, lined with grass
102 open ditches with bushes	206 stonewall, with bushes
103 open ditches with trees	207 stonewall, with trees
104 open ditches with forest	208 forest edge, coniferous
200 road lacking verges	209 forest edge, deciduous
201 roadside, lined with verges, dominated by grass	
202 roadside, lined with bushes	
203 roadside, lined with trees	

Table 1. Classification system with digital codes.

trampling, uneven grazing and the occurrence of small paths. Coniferous plantations on former meadows can be first detected in the photos when the plants are about 1-1.5 m, which is equivalent to about 5 years old. Newly planted meadows cannot be detected. Bushes invading derelict pastures can be detected.

Remnant biotopes such as 'islands' of forest, bushes or bedrock in cultivated fields appear as dark dots on a brighter field. Small ponds, surrounded by trees, can be misinterpreted as being 'islands' of forest. Pond water cannot always be detected among the dark shadows of the trees. It is therefore difficult to make any quantitative estimation of these biotopes. Verges, even only 0.5 m wide, can often be distinguished as dark borders along bright smoothly curved roads. Depending on the time of photography, the smallest verges can be difficult to distinguish from crops that are green and the same height as the verge. Verges have probably been underestimated from the old photos.

Ditches appear as dark, straight lines crossing fields or following roads. Trees and bushes lining ditches make them still easier to distinguish because of the dark shadows they produce. Stone walls and hedges are also seen as straight, dark lines, but since their height is normally 1 m or more it is easy to distinguish them from ditches.

The periods of photography

I have chosen to study the landscape as it was (1) at the beginning of the 1940's; (2) during the middle of the 1960's; and (3) at the end of the 1970's or the beginning of the 1980's. These represent not only 3 occasions, but three different periods. The oldest photos represent the small-scale landscape from the 19th century. We can say that the cultural landscape of the 19th century with its methods and management did not really end until after the Second World War. In the photos from the mid-1960's effects of mechanization and rationalization are first discernable. The most recent photos from the end of the 1970's and the beginning of the 1980's show the landscape of today, where rationalization and mechanization have continued and created a broad-scale, monotonous landscape.

The advantages of air photos for the interpretation of biotopes and of changes in the cultural landscape are that they give (1) new information, not available elsewhere, such as statistics on ponds that have been filled in or on hedges that have been removed; (2) information about small areas, 0.5-1 m; (3) unbiased information from a specific area; and (4) repeated information from several occasions. However, it must be stressed that all interpretation is a qualified 'guess' and that it does not always give a precise classification.

Methods of interpretation

A combination of air-photo interpretation and digital image analysis is a very quick and efficient method for studying changes in areal elements.

Interpretation began with the most recent photos, the only ones for which it is possible to obtain ground truth. The interpretation was made with a Zeiss Jena Interpretoscope with an enlargement capability of 2-16 times. After ground control a second interpretation was made.

Three transparencies were made for each date showing (1) the areal elements, such as land-use, refuges and small biotopes; (2) the size of cultivated fields; and (3) linear elements (verges, ditches, stonewalls, hedges).

The linear elements and field size were completely mapped for all dates. Instead of mapping the areal elements over the whole area for all dates, the total distributions are only mapped once, on the most recent photos. This often means several hundred objects in one test area. By using this transparency as a model for the older photos it is easy to detect changes. The changes often concern only a third or less of the objects and the time needed to interpret this change will thus be very short. The maps showing the total distribution of the areal elements for every date are then made by combining the changes with the total distribution by digital image analysis.

The interpreted data were transferred to a manuscript map, the economical map (scale 1:10 000), before digitalizing.

Computer cartography

Methods

Map information can be digitalized and stored in a computer in two different ways: in vector or raster form. A vector-stored map has good resolution, preserved border lines and needs minimal storage capacity. A raster-stored map has less good resolution, distorted border lines and needs a higher storage capacity.

	1947	1967	1978
Number of fields	394	257	32
Total area (ha)	1352	1358	1375
Median size (ha)	3.4	5.3	10.4
Largest field (ha)	47.3	25.6	86.4
Smallest field (ha)	0.02	0.1	0.4
Total length of border zones (km)	178.5	153.5	115.4
Border zones per area (m ha^{-1})	132	113	84

Table 2. Statistics from the changes in field size at Trolleholm, Scania, from 1947 to 1978.

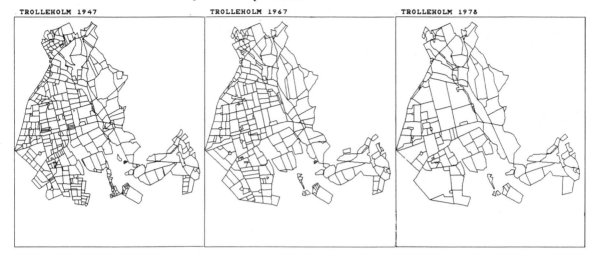

TROLLEHOLM 1947   TROLLEHOLM 1967   TROLLEHOLM 1978

Fig. 2. Field sizes on the Trolleholm estate from 1947, 1967, and 1978 interpreted in panchromatic air photos (scale 1:30 000) and digitalized.

With raster-stored maps it is easy to make computerised comparisons, pixel by pixel, between two maps to detect changes. Maps digitalized in vector form can easily be transformed to raster form by computer.

A combination of these two methods has been used in this study. The manuscript maps were digitalized manually in vector form. The vector map with 'areal elements 1978' and 'changes to 1967' as well as 'changes to 1947' have been transformed to grid maps. The total distributions of areal elements in 1967 and 1947 have been made by logical operation in the image analysing system EBBA. By other types of logical operations betwen the raster maps it is then possible to create new maps showing 'what is new' and 'what has disappeared'.

The program MIDAS, developed at the Department of Physical Geography, University of Stockholm, has been used. This program has routines for digitalizing, transforming to several orthogonal map projections, and computing many statistical functions, such as areas and lengths of border zones, grouping and sorting. MIDAS can also organize the coordinates for printing lists, different types of diagrams and figures, as well as colour maps (Alm & Nordberg 1985).

Field size

Digitalizing every field takes a long time. But compared to conventional areal measurements, digitalizing makes it possible to obtain not only the area of each field but also the length of the boundaries between fields, as well as maps. Changes in each field can be studied by comparing maps covering the different periods. In some areas, more than 20 fields in 1940 have become consolidated into a single field (Fig. 2). Statistics are a useful complement to the maps (Table 2). On the digitalized map it is also easy to group these fields into different

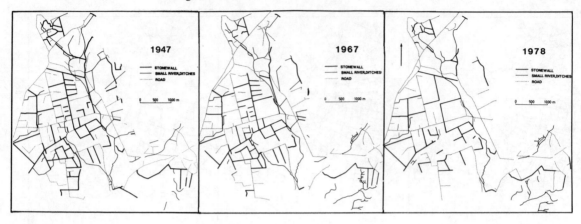

Fig. 3. Linear elements on the Trolleholm estate. The open ditches have diminished from 25.9 km (1947) to 25.6 km (1967) and 15.8 km (1978). The characteristic stone walls have also diminished, from 38.1 km to 31.5 km and to 23.8 km, which means a reduction of nearly 40% between 1947-1978. The road-side verges have diminished by more than 20%.

size classes. In some estates in 1940 two zones of field are clearly seen, one with large fields close to the main building, and one with small fields at the border. Rationalization is seen as larger fields spreading outwards from the main building to the border. The zone with large fields expanded in the middle of the 1960's and by 1978 nearly all the area consisted of large fields.

Linear elements

Open ditches, roadsides, stone walls and hedges are digitalized as linear elements into 14 classes (Table 1). These elements have rapidly diminished as seen in the maps (Fig. 3). Bushes and trees along ditches, roadsides and stone walls are important for stopping wind erosion and for animal protection, as well as in landscape appearance. With computer cartography it is easy to group all the elements with bushes and trees together. They are rapidly diminishing and only 7 of 82 km (±10%) of the linear elements had trees or bushes at the Trolleholm estate in 1978.

Land-use and biotopes are classified here into 24 different classes for special studies of the consequences of change for game (Table 1). Many different types of thematic maps can be made by grouping different classes together, or giving them different colours. Statistics can be given for the total or percentage area of the different classes. Maps can also be made with only one or a few parameters. One example is given in Figure 4, which shows only two biotopes and the changes in them. These biotopes, which are considered 'obstacles for agricultural techniques in the field', are 'islands' of forests and small ponds, which are important refuges for fauna and flora. They have diminished by *ca.* 50% .

The advantages of digitalized maps are (1) ease of obtaining statistics for the areas; (2) ease of making maps in different scales; and (3) ease of obtaining new information by combining the mapped parameters in novel ways.

Habitat changes and its consequences

Modern industrial farming on big estates creates a monotonous landscape. The last remnants of the cultural landscape created over several hundred years have been removed as 'obstacles for farming'.

Does it matter if the open ditches disappear, ponds are drained, or hedges and bushes are cut down? Yes, because these biotopes are required for the survival of many plants and animals. Small ponds give protection and food. 'Islands' of forest are often called 'the nursery of game'. Stone walls, hedges and open ditches are important communication and transport routes for plants and animals including humans. These linear structures are important from an aesthetic point of view as they create 'rooms' in the landscape and also because they inhibit wind erosion. Broad road-side verges provide a habitat for many plant species, which attract many insects, and thus their predators, so that a high diversity of birds and mammals is supported.

All these biotopes have diminished in the areas investigated (Ihse 1984, 1985; Ihse & Lewan 1986). Many modern farms no longer have

Fig. 4. Remnant biotopes, small ponds, and forest islands on the Trolleholm estate in 1947 and 1978.

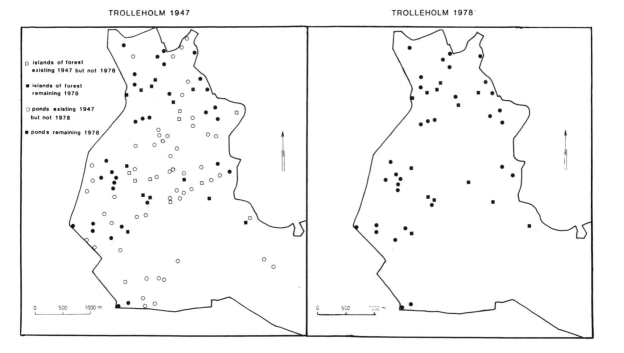

TROLLEHOLM 1947

TROLLEHOLM 1978

□ islands of forest
   existing 1947 but not 1978

■ islands of forest
   remaining 1978

○ ponds existing 1947
   but not 1978

● ponds remaining 1978

cattle and the change in appearance of the former pastures is drastic. Habitat change can be seen in the different land-uses of the 1940's and 1980's (Table 3).

The most important changes can be summarized as follows: (1) the sizes of the fields have greatly increased, 3 to 4 times; (2) 'islands' of forest have been removed or planted with spruce (up to 50%); (3) small ponds have been drained, in some areas more than 50%; (4) open ditch drainage has been converted to subsoil drainage (*ca.* 40%); (5) verges have become smaller or are entirely lacking and crops are grown up to the edge of the road; and (6) meadows have been converted into cultivated fields, planted with spruce, or overgrown by bushes.

How fast will these biotopes disappear? It is both difficult and uncertain to make a prognosis. However, the rate of disappearance of linear elements in the Trolleholm estate gives frightening perspectives (Fig. 5). The rate of annual change was less than 1% between 1947 to 1967, but accelerated between 1967 to 1978 to more than 3% per year. At this rate all linear elements will have disappeared before 2000.

How great then are the consequences? We do not yet know. But some examples from other research projects illustrate the problems. Over 290 of 700 wild, cultural-dependent plants are strongly threatened (Larsson 1986). The wider the verges, the greater the number of plant species found on them (Fogelfors 1979). A study at the University of Lund shows that large, monocultural fields threaten game, especially hares (B. Frylestam personal communication). The potential biotopes suitable for pheasants are calculated to be diminishing by nearly 50% at one estate (G. Gransson 1986). Strong dependence has been shown between the number of bird species and the different number of biotopes available. Cultivated fields without woods had 5-7 species of birds, cultivated fields with small deciduous woods had 12 species, and small woods with water had 16 species (Emanuelsson 1985).

Air-photo studies have illustrated the current development of the cultural landscape in which a broad-scale monotonous landscape only aimed at the production of crops has been created. Animals and

1940's	1980's
fresh meadows	cultivated fields
	deciduous forest
	coniferous forest
wet meadows	reed, bushes, trees
meadows with trees	deciduous forests
deciduous forests	coniferous forests
coniferous forests	clear-cuts

Table 3. Changes in land-uses between the 1940's and the 1980's.

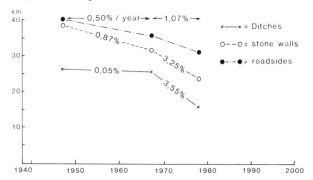

Fig. 5. The rate of disappearance of linear elements on the Trolleholm estate.

plants have nowhere to live. It is important to preserve the small biotopes, as they are small areas of great importance for providing a living landscape with space for animals, plants and humans.

Summary

(1) Modern agricultural techniques have drastically changed the cultural landscape since the Second World War. Small biotopes have rapidly decreased during these four decades.

(2) A method is presented for studying these changes by aerial photo interpretation and computer cartography. Three periods are investigated, *ca.* 1940, 1965 and 1980. Changes have been interpreted using colour photos from *ca.* 1980 (scale 1:30 000) and black-and-white photos from *ca.* 1940 and 1965. The interpretations have been digitalized and computer maps and statistics produced using the program MIDAS.

(3) The biotopes investigated are disappearing biotopes such as meadows, pasture lands and wetlands; remnant biotopes such as small ponds and 'islands' of forest and bushes in cultivated fields; and linear biotopes such as border zones around fields, open ditches, stone walls, hedges and verges.

(4) In addition, the size of cultivated fields has been studied. Some consequences for the flora and fauna of these changes are discussed.

Cultural Landscape Histories in Hungary - Two Case Studies

Dénes Lóczy

Introduction

Hungary consists of flat lowlands, dissected hill regions and low mountains up to 1015 m above sea level. Although the first impression of the lowlands may be of homogeneity, their landscapes are ecologically diverse (Pécsi & Somogyi 1983) with variations in lithology, landforms, soils, climate, vegetation and land-use. While a landscape unit is a fairly clearly delimited area with an individual character and a certain combination of environmental factors not repeated elsewhere, a landscape type is a more general category, poorer in details and only comprising major features, that can occur in a mosaic distribution (Pécsi 1977; Pécsi *et al.* 1971). The individuality of a landscape unit is reflected by its distinctive name. Landscape types are more difficult to define and all their main features must be listed. Landscape-type maps are useful in landscape planning, regional development and nature conservation.

This chapter assesses the extent to which human influence is responsible for the contemporary state of Hungarian landscapes and pinpoints the periods when human activities induced major ecological change. A historical study of the landscape also helps to evaluate future prospects.

As this chapter cannot cover the whole of Hungary, the history of two representative flood-plain landscapes is compared on the basis of their similar setting and parallel evolution from the time of human colonization to river regulation in the last century. Although case studies always imply regional restrictions, here they are assumed to be representative of the degree of cultural influence on the natural landscapes in lowland Hungary.

Geomorphological, archaeological, palynological, palaeohydrographical, ethnographical and historical-geographical evidence is used to reconstruct the landscape evolution over the last 10 000 years. The scarcity of archaeological finds and pollen diagrams has often forced me to rely on analogies from similar areas to provide the missing links in the chain of events.

When the present situation as shown on maps of landscape types

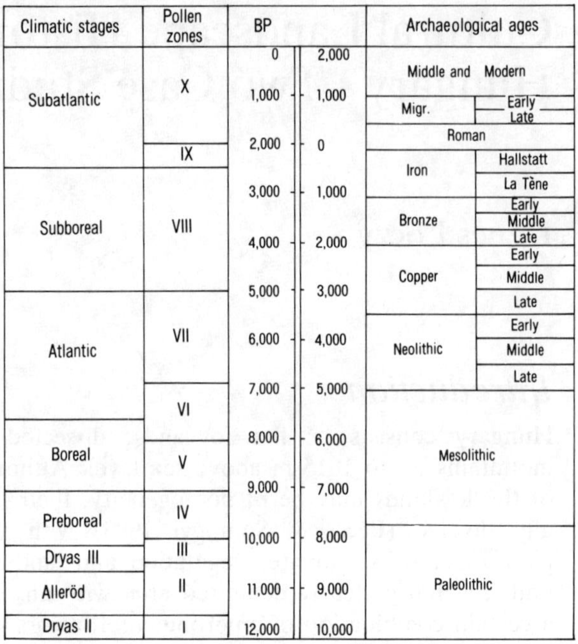

Climatic stages	Pollen zones	BP		Archaeological ages
Subatlantic	X	0 / 1,000	2,000 / 1,000	Middle and Modern
				Migr. — Early / Late
	IX	2,000	0	Roman
		3,000	1,000	Iron — Hallstatt / La Tène
Subboreal	VIII	4,000	2,000	Bronze — Early / Middle / Late
		5,000	3,000	Copper — Early / Middle / Late
Atlantic	VII	6,000	4,000	Neolithic — Early / Middle / Late
		7,000	5,000	
	VI	8,000	6,000	Mesolithic
Boreal	V	9,000	7,000	
Preboreal	IV	10,000	8,000	
Dryas III	III			Paleolithic
Alleröd	II	11,000	9,000	
Dryas II	I	12,000	10,000	

Fig. 1. Historical time-scales for the Carpathian basin, Hungary (after Járai-Kómlodi *et al.* in Tardy 1982; Zólyomi 1980).

and land-use (corrected from satellite images) is viewed against the historical background, the extent of utilization of the environmental potential can be evaluated. Future nature conservation and environmental management should relate to the history and the present state of the landscape.

Absolute dating techniques provide a timescale for ecological change in the Holocene of Hungary (Fig. 1).

The areas investigated

Lowland areas in Hungary occur mostly in two broad regions: (1) the Alföld (Great Hungarian Plain) on fluvial sediments of the Danube and Tisza and the tributaries of the latter; and (2) the Kisalföld (Little Plain) on deposits of the Danube and its right-hand tributaries (Fig. 2).

The Szigetköz ('interfluve of islands') is along the Danube. On the landscape-type map it is shown as a flood-plain that is partly flood-free (protected by dykes) and partly seasonally inundated on alluvial fans with cut-off channels (bar-and-swale terrain). It has alluvial, meadow alluvial or swamp-forest soils, mostly used for agriculture, and a moderately warm, humid subalpine and subatlantic climate. There are a few remnants of forests on the active flood-plain (Pécsi 1968; Ádám & Marosi 1975).

The Hortobágy (the name indicates infilled ox-bow lakes) represents the flood-plain along the second largest river, the Tisza. On the landscape-type map it appears as flood-plain areas separated by natural levées and sandy areas. It has mostly alkaline meadow soils and a moderately continental climate. land-use is pasture, horticulture and arable. There are a few small forest remnants.

Discussion

Since the areas investigated subsided and were partially inundated in the early Holocene, geographers propose that finds older than Neolithic cannot be expected (Somogyi 1984). Among the braided channels which now enclose the Szigetköz, however, some sporadic finds have been recovered from the Late Palaeolithic, indicating temporary settlement in the southernmost sandy parts of the area, which is also closest to the most important trading routes ('Amber Route') of later times (Gábori 1964). The first inhabitants were gatherers, hunters and fishermen; their impact on the natural environment was presumably small and now untraceable. Although the warm and dry climate during Mesolithic times is regarded as favourable for human settlement, no finds have been found. The first agricultural activities near Lake Balaton, 120 km south of the Szigetköz (Zólyomi 1980) were dated

Fig. 2. Location of the regions investigated.

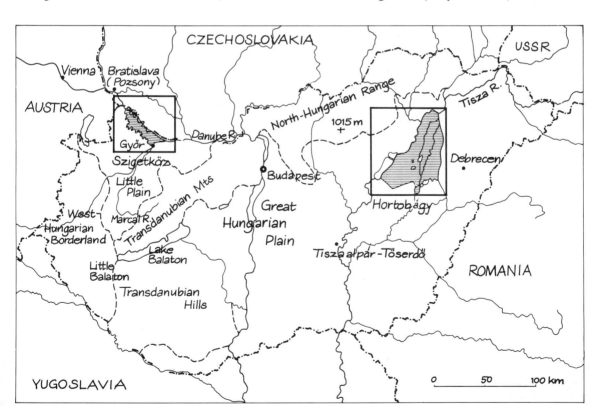

No	Natural type (4000-5000 B.C.)	Cultural (present-day)
1.	Low flood-plain willow and poplar woods (Salicetum albae-fragilis) a. *Rubus caesius* type b. *Cornus sanguinea* type	*Populus euramericana* cultivars plantations (ca. 60%) *Populus alba/nigra/canescens* (10%) *Salix alba* (20%) Others (10%)
2.	Flood-plain meadows *(Deschampsia caespitosa)* and pastures	Degraded into seminatural hay meadows (Arrhenatheretum, Alopecuretum pratensis, Festucetum pratensis)
3.	Wetland successions of cut-off channels, partly filled	Mostly intact
4.	High flood-plain oak-elm groves (Querco-Ulmetum hungaricum)	a. wet type: meadow soils - gardening (vegetables-cabbage) and arable (sugar-beet) b. dry type: chernozem soils - arable (cereals and maize)
5.	Areas of wind-blown sand (Festuco, Convalario-Quercetum roboris)	Sand and gravel pits, built up, gardens and orchards

Table 1. Natural and cultural vegetation in the Szigetköz (after Simon 1975).

No	Natural type (4000-5000 B.C.)	Cultural (present-day)
1.	Low flood-plain willow and poplar woods (Salicetum purpureae), swamps and boggy meadows	Wood remnants in active flood-plain, seminatural meadows
2.	Cut-off channels and backswamps - wetland successions, reed and sedge beds (Phragmitetum, Magnocaricion)	Commercial reed-beds, alkali 'puszta', pasture
3.	Alkali flats ('szikfok') - halophytic grassland (Puccinellio-Salicornetea)	Alkali 'puszta' (Puccinellietum limosae hungaricum), seasonal pasture
4.	Alkali plains with berms ('szikpadka') - halophytic forested steppe (Galatello-Quercetum roboris) with open meadows (Peucedano-Galatelletum)	Alkali 'puszta' a. heavily alkalized (Artemisio-Festucetum pseudovinae), fishponds and ricefields b. more continuous grass (Achilleo-Festucetum pseudovinae) pasture and rice-fields
5.	Loess-mantled terrain - forests (Aceri-tatarico-Quercetum), loess 'puszta' (Salvio-Festucetum-sulcatae) and scrub (Crataego-Prunetum fruticosae)	Cultivation (cereals, sugarbeet, sunflower)

Table 2. Natural and cultural vegetation in the Hortobágy (after Zólyomi & Simon 1969).

palynologically to 4500-4000 B.C., with small-size *Triticum* pollen at the first date and large numbers of cereals from the Middle Neolithic. The easy communication between Little Balaton swamp, from where the pollen was washed into Lake Balaton, and the southern corner of the Szigetköz, around Györ, along the basin of the Marcal river suggest that this dating may be applicable to the whole area. Apart from cereals, the expansion of *Fagus* seems to be closely correlated with cultivation. The first expansion is in the younger Atlantic and the second, with *Carpinus*, in the middle Subboreal, i.e. Middle to Late Bronze Age. As cultural impacts were, at that time, far from decisive in the landscape, the *Ulmus*-decline is probably unconnected with human activity (see Rackham this volume). Early deforestations are indicated by minor rises in NAP (Gramineae, Cyperaceae, *Artemisia*, Chenopodiaceae). Archaeological finds for the Neolithic in northern Transdanubia (Kalicz 1980) suggest large villages with long houses. Slash-and-burn farming and forest cutting are assumed. Direct evidence for settlement in the Szigetköz, however, is lacking. Floods of the shifting Danube channels on the alluvial fan might have destroyed settlements or their remnants (Timaffy 1980). The natural vegetation types on the different landforms from Atlantic times prior to permanent human settlement are listed in Table 1 (after Simon 1975).

Very little palynological evidence exists that can be used to reconstruct the prehistory of the Hortobágy region. The Tisza only adopted its present course at the Pleistocene-Holocene boundary (Somogyi 1975) and delimited the region in the north and west. The first temporary settlement is proposed (Somogyi 1984) in the Boreal with summer drought and steppe vegetation, and when part of the wetlands dried out and primary, natural soil alkalization took place. Pollen analyses by Járai-Komlódi (1966, 1969) were aimed at climatic and vegetation reconstructions. Unfortunately, not all the pollen diagrams cover the entire Holocene. It is well documented, however, that Atlantic times, beginning with the thermal maximum around 5500 B.C., brought a warm and humid climate, which resulted in the Great Plain becoming forested (Fraxino pannoniae-Ulmetum pannonicum association) with areas of steppe, groves and extensive swamp forests (*Salix, Populus, Alnus* - Table 2, after Zólyomi & Simon 1969). The first clear *Ulmus*-decline in the pollen diagram from the Tiszaalpár Töserdö swamp forest (100 km south of the Hortobágy puszta but occupying a similar position along the Tisza river) occurs at 3500 B.P. Unlike in Transdanubia, it is accompanied here by the simultaneous increase of *Fagus, Carpinus* and Cyperaceae pollen. Cultural plants and weeds (*Plantago, Rumex, Polygonum*) indicate widespread slash-and-burn farming at the end of the Neolithic. Archaeological data suggest forest clearance, possibly of lesser extent, since less timber was needed for low houses. The open grasslands were suitable for cultivation. Early deforestation resulted in secondary, man-induced

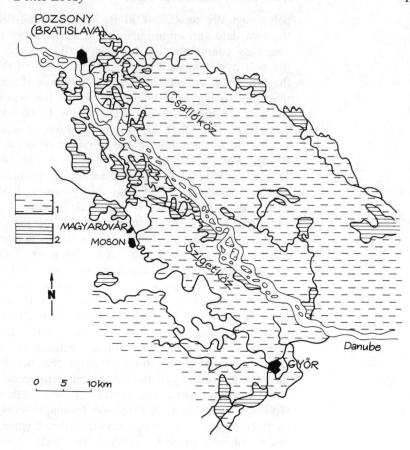

POZSONY
(BRATISLAVA)

Csallóköz

MAGYARÓVÁR

MOSON

Szigetköz

N

Danube

GYŐR

0   5   10km

Fig. 3. Hydrography of the
Danubian alluvial fan before
the flood-control measures of
last century (Magyarország
vízborította... 1938).
1 = areas inundated during
floods;
2 = areas inundated for most
of the year.

alkalization in the Hortobágy. Thus, a major feature of the present
environment (Marosi & Szilárd 1969) dates to this time.

On the loess-mantled margins and natural levées near water, cultivation may have started much earlier, as attested by extended 'tell'
settlements and Neolithic activities of the Alföld Linear Pottery Culture.

Under the gradually deteriorating climate of the Subboreal, reduced
carrying capacity only allowed animal husbandry as a main form of
subsistence. While in the Copper Age the Szigetköz was sparsely
populated between two cultural groups, the Hortobágy was the centre
of the Tiszapolgár culture with cattle, sheep, goats and pigs, in this
order of importance. Twenty four finds (Bognár-Kutzián 1972) indicate
that natural levées, 1.5 to 4 m above flood level, were the most
favourable sites for settlement (Pécsi 1968).

In the Early-Middle Bronze Age, settlement was still sparse in the
Szigetköz, but a second major clearance phase occurred in the Hortobágy (Járai-Komlódi 1966, 1969). Animal herding was suppressed by
the farming of settlers migrating from the east, and the Tisza Region
Tell Cultures flourished (Kovács 1977). Crops cultivated in the Horto-

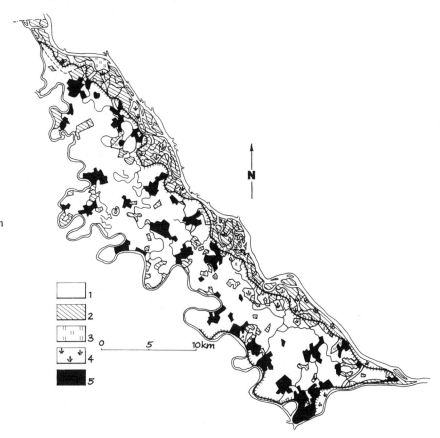

Fig. 4. Map of present-day land-use in the Szigetköz (from topographic maps and field survey revised from LANDSAT TM image for April 4 1984).
1 = arable land;
2 = forest;
3 = meadow and pasture;
4 = wetland (reed and sedge);
5 = built-up area, gardens, orchards, and vineyards.
Barbed line (⊢⊦⊦⊦⊦⊦⊣) shows flood-control dykes and pecked line (------) marks the extension of the projected Hrušov (Dunakrtvélyes)-Dunakiliti reservoir.

bágy included wheat, barley, millet, peas and flax. In the Late Bronze Age (*ca.* 1000 B.C.) the Hortobágy was invaded by the semi-nomadic tumulus Grave Culture people, who introduced horse-breeding into the Carpathian basin (Kovács 1977).

The influence of the Celts, Romans and the people of the Migration Period was manifest in the landscape encountered by the 7 tribes of the Magyar (Hungarian) Conquest (896 A.D.). These, however, were relatively insignificant compared to the changes brought about by the population expansion following the Conquest. The Magyars settled in the Carpathian basin during the so-called 'Little Climatic Optimum'. The Szigetköz, a pre-limes area during Roman times, was first used by the Magyars as a border area ('gyep') of swamp forests, but the first written document, from 990 A.D., tells about the settlement of frontier guardsmen. The first village (Vének - see Fig. 4) dates back to 1093 (Timaffy 1980).

In the Szigetköz proper farming spread from the south after the Conquest. In the Hortobágy puszta most land suitable for two and three field-system agriculture was converted into ploughland. The

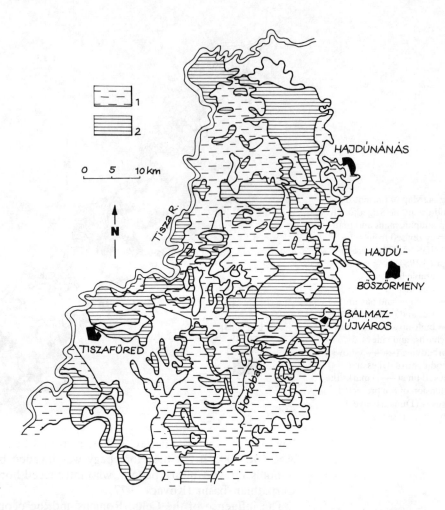

Fig. 5. Hydrography of the
Hortobágy before
flood-control measures. For
source and legend see Fig. 3.

preservation of the Ohat forest (known from the 1220's - Nagy 1976)
indicates that further expansion of arable land had been limited up to
the 17th century. There was even a decline in population and economy
in the wake of the Mongol Invasion (1241-42) coupled with devastation
by cattle-plague and locusts. Forests spread over some abandoned fields.

A common feature of Medieval history in both regions is the disast-
rous floods. The vicissitudes of human history almost totally avoided
the Szigetköz, but the inundations, recorded from 1242, remodelled
the settlement patterns from time to time (Timaffy 1980). For a long
time the main occupation was fishing supplemented by gathering,
hunting and gold-panning, activities with limited environmental impacts.
The early 19th century saw a peak of animal husbandry (partly
fodder-based), when the forest-belt along the Danube and the parallel

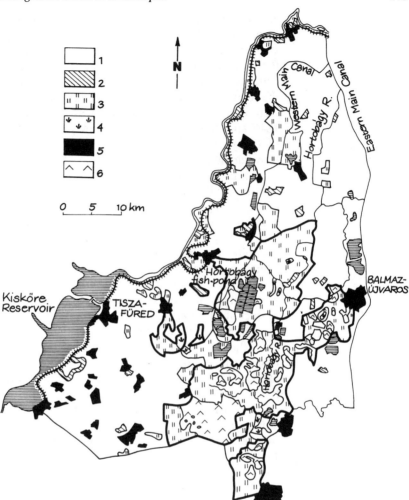

Fig. 6. Map of present-day land-use in the Hortobágy (from topographic maps revised from LANDSAT TM colour composites (bands 2, 3, and 4) for April 22 1985, and September 20 1985. For legend see Fig. 4 with the addition of 6 = alkali puszta. The boundaries of the Hortobágy National Park are indicated by a heavy line.

broad belt of meadows narrowing to the south, supplied large numbers of cattle, horses and sheep. Water birds were also important in the economy of this region with its numerous ox-bow lakes. The hydrography before river regulation is shown in Figure 3 and the present-day drainage in Figure 4.

Livestock breeding was flourishing in the Hortobágy by the 15th century with considerable exports of animals to western Europe (Gaál 1966). This boom was broken by the Ottoman Occupation (second half of 16th century to the end of the 17th century). In peace time, the reduced population resumed cattle breeding to pay the taxes levied by the Turks. For religious reasons, pig breeding was replaced by sheep husbandry, which had a negative effect on plant growth. After the reconquest of Ottoman areas, the population began to rise and arable land spread. The decline of traditional forests paralleled the spread of *Robinia pseudoacacia* L. imported from North America to stabilize sandy areas. *Robinia* forests soon became typical of the Great Plain landscape. The 11 devastating floods from 1730 to 1888, during which part of the discharge of the Tisza river ran to the south along the course of the small Hortobágy river, encouraged flood control. Before river regulation of the Hortobágy section of the Tisza between 1846 and 1857, a third of the area was waterlogged throughout most of the year and another third during floods (Fig. 5; cf. Fig. 6). The growing demand for wheat in the 18th century could not be satisfied by the produce from the limited ploughlands. Swamp drainage became a task of vital importance. The main occupations, however, remained cattle, sheep and horse husbandry (Ecsedi 1914).

River regulation was principally intended to prevent floods by accelerating runoff in channels between embankments (Somogyi 1978). In spite of their negative impact on river-bed sedimentation, the somewhat belated drainage measures and irrigation schemes gave rise to intensive farming over 1.5 million ha of arable land in Hungary. The course of the Danube changed very little in the Szigetköz. The floodcontrol dykes border a relatively wide active flood-plain with forests. Meander cut-offs were more significant along the Tisza river.

In the Hortobágy with summer drought, the fall of ground-water table induced upward cation movement in the soils and secondary alkalization spread over part of the drained areas. A decline in animal husbandry (cattle, sheep, horses, pigs, poultry) ensued. The attempt at converting large portions of public pasture into improved arable land at the turn of the century was not successful. Better use was made of the alkali puszta as fish-ponds and irrigated pastures from the 1910's, and for rice cultivation and reed cutting from the 1930's (Salamon 1976).

In the irrigation fields of the Szigetköz, with no danger of alkalization and with Vienna nearby, horticulture proved to be profitable. Cabbage gardens and orchards are still widespread around villages. In the Hortobágy *Quercus* and *Alnus* forests were planted to restore the natural vegetation, whereas in the Szigetköz *Populus euramericana* (Dode) Guineir plantations (cf. Table 1) marked the advent of commercial forestry.

Post-war social reforms had major impacts on the landscape. Collective farms, introduced in the 1950's and 1960's, had an important ecological effect: the small plots adjusted to the physical units, such as filled meanders or natural levées, merged into large fields consisting of various different habitats. At the same time, the cooperative farms were able to use machines, fertilizers and chemicals on broad scales to achieve higher yields.

Since 80% of the Hortobágy soils are solonetz type, their fertility can be increased through amelioration. Intensive cereal cultivation, however, is restricted to the more favourable margins, and the heart of the puszta is reserved for stock breeding, fodder production and, on the heavily alkalized sites, fishery and rice cultivation. Realizing the attraction and scientific value of typical Hungarian puszta, tourist and nature conservation authorities decided to establish the first Hungarian National Park in the Hortobágy in 1973. The aim was to preserve the alkali puszta as it was then, without reconstructing any of the physical conditions that existed before river regulation. A small portion of the Park is designated for tourists, in some other restricted areas the traditional occupations in the landscape are practised, while most of the puszta is open only for academic and educational purposes. The position of the Szigetköz did not favour similar conservation of the flood-plain forests.

The present use of the regions studied (Figs. 4, 6) primarily reflects changes in hydrography due to human intervention during the last 100 years. A common feature is that flood-control dykes appear as marked boundaries between forests (22% of the total area of the Szigetköz) and wetlands (3%) of active flood-plains and agricultural land. The contrast is best seen in the Szigetköz. Here broad-scale farming more or less obliterated the network of mostly infilled meanders and resulted in an almost homogeneous belt of field cultivation (58%) in the south-western half of the interfluve. Very few remnants of the once contiguous pasture belt (7.5%) persist along the dyke. Developments in the last decade, however, suggest that this zone should have had promising prospects (Góczán *et al.* 1983) if the barrage system (see below) is not built. The land-use belts of the Hortobágy are arranged in a concentric pattern. Very little (1.5%) has survived of the marginal forests. The richest agricultural lands are on sand and loess terrains in both the east and west, while in the centre alkali puszta predominates. Most of it belongs to the Hortobágy National Park including fish-ponds and rice-fields. The amount of pasture of various quality within the National Park is 70%. In contrast with most Szigetköz wetlands, the waterlogged surfaces of the Hortobágy have been converted into commercial reed-beds (1.5%) and only 400 ha are natural wildfowl refuges (Sterbetz 1976).

Parallel analyses of land-use maps and the projects affecting the two regions allow the following predictions. Nature-conservation legislation guarantees that the Hortobágy will not be influenced by either intensification of agriculture or expansion of settlements. Agricultural activities within the Park will be reduced to local-scale horticulture and fodder production for maintaining protected grey cattle, horses and sheep. Poultry will be restricted because of their impact on the grass cover around ponds.

In the Szigetköz the projected Slovak-Hungarian Barrage System, intended to improve navigation on the Danube, is expected to cause major changes in the environment. Most discharge of the Danube will be diverted to a navigation canal on the Slovakian side and the present active flood-plain area will have insufficient water to maintain its vegetation. The ground-water table will rise in the north around the projected reservoir and drop over most of the southern areas, thus influencing ground-water flow. *Populus* stands may be reallocated and the crop pattern adjusted to the altered conditions. The meadows and pastures, which are in ecological balance today, will need irrigation if they are to be maintained.

Summary

(1) Hungarian landscape types have recently been mapped. The history of landscape development has been investigated in two contrasting flood-plain landscape units, the Szigetköz and the Hortobágy puszta.

(2) Geomorphological, palaeobotanical, archaeological and historical-geographical data are used to assess how the natural potential of the landscape has been utilized.

(3) The sparse data available for reconstructing the prehistory of the two regions suggest important differences at the beginning of human settlement. The similar physical setting resulted in similar patterns in environmental utilization. The economies introduced at or soon after the Magyar conquest of 896 A.D. remained the most important until the start of river regulation in the second half of the 19th century. These affected the two landscapes differently. The Hortobágy developed into a landscape of intensive use, still retaining its puszta character. The higher ground of the Szigetköz was intensively cultivated, but seasonally inundated areas are kept intact.

(4) The Hortobágy landscape is now preserved as a National Park. The Szigetköz will be profoundly affected by the proposed Slovak-Hungarian Barrage System on the Danube.

Acknowledgements

As a physical geographer, I am grateful for the advice and source material from related disciplines suggested to me by M. Járai-Komlódi (palynology), M. Domokos (remote sensing) and A. Ringer (archaeology).

Part Two

The Past

Introduction

H. J. B. Birks

Past cultural landscapes cannot be observed or described directly. Their composition, structure and development must be reconstructed from available palaeoecological (e.g. pollen, seeds and fruits, animal remains, etc.) and archaeological evidence. The ecology, functioning and maintenance of a cultural landscape through time must be inferred from such reconstructions. Pollen grains and spores are, by far and away, the most abundant Quaternary fossils, with the result that pollen analysis is the dominant technique in terrestrial Quaternary palaeo-ecology. The chapters in this Part reflect this, as nearly all of them use pollen analysis as a tool for reconstructing and understanding past cultural landscapes over time scales of 1000-6000 years.

Since the classic studies by the Danish ecologist and pollen analyst Johs. Iversen (1941) on landnam and the influence of prehistoric people on vegetation, the detection and interpretation of anthropo-genically induced changes in vegetation by means of pollen analysis has been a major research concern and preoccupation of vegetational historians and palaeoecologists in Europe. Initially, emphasis centred on intensively occupied and utilized areas such as Denmark, southern Sweden, Germany, England and The Netherlands. More recently, interest has developed in marginal areas such as Finland, western and northern Norway, Scotland, Iceland and Greenland. Throughout this type of work, following Iversen (1941), emphasis is placed on the spatial and temporal patterns of (1) selected pollen types (e.g. *Plantago lanceolata, Artemisia,* Cerealia-type), so-called anthropogenic indicators, (2) microscopic charcoal, and (3) particular tree pollen types (e.g. *Tilia, Ulmus, Fagus, Picea*).

In the 1960's and 1970's several important developments occurred in the theory and practice of pollen analysis. These included (1) the development of absolute pollen frequency (APF) techniques, (2) the critical use of surface pollen samples from known modern vegetation types as a tool in interpreting fossil pollen spectra, (3) the application of multivariate data-analytical techniques for detecting and summarizing patterns in pollen data, and (4) renewed interest in pollen represent-ation, through studies on the processes of pollen production, dispersal and sedimentation which has resulted in an increased appreciation of site characteristics and how they affect the spatial and temporal resol-ution of pollen-analytical data. These developments are reviewed by, for example, Birks & Birks (1980).

Pollen-analytical studies on reconstructing cultural landscapes have been comparatively slow to exploit these new developments. Several chapters demonstrate the potential value of these approaches. *Hicks* shows the potential of combining detailed ecological information about land-use practices in Finland with studies on modern pollen deposition to provide a factual basis for characterizing land-use patterns palynologically (see also Hicks 1985c). She develops an interpretative strategy and a key for inferring past local and regional land-use patterns based on defined and hence repeatable palynological criteria. She then applies this approach to the interpretation of two pollen profiles, both with a low intensity of agriculture but one with thinly scattered settlements, the other with a locally concentrated settlement pattern. This study highlights the importance of differences in scale and hence in the location, type and size of pollen site (cf. Jacobson & Bradshaw 1981). The location and size of the source area of 'anthropogenic indicator' pollen and the vegetation between source and site are also shown to be critical in reconstructing cultural landscape history in such marginal areas. Important methodological aspects emerge - the potential value of APF, particularly influx, estimates for characterizing land-use activities involving tree utilization (e.g. Austad this volume); the careful delimitation of anthropogenic indicators on the basis of local ecology rather than analogy with southern Scandinavia or central Europe; and the problems of interpreting very rare occurrences of possible indicator types. The latter aspect arises repeatedly in studies from other marginal areas, particularly the Norwegian mountains (Kvamme this volume; Moe *et al.* this volume).

The use of pollen influx is taken up by *Aaby*. He shows the importance of considering changes in both pollen percentage and influx values in reconstructing cultural landscape development, particularly when the landscape is still primarily forested, as in the Neolithic. Aaby avoids the problems that bedevil almost all influx studies based on lake sediments, such as sediment focusing, by estimating pollen influx from peats. He studied two bogs in contrasting areas today and demonstrates that the long-standing contrasts reflect differences in intensity of cultural activity. Changes in AP influx are primarily a response to changes in forest structure, with AP influx increasing with increased human activity when the landscape is mostly forested. In contrast, AP influx is negatively related to human activity when the landscape is predominantly treeless. The result in the former situation is that relative NAP percentages will be seriously depressed in *early* phases of human activity. Early cultural impact may thus be grossly underestimated compared to later phases if influx data are not considered. Relative representation of many anthropogenic indicator pollen types in percentage diagrams may thus be strongly influenced by changes in AP influx. Aaby's chapter provides one of the first demonstrations of the importance of studying both percentage and influx data in cultural landscape research.

In another primarily methodological contribution, *Birks, Line &
Persson* estimate changing human impact on cultural landscape development by means of correspondence analysis of pollen-stratigraphical
data. Variations in human impact (expansion and regression phases)
in time and space have long been a concern of palaeoecologists and
archaeologists. Previous approaches for estimating human impact have
relied heavily on anthropogenic indicators, the indicator value of which
may vary geographically, temporally and ecologically. Correspondence
analysis provides a tool for detecting major patterns in pollen data and
it has desirable properties when applied to entire pollen assemblages
expressed as relative percentages. Birks *et al.* show close relationships
between inferred human impact and palynological richness which may
reflect changes in the mosaic structure of the cultural landscape through
time. Correspondence analysis is also used to compare the relative
magnitude of human impact at sites in contrasting archaeological
settings. This chapter suggests that there is considerable potential in
applying these and related numerical techniques in cultural landscape
research.

There is more to the study of past cultural landscapes than reconstruction, description and chronology of changing patterns, just as there
is more to present-day cultural landscape ecology than description,
classification and monitoring of patterns. There is the interpretation
of observed temporal and spatial patterns of human impact on landscape
development in terms of underlying ecological and sociological processes. Such interpretations require interdisciplinary studies. *Berglund*
discusses the hypotheses, aims and organization of one such study, the
so-called Ystad project in southern Sweden. This project involves
archaeologists, historians, ecologists and palaeoecologists. It is an
attempt to interpret cultural landscape changes, in particular expansion
and regression phases of human activity, as reconstructed from regional
pollen sequences, in terms of limiting factors such as nutrients, climate
and other environmental parameters as well as sociological factors.
Berglund's chapter links directly with those by Olsson and Emanuelsson
on present-day ecology and with Gaillard & Berglund and Birks *et al.*
on palaeoecology.

Studies of cultural landscape development in time and space can
operate at a variety of spatial and temporal scales. Different hypotheses
about processes are appropriate to explain patterns observed at different
scales (Birks & Moe 1986). The emphasis of many chapters in this part
is on changing patterns over time periods of 1000-5000 years within
specific spatial scales. The chapter by *Edwards* is an exception as it
concentrates on the palaeoecology of one event, the hunter-gatherer
to agriculturalist transition. Although this event is of very great significance in human history and has been extensively discussed by
archaeologists, palaeoecologists have, in recent years, largely ignored
it despite numerous discussions of the *Ulmus*-decline at *ca.* 5000 B.P.

in north-west Europe. Edwards presents evidence for fine-scale local differences in land-use, with hunter-gatherer and agricultural activities apparently existing contemporaneously at nearby sites in Ireland. He uses special sampling and pollen-counting procedures to demonstrate early pioneer Neolithic cereal cultivation prior to the elm-decline. A critical consideration of charcoal stratigraphy shows that no *simple* interpretation in terms of land-use activities and human presence in the Neolithic appears possible, in contrast to the interpretation of charcoal adopted in other chapters. Edwards similarly encourages caution in the uncritical use of 'anthropogenic indicators' because of the wide or even varying ecological amplitudes of many such taxa. A conservative interpretation of current palaeoecological evidence suggests that hunter-gatherer and agricultural practices may well have co-existed and overlapped spatially and temporally, providing an exciting challenge in the interpretation of how this major cultural change occurred.

Turning to the development of the cultural landscape at the broad regional scale, the chapter by *O'Connell, Molloy & Bowler* discusses how a very distinctive marginal landscape, the treeless blanket-bog dominated landscape of Connemara in western Ireland came into being. They show that the area was extensively wooded (mainly *Pinus, Ulmus* and *Quercus*) until blanket-bog developed between 4000 and 3000 B.P., probably as a result of forest clearance and burning. This was the major event in the development of this cultural landscape. Population pressure in the 17th to early 19th centuries led to further woodland destruction, expansion of bog and heath pasturage and construction of cultivation ridges in extreme marginal habitats. Many of these ridges are now being obliterated by further blanket-bog growth, an interesting modern analogue to bog growth over prehistoric fields *ca.* 4000 years ago in western Ireland and Scotland.

Ammann discusses evidence for presumed human-induced changes in forest composition and abundance between 6300 and 2000 B.P. as reconstructed from relative and absolute pollen diagrams from the Swiss Plateau. Rather surprisingly repeated peaks and declines of *Fagus* alternating with peaks of *Alnus, Betula* and *Corylus* pollen can be matched at sites up to 190 km away in the North Alpine Foreland. These changes are interpreted in terms of forest clearance and regeneration, presumably at a regional scale. Interestingly these patterns are not detectable in the pollen record from a large nearby lake, emphasizing the interaction of site type and discernable patterns in palaeoecology (Jacobson & Bradshaw 1981; Birks 1986a). Ammann's chapter illustrates the potential of studying a network of sites of different sizes within an area for reconstructing and interpreting forest dynamics in relation to land-use at a variety of spatial scales.

Küster's work in southern Bavaria also illustrates spatial differentiation. He has analyzed 3 sites in considerable stratigraphical detail around the Auerberg hill on which a Roman town was built. The sites appear to reflect a regional scale (Birks 1986a) and thus his network of sites provides a detailed picture of broad-scale landscape development. Detectable human influence appears to have commenced as early as 6500 B.P., but it was not until Roman times that human impact became widespread and intensive. Interestingly there is strong evidence for fine-scale variation in forest history, with different rates of expansion of montane trees (*Abies, Fagus, Picea*) at sites within a 3 km radius of each other. Present forest patterns appear to mirror past patterns. Montane forest today occupies areas formerly dominated by *Ulmus* 6000-7000 years ago, whereas *Fagus*-dominated forests today are restricted to areas of former *Quercus*-dominance. Pre-Roman phases of human impact are detectable at some but not all sites, whereas Roman and post-Roman phases are universally detected. Such fine-scale differentiation illustrates the inherent complexity of palaeo-ecological data and challenges ideas about regional pollen deposition and sizes of pollen-source areas (e.g. Prentice 1985). An important conclusion from Küster's detailed study is that there is no distinct transition from the natural to the cultural landscape (see Fægri this volume), just as there does not appear to be any distinct change from hunter-gatherer to agricultural practices in Edwards' analysis. As with so much of ecology, there is a gradual continuum from one type to another, whether it be human culture, landscape type or vegetation unit.

A second and finer scale for studying cultural landscape development concentrates on a particular distinctive landscape type (e.g. heathland) and studies its development in several areas. This approach is taken in the chapters by Odgaard, Bos *et al.*, Chambers *et al.*, Kvamme, Nilssen and Fredskild.

Odgaard discusses a very distinctive but increasingly rare cultural landscape, the *Calluna*-dominated heathlands of western Denmark. His approach uses pollen analyses of lake sequences to provide a regional landscape picture against which local-scale pollen analyses of soil profiles yield evidence for local vegetational change. This combination of sites with different pollen-source areas is particularly effective at providing insights into the local ecological processes that produce the regional patterns. Regional pollen diagrams also provide a means of dating landscape change, whereas local diagrams provide a means of testing hypotheses about the mechanisms of heathland formation. Both scales of study are required to reconstruct cultural landscape development in terms of local processes operating at a site, such as grazing, cutting and burning, that were presumably important over the heathland region as a whole.

Bos, van Geel & Pals reconstruct the development of another, very distinctive cultural landscape, the Waterland in the western Netherlands, a former raised-bog area transformed by drainage and reclamation into damp pastures. Their approach is primarily that of archaeology and historical geography and draws on archaeological, documentary, place-name and palaeoecological evidence. They show how landscape changes brought about by drainage have had major effects on the economic and social life in the area which, in turn, have led to further landscape changes. The landscape is dynamic, with a critical balance between peat-shrinkage lowering the ground surface to the water table and the need for further drainage which is, in turn, only possible with further technological developments. Their study illustrates the complex interaction between environmental, human and socio-economic factors in determining the course of cultural landscape development.

The remaining four chapters that consider particular landscape types at this scale all deal with marginal areas, either in the mountains (Chambers *et al.*, Kvamme) or northern latitudes (Nilssen, Fredskild). *Chambers, Kelly & Price* discuss the cultural landscape of upland North Wales with its absence of trees and abundance of stone hut-circles and upland farms. By studying pollen percentages and influx at a mire critically positioned in relation to archaeological sites, they reconstruct the landscape development since Mesolithic times. They suggest, as is commonly proposed in other British uplands, that Mesolithic people influenced vegetational development even though their study area "is devoid of known Mesolithic sites or artefacts". Major forest clearance and the development of the characteristically treeless upland landscape of North Wales occurred in the Early Bronze Age. They evaluate the hypothesis of prehistoric 'transhumance' with summer cattle and sheep grazing in the uplands, and permanent winter sites in the lowlands. Their verdict is not proven. Hypothesis-testing should be more common in palaeoecology as it is a central process in science. Unfortunately hypothesis-testing in a descriptive historical science such as palaeoecology is rarely straightforward, as the available evidence is frequently ambiguous and can be interpreted in several ways. Deducing possible Mesolithic impact on upland vegetation from the available pollen evidence is one such example.

Kvamme considers the development of an important and distinctive cultural landscape in the west Norwegian mountains associated with the transhumance practice of summer farming. Animals are brought up into the mountains for summer grazing (Indrelid this volume). By using local pollen diagrams with a limited pollen-source area, Kvamme demonstrates the various ecological effects of summer farming on local vegetation, on forest composition, density and structure, and on soils. By detecting the onset of these ecological impacts, pollen analysis in conjunction with ^{14}C-dating can date the onset of particular land-use practices. Kvamme makes the important point that "vegetation types can ... turn out to be as informative as archaeological material". This chapter illustrates the value of careful site selection in relation to the land-use of interest in marginal areas such as the west Norwegian mountains, a point re-iterated by Hicks and illustrated by Nilssen, Chambers *et al.*, Moe *et al.* and Bohncke. Problems of interpreting rare and numerically small indicator pollen types and understanding charcoal patterns arise here, just as they do in the lowlands of the boreal forest of northern Finland (Hicks this volume) and at the hunter-gatherer to agriculturalist transition in Scotland and Ireland (Edwards this volume).

Nilssen moves us north to the spectacular Lofoten Islands of northern Norway and shows from local pollen diagrams situated within farming areas today that cultural landscape development began as early as 5500 B.P. By studying several sites in a limited area, Nilssen is able to demonstrate abandonment due either to recession phases or to shifting settlement. The oldest farms studied appear to have been established *ca.* 2800 B.P. but the present landscape developed *ca.* 1800 B.P. following extensive deforestation and expansion of pasture and heath. These interpretations are based primarily on changes in pollen assemblages, charcoal stratigraphy and anthropogenic indicator pollen types carefully selected on the basis of the ecological preferences of the taxa in the area.

Fredskild's chapter concerns Greenland, one of the northernmost cultural landscapes in the world. He considers the past cultural landscape development in 985 A.D. following the Norse landnam, the re-introduction of sheep in this century and the serious threats to the future of the cultural landscape in Greenland. Fredskild's synthesis of the Norse landnam is based not only on pollen analyses but also on plant macrofossils. Seeds and fruits of introduced weeds and apophytes provide invaluable and taxonomically precise information about early land-use, information that cannot be obtained from pollen analysis alone. Coordinated detailed studies of pollen and macrofossils provide a picture of both regional and local flora and vegetation and yet are surprisingly rarely attempted in cultural landscape studies. From an ecological and anthropological viewpoint, it is fascinating that the threats to the present landscape are the same as in Norse times, namely overgrazing, soil erosion, nutrient depletion, climatic change and social pressures. Fredskild's chapter is unique in this volume in considering the past, present and future! It illustrates how essential a historical perspective is, not only in understanding present-day patterns, but also in predicting future trends.

The third and finest scale of spatial resolution possible in reconstructing cultural landscape development is provided by pollen analysis of small, local sites within a *single* component of the landscape mosaic. This approach is represented by the chapters by Andersen, Gaillard & Berglund, Moe *et al.* and Bohncke.

The approach of using sediments in very small (10-30 m diameter) hollows as a means of reconstructing fine-scale vegetational patterns has primarily been developed by *Andersen*. In his chapter, he illustrates this approach as a tool in elucidating cultural landscape development. The hollow was, prior to a prehistoric deforestation phase, situated within *Tilia* forest and Andersen interprets striking changes in the pollen stratigraphy as possibly reflecting shredding of *Tilia* trees in the Early Neolithic, probably one of the earliest indications of this practice in prehistoric times. During the Bronze Age a field was cultivated next to the hollow. Surprisingly the pollen assemblage at this stage barely reflects the cereal or annual weed flora in the adjacent field, presumably because of poor pollen dispersal even at this local scale. This study demonstrates the advantages and the limitations of such sites in cultural landscape research. Such small sites are excellent when the overhanging vegetation consists of trees with high or moderate pollen representation. However, they are less useful when the surrounding vegetation consists of cereals and low-growing entomophilous herbs such as annual weeds, many of which have a very poor representation.

Gaillard & Berglund discuss a detailed pollen diagram from a small (200 m diameter) lake situated within a south Swedish village. The pollen assemblages are particularly rich in land-use indicators. From such a pollen diagram covering the Late Bronze Age to today, Gaillard & Berglund are able to reconstruct the landscape development in some detail and to link pollen-analytical changes with archaeological and historical events within the village. The high diversity of herbs in the pollen assemblages contrasts with the limited assemblage found by Andersen in his minute hollow. Presumably pollen input by surface-water run-off is important in small enclosed lakes and results in good representation of pollen of arable weeds and cereals, whereas in very small hollows with small catchments such run-off is minimal.

The chapter by *Moe, Indrelid & Fasteland* concerns evidence for human exploitation of an extreme marginal area in western Norway, namely at *ca.* 1130 m elevation on the Hardangervidda plateau. They integrate archaeological and palynological evidence. Inevitably, problems of interpretation occur with the scattered, numerically low anthropogenic indicator pollen types, just as in other marginal areas such as northern Finland (Hicks this volume). Moe *et al.* provide evidence for continuous summer-use of the area from at least 4800 to 2800 B.P. and from 2200 B.P. to the present, for hunting, fishing, animal grazing and droving. They suggest that the so-called lowland weed pollen (e.g. *Plantago lanceolata, Urtica, Artemisia*) do not originate locally but that these pollen types are transported by animals driven up from the lowlands and deposited in dung (Moe 1983b). Moe *et al.* use these types as evidence for animal droving whereas changes in locally derived pollen are interpreted as evidence for local grazing. Dispersal of pollen by animals is an interesting hypothesis with important and wide-ranging implications in many areas, for example in discussion of scattered grains of *Artemisia, Plantago* spp. and Chenopodiaceae in pre-Neolithic assemblages in the British uplands (e.g. Edwards & Ralston 1985).

In the last chapter *Bohncke* discusses the development of the cultural landscape near the impressive Standing Stones of Callanish, a massive

Megalithic stone circle on the Isle of Lewis in the Outer Hebrides. Callanish is one of the westernmost large prehistoric monuments in Europe and is situated in a highly exposed treeless setting today. By means of pollen analysis of peat from small hollows, Bohncke demonstrates that small areas of *Betula-Corylus* woodland were present even this far west, and that anthropogenic impact on vegetation may have begun as early as 8400 B.P. and certainly by 5000 B.P. This study in an extremely marginal area again shows the importance of studying small local-scale sites in reconstructing the history and development of elements in the landscape mosaic. A pollen diagram from a nearby mire that samples a regional pollen-source area (Birks & Madsen 1979) suggests that the landscape was predominantly treeless for much of the Holocene, whereas Bohncke shows that, at a finer scale, small areas of woodland occurred locally.

The chapters in this Part cover a range of methodological problems, geographical regions and spatial and temporal scales. The emphasis is almost entirely on reconstruction of cultural landscape development and on detecting patterns of change in time. The chapters are thus primarily descriptive or narrative in character, rather than hypothesis-testing or analytical (*sensu* Ball 1975). This bias largely reflects the nature and underlying philosophy of a predominantly historical science (Birks 1985).

I end this introduction with a warning to the reader and an apology from the editors. The chapters differ in their presentation and use of ^{14}C-dates and so-called corrected or calibrated calendar years. Many authors present 'raw' uncorrected ^{14}C-dates as years Before Present (B.P.) based on a half-life of 5570 years. Others present corrected dates as years B.P., B.C. or A.D. However, different authors use different calibration bases, such as Clark (1975) or the University of Pennsylvania MASCA calibrations (Ralph *et al.* 1973). With the publication of an internationally accepted high-precision calibration curve (Stuiver & Pearson 1986; Pearson & Stuiver 1986) and the ready availability of computer programs to convert ^{14}C-years into calibrated calendar years (e.g. Stuiver & Reimer 1986), it is hoped that palaeo-ecologists will now use dates calibrated by this internationally accepted calibration curve. In this volume the particular way of expressing ^{14}C-dates is given in each chapter.

The Representation of Different Farming Practices in Pollen Diagrams from Northern Finland

Sheila Hicks

Introduction

Northern Finland is a marginal area for human settlement, with only a small area of land being used for agriculture even today. Farming depends mainly on cattle-rearing based on hay grown in permanent fields, and some grain cultivation (Varjo 1977). Historical records show that in the past land-use varied more between regions, being strongly controlled by geographical and ecological factors. This variation is well illustrated in a comparison between two type areas, Kuusamo in the east and the island, Hailuoto, in the west (Fig. 1).

The Kuusamo area, which was deglaciated *ca.* 9500 B.P. and has had a supra-aquatic position since then, is in the northern boreal zone (Ahti *et al.* 1968) midway between the White Sea and the Baltic. It supports forest vegetation in which *Picea* is important. Hailuoto, in contrast, lies just off the west coast, in the middle boreal zone and is dominated by *Pinus*. After deglaciation it remained sub-aquatic until just under 2000 years ago. However, with the present rate of isostatic land-uplift of 9 mm yr^{-1} (Kääriäinen 1953), the island is still rapidly increasing in size (Alestalo 1979).

The earliest occupation of both areas is poorly known. Written sources (Kortesalmi 1975; Ervasti 1978) record that the first Finnish settlers moved to Kuusamo in 1676 A.D., attracted primarily by the fishing, and prior to that the area had two Lapp winter villages. They practised 'huuhtaviljely' a type of slash-and-burn cultivation which was, nevertheless, less intensive than the more familiar 'kaskiviljely' of southern Finland (Heikinheimo 1915) and utilized the moister, richer spruce forests, the main crop being rye. Farming was supplemented by fishing and reindeer herding. By 1800 farmers had changed to growing barley in permanent fields and raising cattle, sheep and reindeer. The animals grazed freely in the forest and winter fodder was gathered from natural meadows, the hay being supplemented with leaf fodder or, in the case of reindeer, by felling spruces, for their

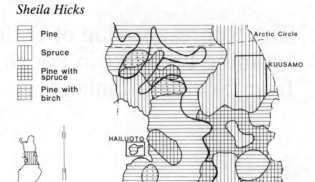

Fig. 1. The main forest vegetation types of northern Finland (Salaminen 1973) and the boundary between the middle and northern boreal zones (Ahti *et al.* 1968), together with the location of the Hailuoto and Kuusamo type areas (see Fig. 2).

lichen cover. The modern situation, with hay fields primarily on drained mires, dates from *ca.* 1950.

Archaeological evidence from Hailuoto shows that the island was settled at least from the Middle Ages. The inhabitants have always relied on fishing and seal hunting, with crops being of minor importance (Virrankoski 1973; Halila 1954; Paasivirta 1936; Suomela 1967). Only at the beginning of 19th century, when state boundaries changed and the best fishing grounds were lost, was any real incentive provided for farming. The forests are on dry infertile, sandy soils unsuited to cultivation and there is no record of slash-and-burn ever having been practised. The silty shore meadows rising from the sea were more suitable for cultivation. Barley was the main crop, together with some rye. In addition to cows and horses, large herds of sheep were kept, with winter fodder coming from the sea-shore meadows and *Betula* and *Alnus* thickets. Again the modern situation of cattle-rearing and sown hay dates from the 1950's.

The cultural landscape of the two areas was, therefore, in strong contrast before 1800 A.D. but has become increasingly similar since then. One major difference still exists. In Kuusamo the small farms and fields are fairly evenly distributed throughout the region at a very low density. On Hailuoto the fields and settlement are concentrated in the centre of the island, which initially was open sea, then a sheltered harbour, and which, with the gradual land uplift, has provided cultivable land. The remaining sandy parts of the island and the mire areas are uninhabited except for the fishing harbour in the extreme west (Fig. 2).

This study aims to delimit pollen criteria which can distinguish between these differences in agricultural activity and their distribution throughout the two regions. If such pollen criteria can be established for the historical period and their reliability demonstrated they could then be used for extending interpretation back into the prehistoric period.

Two major and interrelated problems are apparent. First, because the amount of settlement and agricultural activity, even at the present day, is small and, particularly in Kuusamo, scattered (Fig. 2) the pollen evidence is extremely slight. Second, the number of tree species in the forests is restricted; *Pinus, Picea* and *Betula* in Kuusamo, and only *Pinus* and *Betula* on Hailuoto. *Pinus* and *Betula* are prolific pollen producers, and thus tend to swamp everything else. Any change in taxa associated with humans, including *Picea*, is scarcely recorded in percentage terms.

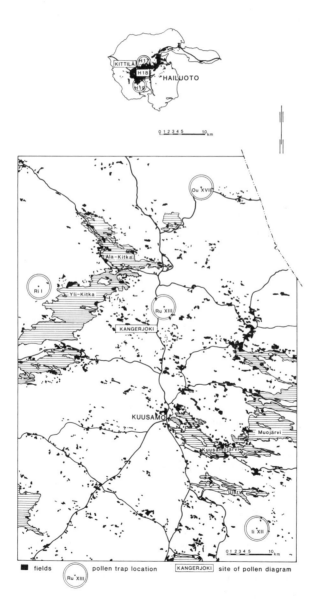

Fig. 2. The island of Hailuoto and the Kuusamo region showing the distribution of fields at the present day, the pollen-trap locations (H 17, H 18, H 19, Ou XVII, Ri I, Ru XII, Ii XII), and the sites of the 2 pollen diagrams, Kittilä and Kangerjoki.

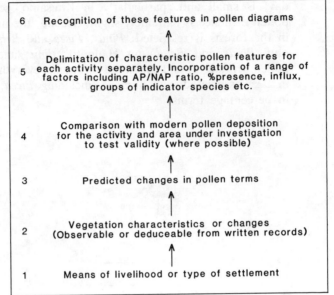

6   Recognition of these features in pollen diagrams
↑

5   Delimitation of characteristic pollen features for each activity separately. Incorporation of a range of factors including AP/NAP ratio, %presence, influx, groups of indicator species etc.
↑

4   Comparison with modern pollen deposition for the activity and area under investigation to test validity (where possible)
↑

3   Predicted changes in pollen terms
↑

2   Vegetation characteristics or changes (Observable or deduceable from written records)
↑

1   Means of livelihood or type of settlement

Fig. 3. A flow chart of the procedure for delimiting pollen characteristics indicative of different types of human presence in a specified region.

Development of the interpretation process

Given this situation a different, more uniformitarian approach to the delimitation of pollen criteria for human presence has been developed. It is summarized in the flow-chart (Fig. 3) as 6 main steps.

Steps 1 and 2 include human-induced situations which can be observed today, or deduced from historical records. Traditional concepts of 'cultural indicators' are initially ignored since many such indicators are either at the limit of or entirely beyond their range. Obviously, as much detailed information as possible is desirable for step 2, in order to make the predictions in step 3 meaningful. Here, however, 4 factors must be born in mind.

(1) Knowledge of vegetation changes associated with an activity or type of settlement which no longer exists may be incomplete or perhaps refers to a single, somewhat atypical situation.

(2) Some types of land-use, particularly if practised at a low density, involve the utilization of natural resources in such a way that no specific pollen changes occur except, perhaps, in terms of influx. Examples include reindeer grazing in forests, temporary settlement and a limited amount of hay cutting from natural meadows.

(3) The pollen type indicated may include taxa which are not characteristic of the activity in question. This is a constant problem with pollen types which are only identifiable to family or genus. A judicious look at the species which could possibly be involved on the basis of their distribution sometimes clarifies the situation but, in any case, such non-specific pollen types should be clearly stated.

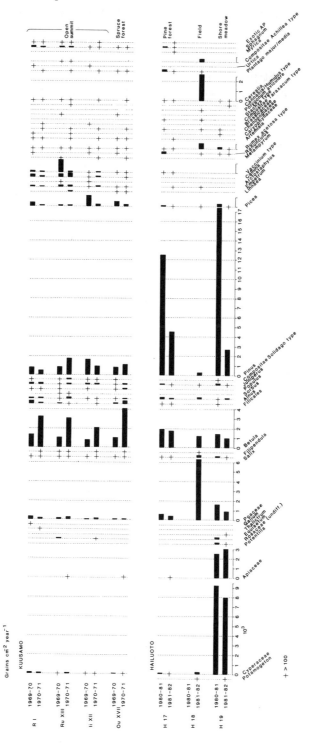

Fig. 4. Modern pollen influx for 2 years (1969-1971 or 1980-1982) for selected sites in Kuusamo and on Hailuoto. The location of each site is shown in Fig. 2.

(4) There are usually several activities occurring simultaneously in an area, so the pollen evidence will be composite, not reflecting any one activity clearly.

Steps 1-3 are shown in Table 1 with just 4 of the many possible human-induced situations being included in order to illustrate the process. The modern pollen deposition records used in step 4 are

Means of livelihood or Type of settlement	Observed or recorded vegetational characteristics and/or changes (main features and species)
Slash-and-burn cultivation of rye. "huuhtaviljely"	Picea removed and the typical ground flora destroyed so that Linnaea and Melampyrum disappear. Secale and Brassica rapa are sown in the clearing. Later the following species spread: Epilobium angustifolium (reaches its greatest abundance in 10-15 years) Trientalis europaea　　　The following may also spread Euphrasia stricta group　after fire: Calluna vulgaris　　　　Deschampsia Antennaria dioica　　　Vaccinium vitis-idaea Solidago virgaurea　　　Empetrum Diphasiastrum complanatum Betula will first colonize the clearing in the succession back to forest.
Permanent settlement	Some local clearing of trees. An increase in, or appearance of the following species: Populus tremula　　　　　　Plantago major Sorbus aucuparia　　　　　Polygonum aviculare Poaceae　　　　　　　　　Stellaria media Trifolium pratense, T.repens　Ranunculus repens Rumex acetosa, R.longifolius　Capsella bursa-pastoris Urtica dioica　　　　　　　Anthriscus sylvestris Tanacetum vulgare (not indigenous in Kuusamo)
Use of natural meadows for winter fodder	In Kuusamo, meadows characterized by:- Juncus　　　　accompanied by: Carex　　　　Thalictrum flavum, T.simplex (neither common) Equisetum　　Filipendula ulmaria Poaceae　　　Veronica longifolia 　　　　　　Ranunculus spp. No changes in forest composition. On Hailuoto a common shore meadow contains:- Schoenoplectus tabernaemontani　In some areas Equisetum may Eleocharis spp.　　　　　　also be dominant and in Calamagrostis stricta　　　others Carex aquatilis. Rumex aquaticus Triglochin spp.　　　　　　All the time the associated Cicuta virosa　　　　　　　shore woodlands are also cut Pedicularis palustris　　　for grass Alnus is preserved Phragmites australis　　　　at the expense of Betula and Lysimachia thyrsiflora　　　grasses expand at the expense Filipendula ulmaria　　　　of woodland herbs. When Galium palustre　　　　　　mowing ceases the meadows Parnassia palustris　　　　become quickly overgrown.
Barley and associated crops grown in permanent fields	Plants characteristic of the barley fields are: Tripleurospermum maritimum　Galium spurium Taraxacum　　　　　　　　Chenopodium suecicum Achillea millifolium　　　C.album (less common) Ranunculus repens　　　　　Spergula arvensis Rumex acetosa, R.acetosella　Galeopsis speciosa Trifolium　　　　　　　　Capsella bursa-pastoris Stellaria media　　　　　Erysimum cheiranthoides Together with various grasses. Polygonum spp. Other species which are or have been cultivated in fields include: Secale, Avena, Brassica, Solanum, Cannabis and Allium.

Botanical nemenclature after Hämet-Ahti et.al. (1984).

Table 1. Vegetational responses and expected pollen changes for 4 different human impacts in the northern boreal-forest zone.

illustrated in Figure 4. Results from 4 Tauber traps (selected from Hicks 1985b, 1986; *q.v.* for location and vegetation details) from Kuusamo are shown. Ri I, Ru XIII and Ii XII are from open summit situations and, therefore, reflect predominantly regional conditions. Ou XVII is within a *Picea*-dominated forest and reflects local forest conditions. Of the 3 Tauber traps from Hailuoto, H 17 is located within *Pinus* forest, H 18 in the field area and H 19 within a shore

Expected pollen evidence		Regionality/comments
Decrease in Picea influx. Presence of charcoal. Disappearance of Linnaeae and Melampyrum. Appearance of Secale and possibly Brassicaceae★ Presence of some of the following: Epilobium★ (may occur later) Trientalis Calluna Compositae Solidago type★ Diphasiastrum complanatum	Later also Poaceae★, Vaccinium type★ and still later increased Betula values.	Commonly practised in Kuusamo from AD 1676 until 1830/50. Unknown on Hailuoto.
Not necessarily any change in AP values. Populus and Sorbus may appear. The presence of several (but not necessarily all) of the following at the same time: Poaceae★ Trifolium type Rumex acetosa type Compositae Achillea type Urtica Plantago major/media	Polygonum aviculare Caryophyllaceae★ Ranunculaceae★ Brassicaceae★ Apiaceae★	Plants associated with settlement show less variation between the two areas than do other groups although some species present on Hailuoto are rare, recent arrivals in Kuusamo.
High values of: Poaceae★ Cyperaceae★ Equisetum★ with possible presence of: Thalictrum, Filipendula, Rhinanthus type★, Ranunculus★ High values of: Cyperaceae★ ±Equisetum★ Poaceae★ Consistent presence of Triglochin Presence in smaller quantities of: Rumex obtusifolius type# Apiaceae★ Pedicularis Lysimachia vulgaris# Filipendula Galium type★ Parnassia	 For nearby shore woodlands Poaceae★ and Alnus increase and Betula decreases. Cessation of mowing in this situation indicated by increase in Betula and Salix and later by re-appearance of Filipendula, Melampyrum and Trientalis and a decrease in Poaceae★	Since these meadows exist naturally it is not really possible to distinguish those which were cut from those which were not, however, in Kuusamo the new meadow areas resulting from flooding or from lowering lake levels in early 1800's should be distinguishable and on Hailuoto changes associated with the cessation of mowing in the associated shore woodlands should also be clearly recorded. Meadows were used for hay from AD 1720-1750 until 1950's in Kuusamo and between around 1800 and 1950's on Hailuoto.
NAP important. Absolute indicators Hordeum, Avena Cannabis. Strong representation of a number of the following: Compositae Taraxacum and Achillea types★ Ranunculus★ Rumex acetosa type Trifolium Caryophyllaceae★ Galium★ Chenopodiaceae	Spergula arvensis Stachys type★# Brassicacae★ Polygonum aviculare Poaceae★	Permanent field cultivation in both the Kuusamo area and on Hailuoto dates from around AD 1800 although small isolated fields certainly existed before that. On Hailuoto the weed flora is generally poorer than on the mainland, partly because some species have not yet spread to the island and partly because edaphic conditions are so unfavourable.

★ Pollen types which could include species other then those regarded as characteristic.

Definition after Moore and Webb (1978).

meadow. These, in contrast to the Kuusamo summit traps, reflect predominantly the local situation. (For the local-regional representativity of Tauber traps see Hicks & Hyvärinen 1986.) The results are for 2 years in each case but a different set of years for the 2 areas so some variation due to periodicity of flowering has to be allowed for.

The results from the 2 regions differ considerably. Despite the dominance of *Picea* in the Kuusamo forests, *Picea* pollen values are low, even in trap Ou XVII (Hicks 1985b). Nevertheless, they are consistently higher than in the Hailuoto traps. *Pinus* values are low in Kuusamo, but dominate the pollen spectra from H 17 in the *Pinus* forest, reflecting the much better pollen production and dispersal properties of *Pinus*. The extremely high *Pinus* values in the shore meadow trap (H 19) for 1980-81 reflect not only a good flowering year for pine but also the exposed situation of this sampling site.

The Kuusamo spectra are fairly uniform, whether from open summits or within forest, but the Hailuoto sites are all quite different. This is a function of regional versus local sources. The open summit spectra of Kuusamo reflect the regional vegetation, namely forest, so a similarity with forest spectra is to be expected. On Hailuoto, however, the field and shore traps reflect specific local situations which are relatively extensive and homogeneous, not scattered and fragmentary (Fig. 2). There are no traps which represent the regional picture in the same way as the Kuusamo summit traps do.

The range of taxa recorded and the representation of forest taxa are greater in Kuusamo, reflecting the greater age and diversity of the Kuusamo forests. The paucity of the Hailuoto *Pinus* forest is also accentuated by the infertile sandy substrate. Often *Empetrum, Calluna, Vaccinium vitis-idaea* and *V. myrtillus* are the only phanerogams present.

Step 4 of the process (Fig. 3) requires examination of the modern pollen data for pollen indicators of known human activities expected from Step 3 (Table 1). Only 'Permanent settlement' and 'Barley and associated crops grown in permanent fields' are relevant today and these occur in both regions. Of the 11 pollen types delimited as indicating permanent settlement, 9 are present in the Kuusamo traps and 8 in the Hailuoto ones, and of the 15 pollen types deemed characteristic of cultivation in permanent fields, 11 are represented in the Kuusamo traps and 10 in the Hailuoto ones (note that some species fall into both categories). The only pollen types not recorded in either area are *Trifolium* and *Spergula arvensis*. The others all occur in extremely low values. Those which are consistently present are *Rumex acetosa* type, Chenopodiaceae and Cerealia (every trap), Ranunculaceae (missing from 1 Kuusamo trap) and *Plantago major/media* (missing from 1 Hailuoto trap). However, the only significantly high values are for Cerealia, Poaceae and *Rumex acetosa* type pollen in H 18 situated actually among the cultivated fields.

Therefore, the pollen-trap evidence agrees with the expected evidence but the quantities involved are minute and would probably be regarded as accidental by workers in areas further south. Nevertheless, the trap results demonstrate that, at the regional level, it is only this sparse evidence which can be expected, and that for higher values the activity in question must take place in the immediate vicinity. The actual pollen types involved appear to be valid.

In step 5 (Fig. 3) the characteristic pollen features of each activity are delimited separately, largely using the predictions in Table 1 but also taking into account modern pollen evidence. The most reliable results are obtained if a combination of factors is used rather than 1 or 2 indicator types. Such a combination should include AP/NAP changes, pollen influx, a group of indicator species, etc. This can be developed further and the characteristic features used as the basis of a dichotomous key which provides a more convenient tool for interpreting pollen diagrams. A preliminary key for the Kuusamo area is provided in Table 2. Similar keys could be constructed for other areas, the definitions and characteristic features being more specific the more precise the initial information is.

Finally, step 6 is the recognition of these features in pollen diagrams. Obviously the difference between local and regional representation is significant. It is observable in percentage diagrams but becomes more obvious when influx values are available. This must, therefore, be taken into account if any attempt is made to relate intensity of pollen representation to distance between source and sampling site or to proportional representation of an activity within a given radius around the sampling site (Hicks 1985a). Moreover, the question of the nature of the vegetation separating the source and the sampling site must also be taken into account.

Results

Using the process outlined above two pollen diagrams, one from each area, are interpreted. These have been selected because they best illustrate the interplay of local and regional representation.

The sites

The Kuusamo Kangerjoki site is in a small mire some 150 m in diameter at *ca.* 288 m elevation formed in a depression between drumlins. It has been partially drained by ditching. At present, fields occur on all sides, the nearest being *ca.* 300 m away. In 1730 there was one dwelling 7 km to the south-south-east and by 1790 3 dwellings *ca.* 2.5-3.0 km to the north. Now the nearest farm is *ca.* 600 m to the north-west.

The Hailuoto Kittilä site is located at one end of a peat-filled

Sheila Hicks
198

```
Provisional key for distinguishing different types of human interference
in pollen diagrams from the northern boreal forest zone

1. a)  AP values high.  Herb pollen values not especially significant but
        some of the following appear or are present in slightly increased
        values, Gramineae, Juniperus, Rumex acetosa type, Compositae
        Achillea and Taraxacum types, Ranunculus, Urtica, Plantago major/
        media.
                                         .........................2

1. b)  AP values lower, either consistently or because of a definite fall,
        and either for all species together or one selectively.  Herb
        pollen values consequently relatively significant, and may be quite
        varied.
                                         .........................3

2. a)  Gramineae important among the herbs.  Some of the following present:
        Rumex acetosa type, Compositae Achillea and Taraxacum types,
        Ranunculus type.  Juniperus values may be higher.
                                         .........................4

2. b)  Plantago major/media present accompanied by some of the following:
        Urtica, Rumex acetosa type, Gramineae, Carduus, Caryophyllaceae,
        Trifolium, Polygonum aviculare, Ranunculaceae, Brassicaceae and
        Apiaceae.
                                         ...........Permanent settlement

4. a)  Possibly increased Juniperus values.  Herb pollen includes a
        selection of the following: Rumex acetosa type, Compositae
        Achillea type, Epilobium, Polygonum aviculare, Caryophyllaceae,
        Linnaea, Trientalis and Lycopodium annotinum.
                                         .......Semi-permanent settlement

4. b)  Herb pollen includes a selection of the following: Compositae
        Taraxacum type, Ranunculaceae, Rubus type.
                                         ....Grazing in forest clearings,

3. a)  Only one tree species, Picea, has lower values.  Herb pollen
        values remain relatively modest.  Presence of Secale.  Possible
        disappearance of Melampyrum and Linnaea and appearance of Epilobium.
        Some of the following may also be present: Trientalis, Rhinanthus
        type, Diphasiastrum complanatum.
                                         ......Slash-and-burn cultivation
                                                (Huuhtaviljely)

3. b)  All AP values lower.  Herb pollen values may be quite high (up to
        80 % ΣP)
                                         .........................5

5. a)  Hordeum and/or Avena and/or Cannabis present.  Great variety of
        herb pollen types including several of the following: Gramineae,
        Compositae Achillea and Taraxacum types, Rumex acetosa type,
        Polygonum aviculare, Spergula type, Stachys type, Brassicaceae,
        Chenopodiaceae, Trifolium, Galium, Ranunculaceae.
                                         ..........Grain crops grown in
                                                   permanent fields.

5. b)  Herb pollen values with a very high proportion of Gramineae
                                         .........................6

6. a)  Gramineae accompanied by high values of Cyperaceae and/or Equisetum.
        Filipendula, Thalictrum, Rhinanthus type and Ranunculaceae may be
        present.
                                         ..............Natural meadows.

6. b)  Gramineae not accompanied by Cyperaceae or Equisetum, instead a
        range of herb pollen types (but not as varied as in 5a) including
        a selection of the following: Elymus repens, Rumex acetosa type,
        Compositae Achillea and Taraxacum types, Trifolium, Chenopodiaceae,
        Caryophyllaceae and Ranunculaceae.
                                         ..............Sown hay fields.
```

Table 2. Provisional key for distinguishing different types of human interference in pollen diagrams from the northern boreal-forest zone.

depression *ca.* 600 m long and 50 m wide lying between sandy ridges. It once formed part of the coastline but now lies at 9-10 m elevation on the north-west margin of the Hailuoto village. The peat supports *Betula-Alnus* woodland and has been partly ditched. At present, fields occur 250-350 m away to the south and east and the Kittilä farm is *ca.* 500 m away. Using the known rate of land uplift of 9 mm yr^{-1} it can be calculated that the base of the peat was at sea level at *ca.* 975 A.D. The position of the site relative to the ever-changing coastline is illustrated in Figure 5 which shows how, during 1000 years, the change from a coastal locality to one isolated at the periphery of the village has taken place. Maps constructed for taxation purposes show that in 1766 A.D. the edge of the cultivated land was *ca.* 600 m to the south-east and by 1866 there were some additional fields *ca.* 400 m to the east.

Interpretation of the Kangerjoki pollen diagram

The Kangerjoki pollen diagram (Hicks 1975, 1985c) is redrawn here (Fig. 6) with the taxa grouped ecologically. The uncalibrated ^{14}C-date suggests that the 14 cm horizon is between 1640 and 1780 A.D., thus providing some chronological control. Tree pollen dominates the diagram. *Picea* values are generally low in keeping with the lower and less frequent pollen production of *Picea*. Using the key (Table 2) and the pollen evidence in Table 1 the following phases can be distinguished:

K1 15-25 cm

Betula and *Pinus* values are high, *Picea* is generally more than 10% Σ P. Both mire (Cyperaceae + *Rubus chamaemorus, Potentilla, Equisetum*) and forest (*Lycopodium annotinum, Vaccinium* type, *Empetrum, Calluna, Melampyrum*) taxa are represented. There are two occurrences of *Rumex acetosa* type and one of *Plantago lanceolata*.

This is interpreted as representing the normal regional forest type. The presence of *Rumex acetosa* type pollen could indicate settlement somewhere in the region but there are no other indicative pollen types present to support this interpretation. Since *P. lanceolata* is beyond its ecological range in Kuusamo, the pollen is regarded as long-distance transported.

K2 14-15 cm

This phase, of one sample, is only weakly substantiated. It exhibits slightly lower values for *Betula* and *Picea* corresponding with a higher value of *Pinus*. Mire taxa are well represented but forest taxa are fewer. There is one occurrence of *Rumex acetosa* type and one of

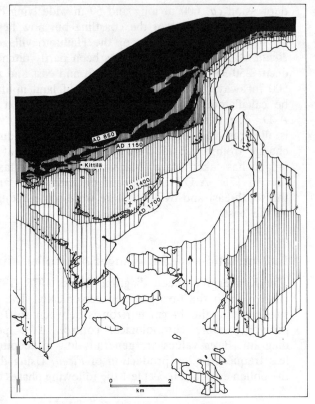

Fig. 5. The central part of the
island of Hailuoto showing the
position of the shoreline at
different points in time as
calculated using the 10 m,
7.5 m, 5 m, and 2.5 m contours
from the 1981 map (based on
air photos flown in 1975) and
a rate of land uplift of
9 mm yr⁻¹ (Kääriäinen 1953).
The site of the Kittilä pollen
diagram and the church are
also indicated.

Cerealia. It is on the basis of these, particularly the latter, that a separate phase is suggested (Hicks 1985c).

This is interpreted as very slight evidence of settlement and slash-and-burn (huuhtaviljely) cultivation (point 3a in the key, Table 2) somewhere in the region.

KK3 8-14 cm

Betula values increase and those of *Pinus* decrease. *Picea* values remain constant. Forest taxa are well represented but mire taxa are fewer. *Ranunculus*, *(Artemisia)*, *Rumex acetosa* type, Brassicaceae, *Galium*, *Cannabis/Humulus* type, Cerealia, *Urtica* and (*Plantago lanceolata*) occur. Like *P. lanceolata*, *Artemisia* is beyond its range in Kuusamo and the pollen is regarded as long-distance transported.

This represents a continuance of slash-and-burn cultivation together with some permanent field cultivation (5a in the key) and more permanent settlement (2b in the key).

KK4 1-8 cm

Cyperaceae shows high values. There are increased values of Poaceae and *Juniperus*, but lower values of *Betula* and *Picea*. Mire taxa are more abundant. *Melampyrum* is absent from the forest taxa. *Ranunculus*, *(Artemisia)*, *Rumex acetosa* type, Chenopodiaceae, *Cannabis / Humulus* type, Cerealia and *Urtica* are all present.

The high Cyperaceae and increased Poaceae indicate the flooding of the mire surface to increase its productivity as a natural source of hay (6a in the key). The other taxa indicate permanent settlement and cultivation in permanent fields.

KK5 0-1 cm

Betula, *Pinus* and *Picea* have increased values while those of the mire taxa fall dramatically. The forest taxa are well represented. *Rumex acetosa* type is present.

This indicates a return to forest conditions. The high *Pinus* values reflect the spread of *Pinus* on to the mire surface following ditching. Using the known history, the ^{14}C-dates and the rate of peat accumulation, the following chronology can be established.

Fig. 6. Relative pollen diagram for Kangerjoki, Kuusamo. The numbers KK1-KK5 refer to the different phases of human presence distinguishable on the basis of the pollen criteria delimited in Tables 1 and 2.

KK1 pre-1670/80 A.D.
KK2 1670/80-1750/60 A.D.
KK3 1750/60-1820/30 A.D.
KK4 1820/30-1950 A.D.
KK5 1950 A.D.-present

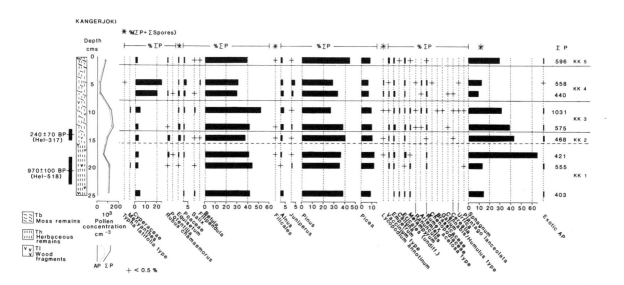

In phases KK1-3 the regional situation is clearly recorded. Pollen indicating human presence is travelling some distance and the intervening shrub and forest vegetation is, no doubt, filtering a lot of it out. Wood remains in the peat also indicate a higher density of *Pinus* on the bog itself than during phase KK4. This would both filter out regional pollen and contribute additional local *Pinus* pollen. Thus only when the level of activity in the region as a whole reaches greater proportions do the number of indicators increase, as in phase KK3. Phases KK4 and KK5, in contrast, reflect local conditions more strongly. The dramatic increase in Cyperaceae clearly shows changes on the mire itself while the smaller evidence of settlement and cultivation continues to reflect the regional situation. In phase KK5, local drainage plus the spread of *Pinus* on to the mire surface combines to prevent the regional pollen indicative of farming from being preserved in the surface peat at all.

Fig. 7. Relative pollen diagram from Hailuoto, Kittilä. The numbers HK1-HK5 refer to the different phases of human presence distinguishable on the basis of the pollen criteria delimited in Table 1.

Interpretation of the Kittilä pollen diagram

The taxa in the Kittilä diagram (Fig. 7) are in the same order as in the Kangerjoki one. *Picea* values are minimal, conspicuously lower

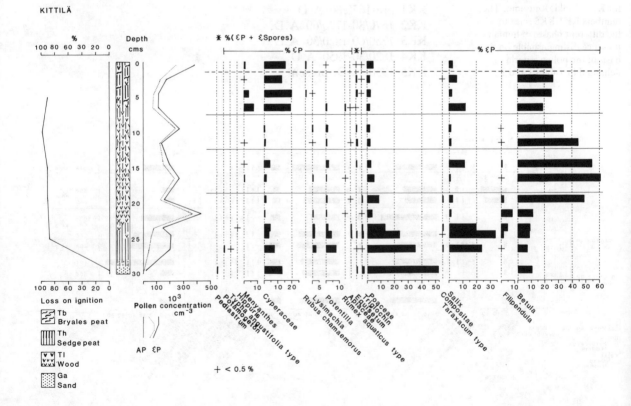

than in the Kangerjoki diagram, in keeping with the difference in forest type in the 2 areas. Using the pollen evidence in Table 1, the following phases are distinguished:

HK1 19-30 cm

High values of Poaceae and relatively high values of Cyperaceae are followed by high values of *Salix*, *Potentilla* and *Lysimachia*, then by peaks of *Alnus* and *Filipendula*, and finally by sharply increased values of *Betula*. Aquatic and wetland taxa (*Pediastrum*, *Potamogeton*, *Typha angustifolia* type, *Hippuris*, *Rumex aquaticus* type, Apiaceae, *Equisetum*, *Triglochin*) are well represented. *Pinus* values are variable and relatively low while the forest taxa are relatively rare. Caryophyllaceae, (*Artemisia*), *Stachys* type, *Galium*, *Elymus repens*, *Hordeum*, Chenopodiaceae, *Rhinanthus* type, Brassicaceae, *Urtica*, (*Plantago lanceolata*) and Compositae *Taraxacum* type are all present. *Artemisia* and *P. lanceolata* pollen are regarded as being long-distance transported. Although Hailuoto is just within the range of *Artemisia* the plant is much rarer on the island than on the mainland (Kaakinen & Saari 1977).

The interpretation is of a fairly rapid successional development from

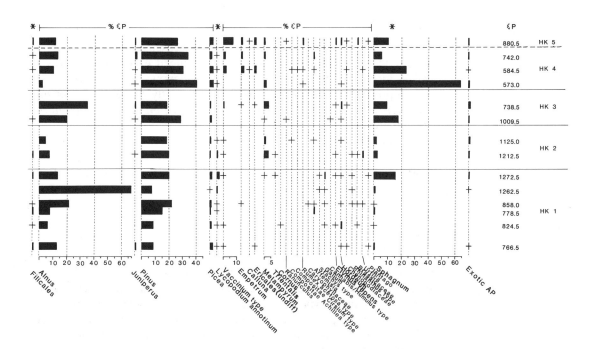

a typical shore meadow with *Phragmites*, through *Salix* scrub and *Alnus* thickets to *Betula* woodland, a conclusion confirmed by the stratigraphy and plant-macrofossil remains. This coastal meadow and developing woodland must have been sufficiently dense and local to filter out effectively much of the *Pinus* pollen from the dry *Pinus* forests which undoubtedly existed on the sandy areas inland, as well as contributing itself to the pollen rain. Overlying this natural succession there is evidence, probably equally local, of crop production and settlement. In its early stages the surrounds of this site must have been highly suitable as a dwelling site for people engaged primarily in fishing.

HK2 12-19 cm

Betula is dominant. The successional shore taxa typical of HK1 are generally much reduced. *Pinus* values are unchanged. Forest taxa are poorly represented except for *Melampyrum*. Compositae *Achillea* type, (*Artemisia*), *Elymus repens*, *Hordeum*, *Rhinanthus* type, Brassicaceae and *Urtica* all have single occurrences.

The evidence shows that *Betula* woodland has developed at the site. The indicators of human presence have diminished but not disappeared perhaps because the centre of activity (the coast) is now farther from the sampling site.

HK3 7-12 cm

Betula values decrease as *Alnus* values increase, and *Pinus* values are marginally higher. Wetland taxa are unchanged except for *Sphagnum*, which increases. Of the forest taxa, *Melampyrum* continues to be well represented. *Juniperus* is more consistently present. *Ranunculus*, *Cannabis/Humulus* type, *Elymus repens*, *Hordeum* and Chenopodiaceae are all present.

The indicators are of local changes to wetter conditions favouring *Alnus* at the expense of *Betula* and allowing the spread of *Sphagnum*. Evidence for human presence in terms of cultivation indicators continues to be clear but slight.

HK4 1-7 cm

There are high Cyperaceae values accompanied by *Menyanthes*, *Salix* and *Sphagnum*. *Betula* and *Alnus* values are lower, but *Pinus* and *Picea* values are higher and *Juniperus* is consistently present. The forest-heath taxa expand. Compositae *Cirsium* type, *Rumex acetosa* type, (*Artemisia*), *Hordeum*, Chenopodiaceae and *Urtica* are present.

The local *Betula-Alnus* woodland has been opened up considerably and mire plants have become important. This change is also evident in the stratigraphy which consists of sedge peat with some bryophytes in which the wood remains so abundant lower down the profile become quite rare. The opening of the local canopy allows for the better representation of the island's regional forest type, namely pine with a ground flora of heaths.

The pollen types indicative of settlement and cultivation remain few in number but the taxa involved are more diagnostic. The local change could be caused by the intensified use of the wooded mire, now marginal to the main centre of settlement, for grazing.

HK5 0-1 cm

The distinction of 1 sample as a phase is questionable. Most values remain similar to those in HK4 but Poaceae and the forest heaths increase, and it is on this slight evidence that HK5 is tentatively separated. *Ranunculus, Rumex acetosa* type, *Elymus repens, Hordeum*, Chenopodiaceae, Brassicaceae and *Plantago major/media* are present.

The local situation of open woodland conditions continues, along with the good representation of the regional *Pinus* forest. Indicators of settlement and cultivation (permanent hay fields) are stronger, in keeping with the closer proximity of cultivated land.

The exact dating of the phase boundaries is problematical. As yet no ^{14}C-dating is available so the only fixed point is the inception of peat growth at *ca.* 975 A.D. The stratigraphy indicates that the rate of peat accumulation has varied, being most likely faster between 29 and 23 cm and again from 5-1 cm, and slower between 23 and 5 cm and at the surface. This is also indicated by changes in pollen concentration. Taking this variation into account and correlating the pollen evidence with the known history the following time-scale is tentatively proposed:

    HK1 950/1000 A.D.
    HK2 1300/50-1700/50 A.D.
    HK3 1700/50-1800/30 A.D.
    HK4 1800/30-1950 A.D.
    HK5 1950 A.D.-present

This diagram records the local situation very strongly in contrast with Kangerjoki. Thus it can be concluded that the early indicators of settlement and cultivation (HK1) are also local. The map (Fig. 5) shows that this must be so. Up until 1150 A.D., at least, the area south of Kittilä was open sea and to the north was infertile sand, so any settlement would have been just at the coast. As the sand spit, on which the church has stood since the Middle Ages, gradually emerged and the sea between it and the Kittilä site gradually shrank to the small lake that it was in 1400 A.D., so the Kittilä locality became less and less desirable as a dwelling site for those engaged primarily in fishing. This explains why the pollen evidence for human activity becomes weaker; the centre of settlement moves closer to the church and the Kittilä area is temporarily abandoned. This is confirmed by the 18th and 19th century taxation maps which show Kittilä as being well beyond the limit of the cultivated land area.

Conclusions

The evidence presented demonstrates that the interpretive process employed here is viable and helps in differentiating types of human activity. However, in boreal forest regions, where cultivated land has only ever occupied a small proportion of the total land area, the pollen evidence for human presence is extremely sparse. For example, in the top samples at both Kangerjoki and Kittilä, representing the last 20-30 years, the occurrence of those pollen types recorded in Table 1 as indicating permanent settlement and cultivation in permanent fields is even fewer than those recorded in the pollen traps, 5 out of 11 and 6 out of 15, respectively, for Hailuoto and 2 out of 11 and 2 out of 15, respectively, for Kuusamo. Therefore even the occurrence of one or two characteristic grains may be significant. Nevertheless, such sparse occurrences should be interpreted all the more strictly in terms of the ecological and edaphic restrictions imposed by the location of the area in question and reliance should be placed, in the first instance, on a group of pollen characteristics rather than on a single pollen find. Additionally, because the evidence is so sparse and the whole range of pollen types characteristic of a specific activity is rarely present, the separation of a number of concurrent activities cannot always be made with certainty.

A second point which emerges clearly is the importance of being aware of differences in scale. Some features reflect the immediate local situation and some the general regional one. This largely depends on the type and position of the sampling site. For example, the Kittilä site was chosen with the specific intention of finding out whether any settlement had ever existed in that locality, which, on general geographical grounds, appeared to be an ideal one for early fishing communities. The larger and more open the mire (lake) is, the greater is the regional representation relative to the local (Tauber 1977) and conversely a small hollow within a forest (Andersen 1984, this volume) reflects primarily local changes.

However, the nature and extent of the pollen source of human activity and the vegetation between it and the sampling site must also be taken into account. Boreal forests present a particular problem, because the major tree taxa, *Pinus* and *Betula*, are prolific pollen producers. *Picea*, which may be influenced by human activity (e.g. huuhtaviljely) only produces pollen sparsely and infrequently. Pollen-influx measurements may help to disentangle these factors, as indicated by modern pollen deposition values recorded by pollen traps.

A more detailed investigation of a restricted area using the process outlined here could prove particularly fruitful in pinpointing actual centres of human presence and the type of activity undertaken in these, otherwise, somewhat difficult marginal areas.

Summary

(1) Northern Finland is a marginal area for human settlement. Comparison of two areas, Kuusamo and Hailuoto, shows that, although they are rather similar today, their agricultural histories are quite different. This reflects their contrasting ecological situations.

(2) This investigation aims to determine to what extent these contrasts can be distinguished in pollen diagrams.

(3) A process is outlined which uses historical documentation and present-day observations of vegetational changes together with modern pollen deposition records to deduce a set of pollen characteristics for the range of human activities practised.

(4) These pollen characteristics are applied to the interpretation of one pollen diagram from each area with attention being focused on whether the representation of an activity is local or regional.

(5) The process proves to be viable. The differing histories of the two areas can be traced but the evidence is extremely slight. Only those activities occurring in the immediate vicinity of the sampling site are strongly recorded.

(6) Boreal forests present a particular problem in investigating human presence since they consist of a few tree taxa which are mostly profilic pollen producers. Not only does the distance from the activity to the sampling site affect its palynological representation but also the nature of the intervening vegetation. These two aspects can be partly overcome if influx rather percentage values are used.

Nomenclature

Plant nomenclature follows L. Hämet-Ahti, I. Suominen, T. Ulvinen, P. Uotila & S. Vuokko (1984) *Retkeilykasvio*, Suomen Luonnonsuojelun Tuki Oy, Helsinki.

The Cultural Landscape as Reflected in Percentage and Influx Pollen Diagrams from Two Danish Ombrotrophic Mires

Bent Aaby

Introduction

Pollen-analytical reconstructions of past floras and plant populations are mainly based on percentage diagrams. They meet most demands, but the limitation that the values of each pollen type depend on other types cannot be solved. To overcome this inherent problem, concentration and influx pollen diagrams are used to examine more realistically the composition and abundance of past vegetation and how it has changed.

The advantage of independently varying pollen frequencies is gained at the expense of a much higher data uncertainty. Errors originate from, for example, sediment weight and volume measurements, exotic pollen added and the number of exotic pollen counted. Futhermore, estimation of pollen influx involves calculation of a matrix accumulation rate.

In addition, depositional and post-depositional processes have considerable influence on the observed absolute pollen frequency (APF). The shape and size of the pollen recruitment site are particularly important factors, as shown by investigations of pollen influx in lake sediments (see Birks & Birks 1980). Post-depositional processes that occur in lakes are insignificant in peat bogs, although some vertical and horizontal movement of pollen occurs in the unconsolidated surface layer (Rowley & Rowley 1956; Salmi 1962; Clymo & Mackay 1987). *A priori*, mires should be suitable for APF studies.

Ombrotrophic mires are preferable because of rather stable environmental conditions over extended time periods. But, even here, several factors can contribute uncertainty. Autocompaction and heterogeneity in peat structure cause short-term variations in matrix accumulation rate (Aaby & Tauber 1975) which is likely to be the main factor determining pollen concentration (cf. Middeldorp 1982; Dupont 1985). Vegetation structure also affects pollen deposition rate; a rough

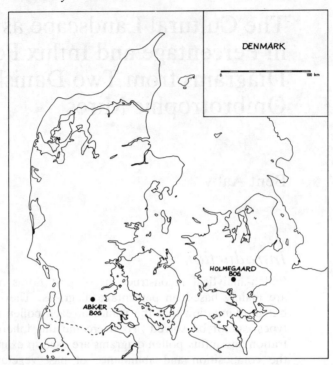

Fig. 1. Location map.

hummock vegetation traps pollen more efficiently than a smooth *Sphagnum* lawn in hollows (Dupont 1985). Influx data obtained from ombrotrophic mires should accordingly be interpreted with caution.

Two extensive ombrotrophic mires have been investigated to reconstruct the regional vegetational development from pollen analysis. Special attention has been paid to the cultural landscape as reflected in percentage and influx diagrams. Preliminary results from these sites have already been published (Aaby in Andersen *et al*. 1983; Aaby 1986a, 1986b).

The sites and methods

The Sites

Holmegaard Bog is situated in southern Zealand, *ca*. 15 km from the coast (Fig. 1). Clayey till deposits dominate the glaciogenic landscape around the bog and only *ca*. 12% of the region has sandy deposits. The region is one of the most fertile farming areas in Denmark.

Holmegaard Bog is part of an extended mire complex, consisting of ombrotrophic mires, fens and carrs, which formerly occupied a wide valley. The investigation area is preserved in a natural state and is *ca.* 500 m from the narrow minerotrophic lagg surrounding the bog. The site was formerly a lake which became overgrown in middle Atlantic time, and since *ca.* 5300 B.P. (4100 B.C.) more than 3 m of ombrotrophic peat has accumulated. The present vegetation at the sampling site is dominated by *Calluna vulgaris* and *Eriophorum vaginatum*.

Abkær Bog is located in southern Jutland, *ca.* 10 km from the east coast (Fig. 1). The hilly landscape east of the bog was glaciated in the Weichselian and the soil is dominated by clayey till and fluvioglacial sand deposits. The area west of the bog has quite another physiognomy, being more or less flat, due to the presence of extensive sandur plains and gently undulating low hills consisting mainly of sandy deposits. The 3 km² bog occupies a depression formed by melting ice at the end of the Weichselian. It was a shallow lake in the early Holocene and became overgrown by fen peat in the Boreal. Shortly afterwards ombrotrophic peat accumulated and since Late Boreal time *ca.* 5.5 m of *Sphagnum*-dominated peat has accumulated in the central part of the bog which remains in a natural condition. The sampling site is *ca.* 700 m from the surrounding upland. *Calluna vulgaris, Eriophorum vaginatum* and *Erica tetralix* dominate the present vegetation. Small hollows with *Sphagnum cuspidatum* Hoffm. and *Rhynchospora alba* are also present on the bog dome.

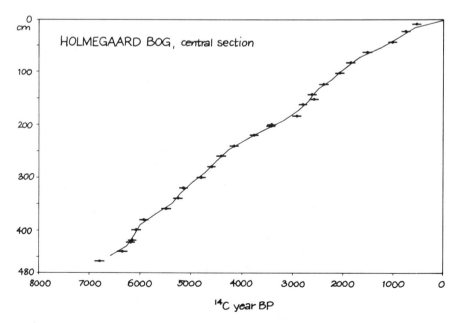

Fig. 2. Age-depth curve for Holmegaard Bog.

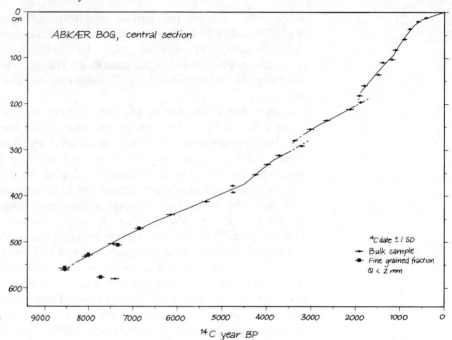

Fig. 3. Age-depth curve for Abkær Bog.

Chronology

At Holmegaard a chronology extending back to the early Atlantic has been established based on 27 ^{14}C-dates at *ca.* 20 cm intervals throughout the 4.4 m deep peat deposit (Fig. 2). At 2 levels, peat collected from the same level in different cores has been dated to compare sample correlation. Both sets of dates agree.

The ^{14}C-dates were not used directly for constructing an age-depth curve. Instead, weak curve-smoothing was done by joining average age and depth for 2 successive dates. This procedure was chosen because of the close ^{14}C-dates. The age-depth curve lies within one standard deviation of the individual dates, except for the uppermost one which is probably too young due to infiltration by modern roots.

Thirty two ^{14}C-dates have been obtained from Abkær Bog (Fig. 3). The construction of an age-depth curve was more complicated because of secondary compaction during coring of unconsolidated *Sphagnum*-peat. A separate age-depth curve was established for core 2 based on its 5 dates. Joined curves for cores 0 and 1 and for cores 3-6 could be constructed, because correlation of cores was possible with a high degree of certainty. The same procedure was used as at Holmegaard Bog. The curves show only slight variations and are within one standard deviation of the individual dates except at 3.78 m and 3.93 m where the samples have similar ^{14}C-ages.

Dates are given as ^{14}C-years B.P. and as calendar years B.C./A.D. Calibrations follow Clark (1975).

The percentage diagrams

The pollen diagrams from Holmegaard and Abkær reflect vegetational development since the late Atlantic (Figs. 4, 5). About 750 AP grains were counted in every sample in addition to non-arboreal pollen types. Pollen assemblage zones (p.a.z.) are established using Jessen's (1935) system. Subzones of regional validity only are indicated by letters.

The cultural landscape at Holmegaard

The percentage diagram reflects the vegetational development in the fertile landscape around Holmegaard Bog (Fig. 4).

Pollen assemblage zone VII, Atlantic time

The landscape was densely forested in the late Atlantic as seen from the dominance of AP and the few herbaceous pollen grains originating from upland areas. *Tilia* was the dominant tree on well-drained brown earths as shown from local diagrams in the Holmegaard area (Aaby unpublished) and elsewhere (Iversen 1969; Andersen 1973, 1984, 1985, this volume; Aaby 1983), whilst *Quercus*, *Ulmus*, *Fraxinus* and *Corylus* were probably growing mainly on damp ground, and *Alnus* exclusively on peaty soils (cf. Iversen 1960).

The vegetation remained stable until *ca*. 5000 B.P. (*ca*. 3850 B.C.) when *Corylus* and *Fraxinus* became more prominent and the frequency of *Pinus* decreased. The *Ulmus*-decline is distinct in the pollen diagram and dated to 4800 B.P. (*ca*. 3630 B.C.). This date is *ca*. 200 years later than the *Ulmus*-decline elsewhere in Denmark (Andersen 1978) and southern Sweden (Nilsson 1964). The occurrence (2 samples) of the indigenous weed *Plantago lanceolata*, just prior to the *Ulmus*-decline, may be due to long-distance transport from Neolithic clearances, or may be of local origin.

Pollen assemblage zone VIII, early and middle Subboreal time

Human interference with the vegetation became evident early in the Subboreal when the representation of *Ulmus*, *Fraxinus* and, later, *Tilia* gradually decreased over a period of 300 years, whereas *Betula*, *Corylus* and *Alnus* were favoured. A single pollen grain of *Hordeum*-type suggests that cereal-growing was practised from the beginning of p.a.z. VIII. Larger open areas originated in the middle of p.a.z. VIIIa, as seen from the relatively high frequency of light-demanding herbs in the pollen diagram. Pastures were present (Iversen 1941, 1973) and *Corylus* became prominent in the landscape during the second part of the subzone.

Fig. 4. Percentage pollen
diagram from Holmegaard
Bog. Pollen from trees, shrubs
and dry-soil herbs are
presented only.

The Holmegaard area was abandoned after *ca.* 700 years of cultural activity. The field flora was sparsely represented whilst *Ulmus*, *Tilia* and *Fraxinus* became frequent again *ca.* 4200 B.P. (2900 B.C.) at the expense of *Corylus*. The tree flora in p.a.z. VIIIb closely resembled

HOLMEGAARD BOG, CENTRAL SECTION

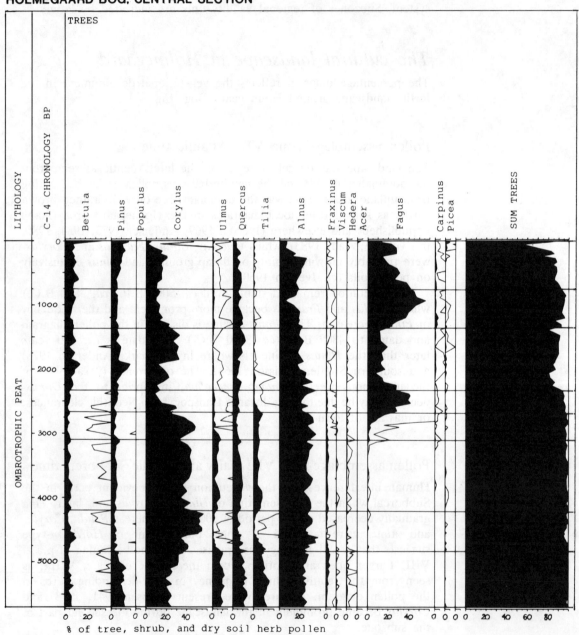

% of tree, shrub, and dry soil herb pollen

that of the Atlantic (VII). The ecosystem equilibrium thus remained stable, although there had been strong human interference with the vegetation over several centuries.

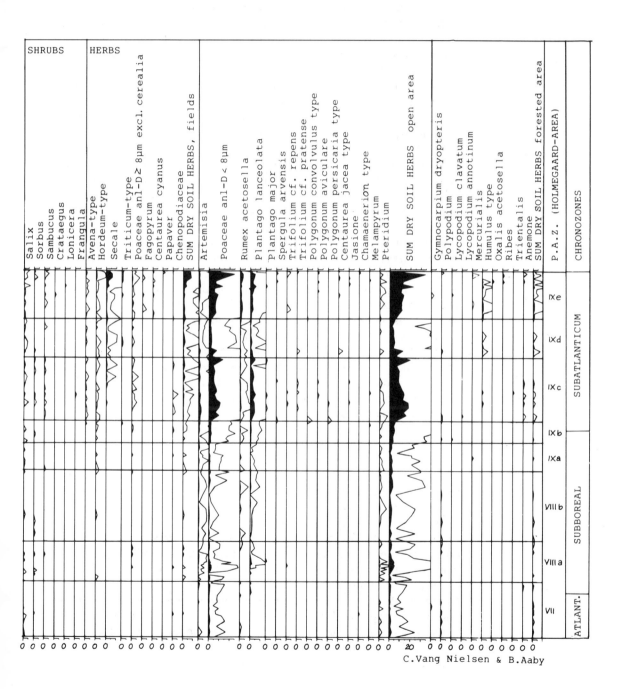

C. Vang Nielsen & B. Aaby

Pollen assemblage zone IX, late Subboreal and Subatlantic time

The *Ulmus* pollen curve declines again at the beginning of p.a.z. IX, reflecting instability in vegetation caused by new cultural activities in the Bronze Age (Aaby 1986b). The subsequent vegetational changes differ from those in the early Subboreal. *Tilia* remains an important tree in p.a.z. IXa and no *Betula* expansion occurs. The dry-soil herbs do not respond to the changes in the arboreal flora, suggesting that there was only interference with the vegetation on damp soils, where *Ulmus* grew.

Fagus immigrated *ca.* 3100 B.P. (1500 B.C.) and subsequently its pollen was continuously recorded. It was rare initially, but after 200-300 years it expanded dramatically and became a dominant tree in p.a.z. IXb. The expansion of *Fagus* seems to have developed unhampered and was probably favoured by moderate cultural activity on dry ground, as seen from the small but distinct increase in pollen from the weed flora. The significant recession of *Tilia* may be due to human exploitation, as it is unlikely that *Tilia* was removed by natural competition from *Fagus* (Aaby 1986b).

Fagus inhabits various soil types, brown earths as well as podzols, but it always occurs on well-drained soils. Thus it is restricted to potential farmland habitats, and is a valuable anthropogenic indicator which is negatively correlated with agricultural activity (Aaby 1986b). The high frequency of *Fagus* pollen (40%) indicates that it occupied most of its potential growing habitats in late Subboreal time. The *Fagus* forest was greatly reduced and extensive open areas appeared for the first time in p.a.z. IXc, *ca.* 2300 B.P. (450 B.C.). Perennial weeds, such as *Plantago lanceolata* and *Artemisia*, and grasses were common, indicating extensive land-use. Cereals were grown and the first *Secale* pollen appeared *ca.* 200 B.C.

Corylus pollen shows a distinct maximum and *Quercus* a smaller one in the latter part of p.a.z. IXc. Similar trends are demonstrated in a local pollen diagram from a small hollow, *ca.* 10 km north of Holmegaard, where an abandoned field was invaded by *Quercus* and *Corylus* and grazed (Andersen 1985, this volume). Changes in farming activity may also account for the observed variations in AP composition at Holmegaard.

The landscape changed again *ca.* 1400 B.P. (600 A.D.) and former open farmland was abandoned and became forested. The distinct pollen maximum of, for example, grasses at the end of p.a.z. IXc may reflect this event. This allowed herbaceous vegetation to flower vigorously without being grazed (cf. Andersen 1978). *Carpinus* and *Ulmus* were components in the developing forest vegetation, which quickly overshadowed *Corylus* and prevented it from flowering. The forest gradually became denser and the pioneer flora disappeared within a few centuries.

There is good evidence that the Holmegaard area was settled again in Early Medieval time. *Fagus*-dominated forest areas were cleared and the heliophilous ground flora again became prominent in the cultural landscape. *Avena, Hordeum, Triticum* and *Secale* were all grown in the region and in p.a.z. IXe, *Fagopyrum* was cultivated. *Centaurea cyanus* appeared in fields, mainly of *Secale* crops (Behre 1981). The farming seems to have been almost stable throughout the Middle Ages, whereas a significant increase in agricultural activity took place 1600-1700 A.D. and has lasted ever since.

The percentage diagram shows that the level of agricultural activity has varied considerably during the last 6000 years. Prehistoric people were active in relatively short episodes, especially in the early Subboreal and in the early Subatlantic. This inland region was probably a marginal area which served as a potential expansion area, for example, for the population living in the coastal region, which was more permanently occupied (Mikkelsen 1949).

The cultural landscape at Abkær

Like Holmegaard, Abkær represents an inland locality, but the surrounding topography and geology are much more variable, with acid sandur plains and fertile, glaciogenic hills. These differences are also reflected in the vegetation which has developed in the 2 areas.

Pollen assemblage zone VII, Atlantic time

Tilia and *Ulmus* were common in the Atlantic forest vegetation, and *Ulmus*, an edaphically demanding tree, was just as frequent in the pollen spectra as in diagrams from eastern Denmark. The less fertile soils supported a more open forest where *Betula* and especially *Corylus* were able to flower. Local diagrams show that *Corylus* was present on sandy upland soils in the Atlantic (Andersen 1978, 1984) and regional diagrams from areas with similar soils in western Jutland also have relatively high values of *Corylus* and *Betula* (Odgaard 1985). Thus, vegetational features typical for both fertile and less fertile areas are reflected in the Atlantic pollen spectra from Abkær (Fig. 5).

Pollen assemblage zone VIII, early and middle Subboreal time

The decline of *Ulmus* began *ca.* 5000 B.P. (3850 B.C.) and the frequency of *Tilia* and *Fraxinus* also diminished in early Subboreal time, whereas *Alnus* and, later, *Betula* were favoured. *Corylus* and *Quercus* seem to have been unaffected by cultural disturbance. The landscape was gradually opened, as seen from the increasing amount of dry-soil herbs in spectra from p.a.z. VIIIa.

Grasses increase *ca.* 4000 B.P. (2600 B.C.) together with *Plantago lanceolata* and *Artemisia*. These increases are probably caused by intensified grazing. *Betula* is less frequent in p.a.z. VIIIb, contrary to expectation, because the landscape attains a more open character that, in theory, should favour this tree, as in the previous subzone. The vegetation in p.a.z. VIIIb was probably managed in a slightly different way than in the previous subzone.

Fig. 5. Percentage pollen diagram from Abkær Bog. Pollen from trees, shrubs and dry-soil herbs are presented only.

Pollen assemblage zone IX, late Subboreal and Subatlantic time

Corylus was widespread until *ca.* 3100 B.P. (1500 B.C.) when it gradually declined. The *Corylus* decline in north-west Europe is

ABKÆR BOG, CENTRAL SECTION

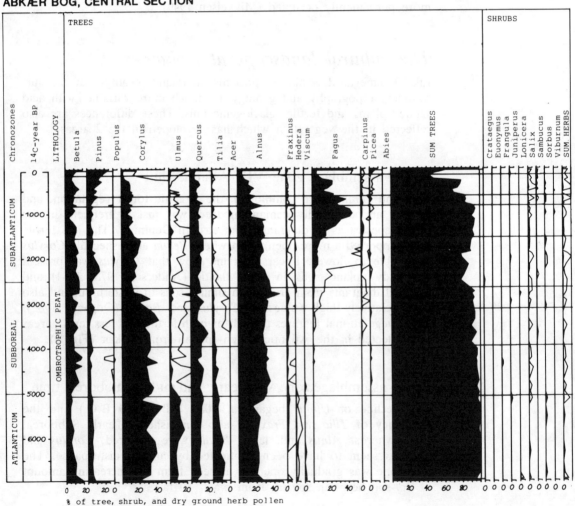

% of tree, shrub, and dry ground herb pollen

generally interpreted as a response to climatic change (Gross 1931; Straka 1970; van Geel 1978; Dupont & Brenninkmeijer 1984). However, in Denmark, it is more likely an effect of intensified anthropogenic activity (Aaby 1986b). The reduction of *Corylus* may reflect increasing agricultural activity in p.a.z. IXa. *Fagus* immigrated *ca.* 3100 B.P. (1500 B.C.), but remained rare, although climatic and edaphic conditions were favourable for it at this time, as demonstrated by the Holmegaard diagram. Its potential habitats were probably cultivated or grazed. This suggests a rather intense cultural activity in the region in the latter part of p.a.z. IXa and in p.a.z. IXb. The abundance of heliophilous dry-soil herbs also demonstrates that the landscape was rather open. The higher frequency of *Corylus*, *Fagus*

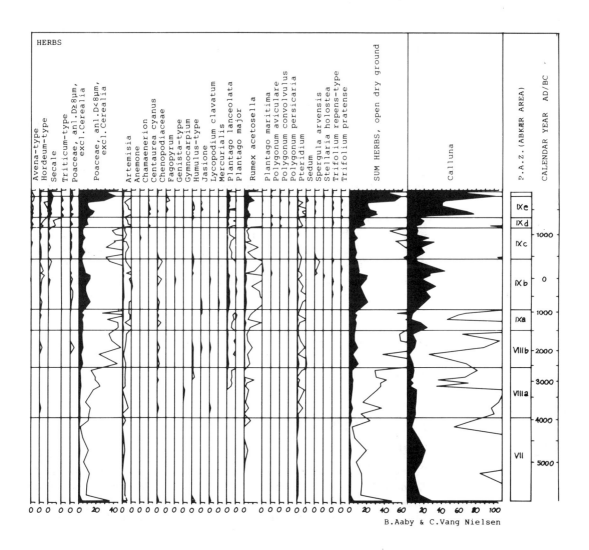

B. Aaby & C. Vang Nielsen

and *Quercus* at the expense of herbs in the latter part of p.a.z. IXb, may indicate land-use changes, allowing forests to expand on dry soils.

Distinct vegetational changes occurred *ca*. 1500 B.P. (500 A.D.). *Fagus* expanded whereas weed and cereal types diminished in the pollen spectra. It is evident therefore that cultural activity decreased and abandoned farmland was invaded by trees. *Fagus* was dominant, but *Ulmus*, *Fraxinus* and, a little later, *Carpinus* were part of the pioneer flora. *Corylus* quickly lost its importance due to shading. *Fagus*-dominated vegetation became widespread, and by 1200-1000 B.P. (750-1000 A.D.), it was probably abundant in most of its potential habitats.

The Abkær region remained forested until the 12th century when a new cultural expansion took place and the *Fagus*-dominated forests were cleared for agriculture. A further expansion in the 14th century is shown by the distinct rise in pollen frequencies of crops and weeds. The exceptionally high amount of *Calluna* pollen at that time probably originated from local bog vegetation as well as from extensive heath areas that mainly developed after Medieval forest clearances.

A cultural decline began *ca*. 1400-1450 A.D. and lasted for *ca*. 200 years. *Fagus* and *Quercus* were favoured, whilst dry-soil herbs and *Calluna* decreased. Thus, heath areas may have been invaded by trees as seen today in abandoned heath areas. This dereliction is recorded in historical documents (Gissel *et al*. 1981; Gregersen 1974, 1977). Cultural activity was again high during the last 300 years, and the uppermost pollen spectra represent the modern open landscape in which *Picea* plantations are a new, characteristic feature.

The percentage diagram shows generally increasing cultural activity since early Subboreal time, only interrupted by one long-lasting period of decreased exploitation. Cultural expansion was restricted mainly to the early Subboreal (3850 B.C.), middle Subboreal (2600 B.C.), late Subboreal (1500-1000 B.C.) and middle Subatlantic (1150 A.D.). Thus, shorter periods of vegetation instability were succeeded by longer periods with almost stable floristic composition.

The Abkær region was more permanently occupied compared to the Holmegaard region and differences in floristic composition are obvious in middle and late Subboreal times, whereas the 2 regions have followed broadly similar vegetational trends over the last 2 millennia.

Pollen influx

Methods

Pollen influx at Abkær and Holmegaard was estimated by adding *Lycopodium* tablets (Stockmarr 1971) to a known sediment volume and counting the exotic spores in routine pollen analysis (Birks & Birks 1980). Peat-accumulation rate was calculated by means of 2

HOLMEGAARD BOG

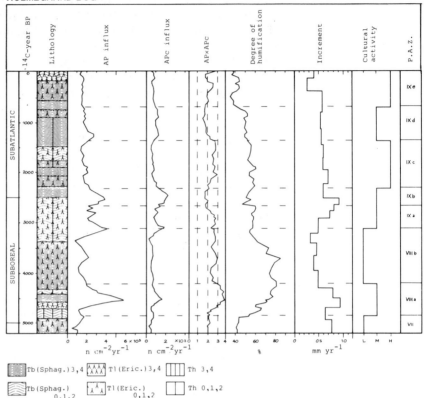

Fig. 6. Total tree pollen (AP) influx, transformed tree pollen (APc) influx, ratio APxAPc⁻¹, degree of humification, increment, and tentative level of cultural activity in Holmegaard Bog. The tentative levels for cultural activity are L = low; M = medium; H = high. Sediment stratigraphy according to Troels-Smith (1955) and Aaby & Berglund (1986).

successive ^{14}C-dates as described above.

Only the influx of total arboreal pollen (AP) and dry-soil herb pollen is considered, as the aim is to describe overall variations in pollen influx and demonstrate how they are related to human activity, as reflected in pollen percentage diagrams. Furthermore, AP is the dominant component in the pollen sum used for calculating percentage values, and variations in AP influx therefore influence the frequencies obtained in relative pollen diagrams.

Tree-pollen influx at Holmegaard

AP influx varies between *ca.* 1000 and 6000 grains cm^{-2} yr^{-1}, but generally it was *ca.* 2000-2500 grains cm^{-2} yr^{-1} since late Atlantic time (Fig. 6). High influx values occur in p.a.z. VIIIa, IXa and IXb.

Results from Holmegaard are comparable with data from similar sites. Dupont (1985) found *ca.* 2000-4000 AP cm^{-2} yr^{-1} in shallow peat and 4000-5500 AP cm^{-2} yr^{-1} in hummock peat. Low pollen producers such as *Tilia* and *Fagus* were frequent at Holmegaard while taxa with a higher pollen production dominated the Dutch locality. By allowing for differential pollen production (Andersen 1970, 1980), almost similar influx values are obtained from Holmegaard and from hollow peat at Bergerveen (Dupont 1985). Mack *et al.* (1979) mention that pollen influx (AP + NAP) in Holocene peat is generally *ca.* 5000 grains cm^{-2} yr^{-1}, with a minimum of 1000 grains cm^{-2} yr^{-1}, and similar influx values have been reported from raised bogs in south-west England (Beckett 1979).

As mentioned earlier, some of the variations in AP influx may be due to errors in estimation. Thus, the local *Calluna-Eriophorum vaginatum* vegetation in p.a.z. VIIIa may contribute to the influx peak recorded from this subzone. The time-scale used may influence the result, having a maximum incremental rate in p.a.z. VIIIa.

Assuming a peat increment similar to that below (0.6 mm yr^{-1}), and a reduction of pollen influx by 30% because of a higher pollen filtration efficiency by the hummock vegetation (cf. Dupont 1985), there still remains an influx maximum of *ca.* 2500-3000 AP cm^{-2} yr^{-1}. The maximum therefore seems to reflect actual changes in AP influx. Variations in peat composition and peat structure may also have influenced influx values at other levels, but there is no obvious relationship between AP influx and increment, degree of humification or macrofossil composition in p.a.z. IXa and IXb, where influx maxima also occur (Fig. 6). Accordingly, it is assumed that the main trends in AP influx display general changes in annual pollen deposition (Fig. 8).

Tree species vary considerably in pollen production. Changes in species composition will accordingly influence AP influx. After transforming for differential pollen production within forests (Andersen 1970, 1980), there still remains considerable variation in the transformed tree pollen (APc) influx curve (Fig. 6). It shows similar trends to the AP influx curve, although they are not congruent - see the variation in the AP x APc^{-1} ratio. This ratio is dependent on species composition and is high when *Alnus* and other prolific pollen producers are common and low when *Tilia*, *Fraxinus* and *Fagus* dominate. Variations in AP influx are therefore determined by the tree vegetation, but more than the species composition could alone account for.

AP influx was low in the late Atlantic and increased considerably in p.a.z. VIIIa when Neolithic people strongly interfered with the vegetation. AP influx fell again with the cultural cessation *ca.* 4200 B.P. (2900 B.C.) and remained stable at *ca.* 1500-2000 AP cm^{-2} yr^{-1} as in the late Atlantic. A new rise in influx occurred in p.a.z. IXa and it was high until the end of p.a.z. IXb. The level of cultural activity was moderate in these zones and obviously higher than in the previous p.a.z. VIII (see Fig. 6).

Variations in AP influx suggest that human activity increased the forest's ability to produce and disperse pollen in p.a.z. VII-IXb. This relationship was reversed at the beginning of p.a.z. IXc, when intensified cultural activity resulted in a decrease in AP influx. A small temporary peak in AP influx occurs at the beginning of p.a.z. IXd, when *Fagus* and other pioneer taxa colonized abandoned farmland. AP influx decreased again as the forest structure became denser, and Medieval forest clearances had an insignificant effect on influx values.

The positive relationship between level of cultural activity and AP influx in early and middle Subboreal time is explained by structural changes in the forest vegetation. AP influx was low in the Atlantic, when the forest was dense (cf. Iversen 1973). Although Neolithic people felled trees and created openings, the forest vegetation had a higher pollen production capacity because the open-structured forest enabled a larger part of the tree's crown to flower. Furthermore, solitary trees often flower vigorously every year, in comparison to trees in closed stands which normally have a 2-year flowering cycle (Andersen 1974).

AP influx declined in p.a.z. IXc when the nemoral landscape was changed into an open landscape. The remaining trees were unable to compensate for the loss in pollen production of felled trees. The expanding, open-structured pioneer forest increased the AP influx at the beginning of p.a.z. IXd, until a continuous canopy was established in the middle of the zone.

The influence of forest structure on AP influx was also mentioned by Dupont (1985). To test my Holmegaard interpretation similar influx investigations were undertaken at Abkær, where the cultural development was different, especially in the Subboreal (see Figs. 7, 8).

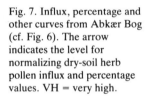

Fig. 7. Influx, percentage and other curves from Abkær Bog (cf. Fig. 6). The arrow indicates the level for normalizing dry-soil herb pollen influx and percentage values. VH = very high.

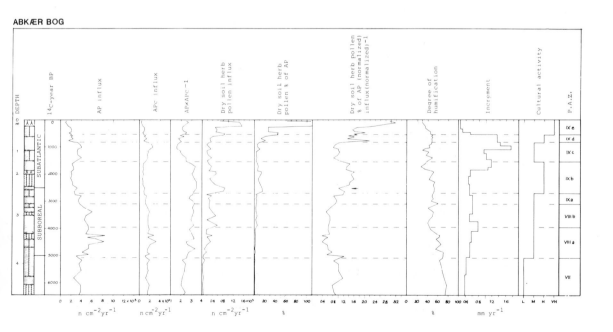

ABKÆR BOG

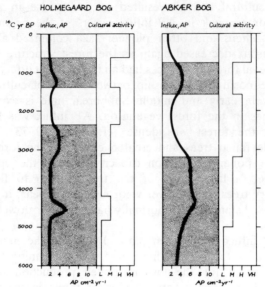

Fig. 8. General trends in total AP influx and levels of cultural activity (see text) based on information from the percentage diagram from Holmegaard Bog and Abkær Bog. For explanation of letters, see Figs. 6 and 7.

Tree-pollen influx at Abkær

AP influx varied between *ca.* 1000 and 8000 AP cm^{-2} yr^{-1} since late Atlantic time with an average of *ca.* 3500-4000 AP cm^{-2} yr^{-1} (Fig. 7). These values are comparable to influx data from Holmegaard, although AP influx is a little higher at Abkær. AP influx was generally higher in p.a.z. VIIIa and VIIIb than in the preceeding and following zones. Relatively high values also occur in p.a.z. IXc. These variations all occur in *Sphagnum* peat, suggesting that they are not affected by peat composition, humification or accumulation rate. The main trends in AP influx are therefore considered to be controlled by the tree vegetation.

Some of the AP influx changes were caused by changes in the composition of the arboreal flora. Transforming for differential pollen production, there was an almost constant APc influx from p.a.z. VII to the end of p.a.z. VIIIb, although a small maximum still remains in p.a.z. VIIIa (Fig. 7). The late Subboreal and Subatlantic changes in APc influx are almost identical to the AP influx changes, but the curves are not congruent. Therefore only some of the trends in AP influx can be attributed to floristic features other than species composition.

The Atlantic AP and APc influx at Abkær is distinctly higher than at Holmegaard. This suggests a more open forest structure at that time around Abkær. This is supported by the higher representation of *Corylus*, which only flowers abundantly when it is unshaded (Andersen

1980), suggesting that it formed part of the forest canopy. Less fertile soils, especially in the sandur plains west of the bog, may have supported a more open forest type where *Corylus* was probably able to flower abundantly (cf. Odgaard 1985). This abundance may also explain the absence of any expansion of *Corylus* in p.a.z. VIIIa, when Neolithic people cleared the forest.

The higher AP influx in p.a.z. VIIIa and VIIIb was caused mainly by changes in species composition and the moderate forest clearances contributed only a small additional increase in AP influx, which was already high.

AP influx gradually decreased from the end of p.a.z. VIIIb and this trend cannot be explained by variation in species composition alone, as APc influx has a similar trend. *Corylus* also diminishes from that time, reduced by intensified cultural exploitation in p.a.z. IXa (Fig. 5).

A distinct minimum in AP influx in p.a.z. IXb probably results from extensive clearance and the formation of an open landscape, as indicated by the abundance of open-ground herbs and the scarcity of *Fagus*.

The AP and APc influx curves rise again when human activity decreased and forest vegetation occupied abandoned farmland. They remain high until the next period of forest clearance.

Variations in AP influx at Abkær are correlated with the level of cultural activity, as reflected in the percentage diagram and they can be explained by changes in the forest structure. The AP influx remained high when humans had a moderate influence on the landscape with an open-structured forest. Conversely, low AP influx values are found when cultural activity was high.

A comparison of changes in AP influx at Holmegaard and Abkær shows that they were asynchronous in the Subboreal (Fig. 8), suggesting that regional climate or other environmental parameters had no significant influence on the observed influx variations.

The explanation given for the varying AP influx at Holmegaard is valid also for the Abkær site, which supports the assumption that the total AP influx was determined mainly by the forest structure in the past 6000-7000 years, although species composition also caused changes.

The herbaceous pollen flora in percentage diagrams

The considerable variations in AP influx influence the herb pollen frequency in percentage diagrams, since NAP is usually expressed relative to AP, and AP is often a major constituent of the pollen sum.

The effect of varying AP influx on dry-soil herb pollen (DSP) percentages has been analysed (Fig. 7). Only the Abkær results are discussed as they are similar to those at Holmegaard.

The DSP influx values are normalized relative to the highest influx values obtained in prehistoric times (indicated by an arrow in Fig. 7).

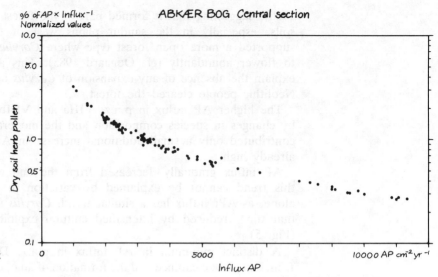

Fig. 9. Relationship between
the dry-soil herb pollen ratio
% of AP (normalized) x influx
(normalized)$^{-1}$ and AP influx.

Also, DSP percentages, calculated as % AP, are normalized using the
% value at the same level. The ratio DSP % (normalized) x DSP
influx (normalized)$^{-1}$ shows that the relationship between DSP % and
DSP influx has varied considerably since late Atlantic time. The ratio
is high in p.a.z. VII, IXb, IXd and IXe, suggesting that the normalized
DSP % are relatively higher than the influx values and *vice versa* in
periods when the ratio is low. Variations in the ratio are related to
changing AP influx. The ratio is high with low AP influx and decreases
with rising AP influx (Fig. 9), and varies by a factor of 10. The
representation of the herbaceous pollen flora in percentage diagrams
is therefore strongly linked to AP influx. This is also valid if AP is
a dominant part of the pollen sum, as in the percentage diagrams
from Abkær and Holmegaard.

A similar relationship between the above ratio and AP influx is
found when the AP values are transformed for differential pollen
production (Fig. 10).

The Abkær investigations shows that the herbaceous pollen flora is
underrepresented in percentage diagrams during periods of high AP
influx compared to periods with a low AP influx. Accordingly, early
and primitive cultural phases may easily be underestimated compared
to the later ones if the effect of different forest structure is neglected.

Summary

(1) Data from 2 extensive ombrotrophic mires demonstrate spatial
and temporal variations in the regional forest and open ground flora
based on percentage and influx pollen diagrams.

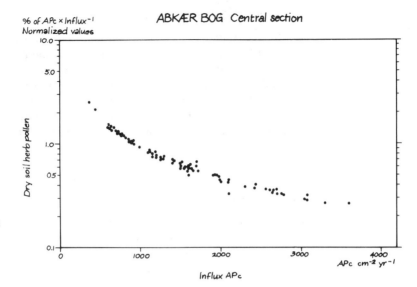

% of APc × Influx⁻¹
Normalized values

ABKÆR BOG Central section

Fig. 10. Relationship between the dry-soil herb pollen ratio % of APc (normalized) x influx (normalized)⁻¹ and the transformed AP influx.

(2) The relative pollen diagrams show a different development in the 2 regions since the late Atlantic due mainly to differences in the intensity of cultural activity.

(3) It is emphasized that the Subboreal and Subatlantic variations in *Corylus* and *Fagus* pollen percentages are determined primarily by human activity, and *Fagus* is a particularly valuable indicator of dry-soil habitat exploitation.

(4) The total tree-pollen (AP) influx varies between 1000 and 6000 grains cm^{-2} yr^{-1} at Holmegaard and 1000 to 8000 grains cm^{-2} yr^{-1} at Abkær. These values are comparable with results from other peat sites in north-west Europe.

(5) The main trend in AP influx is determined by tree vegetation, but variations in peat composition, structure, increment, etc. may cause small irregularities.

(6) The considerable variation in total AP influx is explained mainly by changes in the forest structure caused by human activity. Tree species composition also caused variations.

(7) AP influx responds positively to increased cultural activities in forested regions and responds negatively when the landscape is open.

(8) If AP dominates the pollen sum, then the herbaceous pollen flora is underrepresented in percentage diagrams during periods of high AP influx compared to periods of low AP influx.

(9) Early cultural phases may easily be underestimated compared to later ones if the effect of different forest structure is neglected.

Acknowledgements

O. S. Jacobsen, C. Vang Nielsen, B. V. Odgaard and M. Thelaus took part in the field work. Coring equipment used at Holmegaard was put at my disposal by B. E. Berglund. C. Vang Nielsen also assisted with pollen analysis. B. Stavngaard and H. Bahnson determined physical properties and peat samples were ^{14}C-dated by H. Tauber. R. Flensborg and H. J. B. Birks made linguistic improvements. R. Flensborg typed the manuscript. I am grateful to all for their skilful assistance.

Nomenclature

Plant nomenclature follows *Flora Europaea* except for Poaceae (= Gramineae) and *Chamaenerion* (= *Epilobium* Sect. Chamaenerion).

Quantitative Estimation of Human Impact on Cultural Landscape Development

H. J. B. Birks, J. M. Line and T. Persson

Introduction

Palaeoecological studies in north-west Europe clearly demonstrate that the magnitude of human impact on local and regional vegetational history and hence on cultural landscape development has varied greatly in time and space over the last 5000 years. At particular times, some areas have experienced intense human impact, whereas others have been subjected to little or no human disturbance. Alternatively at one point in space the extent of human impact has changed considerably through time. Human impact has thus varied in time and space, just as it has in the recent past.

Such variations in human impact are commonly termed expansion and regression (or recession) phases (e.g. Berglund 1985a, this volume) or clearance (or interference) and regeneration phases (e.g. Edwards 1979). The underlying ecological and archaeological causes of these variations are subjects of much speculation (e.g. Aalen 1983; Berglund 1969; H. Göransson 1986, 1987; Gräslund 1980; Welinder 1975, 1977, 1983a). These include changes in human settlement and land-use patterns, population density, climate, vegetation composition and structure, local or regional hydrology, and soil fertility, as well as socio-economic, demographic, and cultural factors, and the complex but poorly understood interactions between these components (Berglund this volume; Emanuelsson this volume). However, such interpretations are not our concern here. Instead we consider how to derive quantitative estimates of temporal and spatial changes in the magnitude of human impact from pollen-stratigraphical data.

The structure of this chapter is as follows. First, we review briefly previous approaches to estimating quantitatively from pollen-analytical data the extent of human impact in time and space. Second, we discuss the suitability of correspondence analysis for estimating the magnitude of vegetational change from palynological data, and of rarefaction analysis for estimating palynological richness or diversity. Third, we apply these techniques to pollen data from Diss Mere in south-east England to derive quantitative estimates of vegetational change and richness over the last 7000 years. Fourth, we use correspondence analysis to compare the extent of vegetational change over the last 7000 years at 2 sites in the Ystad area of southern Sweden.

Approaches for estimating the magnitude of human impact

Previous approaches at estimating the magnitude of human impact on vegetation have involved summing various 'anthropogenic indicator' pollen types and plotting this sum through time to produce so-called human-influence or impact diagrams (e.g. Berglund 1969, 1985a; Gaillard 1984). Pollen indicators commonly used include anthropochors and selected apophytes and archaeophytes (Birks 1986a). There are many problems in selecting and using 'anthropogenic indicators' (Behre 1981; papers in Behre 1986). These include taxonomic problems resulting from current pollen-morphological limitations (e.g. Chenopodiaceae, Gramineae, Liguliflorae) and hence problems in deciding what is anthropochorous; representational bias due to differential pollen production and dispersal; problems in finding modern analogues for early land-use; differences in dependence on human influence of a species in different geographical areas; and different ecological tolerances in different parts of a species range (centre *vs* margin) (e.g. Holzner 1978). Berglund (1985a) discusses these and related problems in the context of Scandinavia.

Human impact can influence most, if not all, of the vegetation at the landscape scale of 10^8-10^9 m^2, the presumed source area for pollen assemblages deposited in lakes or bogs of >500 m diameter (Birks 1986a). It is relevant when estimating human impact to consider the whole pollen assemblage, rather than selected 'anthropogenic indicators', the indicator value of which may be unclear. Thus attempts at quantifying human impact from pollen data should use the entire pollen assemblage at a site. As these data are multivariate, comprising quantitative counts of many pollen taxa in numerous samples, approaches to quantification involving multivarate data-analytical procedures are needed.

Numerical approaches

Introduction

The problem is to analyse the total pollen assemblage from a site of interest and to extract and display the major underlying patterns of change or 'directions of variation' within these data. If we restrict our analyses to the time period when human impact was the major determinant of vegetational and hence palynological change in the study area (the so-called *Homo sapiens* phase, Birks 1986a), we can assume that the *major* patterns of change detected by numerical analysis are a reflection or 'proxy-record' of the intensity of human impact on the pollen-source area of the study site.

Within all pollen data there are invariably pollen types of high and low relative representations. Some of these may reflect human impact either positively or negatively, whereas others may be indifferent to human influence. There are usually only a few numerically abundant pollen types but many numerically rare taxa, often of considerable interpretative importance. The data are usually expressed as relative percentages and are always recorded in a stratigraphical and hence temporal sequence. Any numerical approach used should take into account these important palaeoecological properties of the data.

Multivariate techniques involving correlations and magnitudes or 'sizes' of different pollen curves, such as principal components analysis (PCA) are not appropriate for studying human impact from percentage pollen data. The numerically rare but ecologically important minor pollen types are important and should be included and analysed appropriately. PCA, in its conventional implementation, is heavily size-orientated and its results are strongly influenced by differences in relative pollen representation between taxa, with numerically abundant pollen taxa dominating. Moreover, conventional PCA is not strictly applicable to 'closed' percentage data (Aitchison 1986). The quantitative contribution of numerically rare pollen types can be increased by standardizing each to unit variance (e.g. Turner 1986). However, such standardization is "illogical" (Prentice 1980) when applied to percentage pollen data, and commonly has deleterious effects on the 'signal' to 'noise' ratio within the data. A further limitation is that the close interrelationships between pollen types and samples can be represented only approximately by either covariance or distance biplots (ter Braak 1981; Gordon 1982) and thus PCA results cannot be readily and accurately presented in a form comprehensible to pollen analysts. An alternative approach is required.

Correspondence analysis

Correspondence analysis (CA) provides an attractive alternative. It is a powerful numerical technique for detecting major patterns in complex data and for displaying and quantifying inter-relationships between samples and variables. In addition it overcomes many of the disadvantages of PCA when applied to percentage data that include both abundant and rare pollen types. The underlying theory is discussed, for example, by Greenacre (1984), Gower (1984) and Hill (1974, 1982). Palynological applications include Gordon (1982), Peglar *et al.* (1984), Jacobson & Grimm (1986) and Berglund *et al.* (1986).

For our purposes CA can be thought of as a variant of PCA that attempts to simplify complex data into a few major axes of variation and to represent the data geometrically. However, CA differs from PCA in several mathematically and palynologically critical ways. These include the following:

(1) As a result of data transformations and scalings, the implicit dissimilarity measure between samples and between pollen types is symmetric (Greenacre 1984), allowing simultaneous analysis of samples and taxa and their joint representation on the principal axes of variation. This allows patterns between samples to be interpreted in terms of patterns in the pollen types, and *vice versa*, thereby permitting easier and more detailed interpretation than is possible with PCA (Gordon 1982; Birks 1985).

(2) Numerical properties of the implicit distance measure ensure maximization of the ratio of 'signal' to 'noise' in percentage pollen data (Prentice 1980).

(3) Because CA considers resemblances between profiles of data rather than absolute magnitudes, problems of percentage 'closure' do not arise (Marcotte 1986).

(4) Because the shape rather than the absolute size of the pollen curve is important, effects of pollen representation are not important, and CA is thus a 'representation-insensitive' technique (Gordon 1982).

(5) CA maximizes the correlation between sample and taxon scores on the first non-trivial CA axis (Hill 1974, 1982). This results in an optimal diagonal structure and concentration of data values when the columns and rows of the original data matrix are arranged in the order of the sample and taxon scores on the first non-trivial CA axis, rather than the block structure produced by PCA (Gauch 1982). Thus CA provides an easy and rapid solution to seriation problems that seek to arrange data in diagonal structure, with the most appropriate sample order according to the observed values for the associated variables (Hill 1974, 1982; cf. Bonham-Carter *et al.* 1986). In palynology, sample order is known *a priori* because of stratigraphical ordering. Sample scores on the first non-trivial CA axis thus reflect major patterns between pollen spectra and curves and, when plotted in stratigraphical order, provide a quantitative estimate of the magnitude of palynological and, by inference, vegetational change between samples (Jacobson & Grimm 1986).

(6) When a human-influence diagram (e.g. Berglund 1969) is constructed using particular indicator pollen types, a simple one-dimensional ordination is performed. A score is assigned to each sample based on the sum of the anthropogenic indicator value for each pollen type (e.g. 1 for anthropogenic indicator, 0 for non-anthropogenic indicator) weighted by its pollen percentage. Weighted averaging is also central to CA (Hill 1974; ter Braak & Barendregt 1986). Pollen types are ordered along the major axis of variation, and are weighted by how much they contribute to the gradient. For example, *Tilia* and *Viscum* may contribute little to the gradient of increasing human impact,

whereas *Cannabis sativa* and *Secale cereale* contribute much to this gradient. Such weights are approximate indicator values and a score for human influence can thus be calculated for each sample. In this context, CA can be regarded as a multivariate extension of the use of selected anthropogenic indicators to construct human-impact curves.

The mathematical properties of CA thus closely parallel important palynological requirements, suggesting that CA is an appropriate method for deriving quantitative estimates of human impact from pollen data. As attention here is centred on the first non-trivial axis, questions of detrending to minimize high-order dependence of the second (and subsequent) axes on the first axis (so-called 'arch' or 'horseshoe' effects) do not arise, fortunately considering current debates about the desirability (or otherwise) of detrended CA (Hill & Gauch 1980; Wartenberg *et al.* 1987).

Rarefaction analysis

An interesting and important feature of pollen assemblages is their total richness, perhaps one of the most useful and reliable measures of biotic diversity (Hurlbert 1971). Diversity indices that consider richness *and* equitability (evenness with which individuals are distributed among species) such as the Shannon-Wiener information index (Moore 1973; Küttel 1984) are not suitable for pollen data because the values of such indices are inevitably influenced by differences in pollen representation between taxa and by the effects of percentage constraints on the relative representation of taxa.

The number of species recorded in a sample depends on the number of individuals (in our case pollen grains) counted in the sample. As pollen counts in a stratigraphical sequence are usually of different sizes and as the relationship between richness and count size is not linear, we need to estimate how many taxa would have been found if all sample counts were the same size. The robust technique of rarefaction analysis, developed in biogeography and ecology, provides unbiased estimates, for a standard count size, of the expected number of taxa (E(Sn)), the so-called rarefaction diversity or richness index. E(Sn) permits richness estimates to be compared even when the samples on which they are based are of different sizes. The theory of rarefaction analysis is presented by Simberloff (1979) and Tipper (1979). Examples of palaeoecological applications include Malmgren and Sigaroodi (1985) and Berglund (1986). Like CA, rarefaction analysis is largely 'representation insensitive' when applied to pollen data. CA and rarefaction analysis provide quantitative summaries of the magnitude of change between samples and total richness, respectively.

Computing

CA was implemented using the FORTRAN program BENZECRI (Birks 1986b). Rarefaction was done using the FORTRAN programs RAREFORM and RAREPOLL developed by JML, the latter based, in part, on Simberloff's (1978) SIM program for rarefaction.

Diss Mere, southern England

The use of these numerical methods for quantifying past human impact on vegetation from pollen data is illustrated by analyses of data from Diss Mere, Norfolk. It is a small, deep lake whose pollen stratigraphy has been studied in considerable detail by Sylvia Peglar (Peglar *et al*. 1988; Peglar & Birks unpublished). The stratigraphy suggests continuous human impact on the vegetation since at least Bronze Age times. Extensive forest clearance occurred in the Iron Age, and the town of Diss developed around the Mere in Anglo-Saxon times. Numerical analyses were restricted to 95 samples covering 0- *ca*. 7000 B.P. (616 cm-1950 cm in the 1979 core series; Peglar *et al*. 1988). CA was confined, for computational reasons, to the 58 pollen and spore types included in the basic pollen sum with values of 1% or more in any sample. Rarefaction analysis involved all taxa in the pollen sum and E(Sn) was calculated for every sample with a constant sample count (n) of 626 grains, the smallest single actual sample count.

The first non-trivial CA axis is large (λ_1 = 0.511) compared with axes 2 (λ_2 = 0.122) and 3 (λ_3 = 0.111) and represents 41% of the total inertia. Its large relative size suggests that there is one major direction of variation within the data, and that for our purposes axis 1 provides an efficient summary of the major trends in the Diss Mere pollen record. Sample scores on axis 1 are plotted in stratigraphical order (Fig. 1) with the vertical axis drawn, not through 0, but at the largest negative score. This graphical presentation- allows CA results to be represented in a form familar to pollen analysts and conceals the concept of negative and positive scores. The pollen types most strongly associated with the lowest sample scores are *Tilia, Taxus baccata* and *Viscum*, whereas those associated with the highest scores include *Vicia* cf. *V. faba, Cannabis sativa, Papaver* and *Echium*. Axis 1 (Fig. 1) reflects vegetational patterns that are most simply interpreted as resulting from the changing intensity of human impact with time, and is thus a 'proxy record' and an approximate estimate of the extent of human impact through time. Rarefaction estimates of richness E(Sn), are also plotted. The curves closely parallel each other. If the patterns of right and left swings in each curve are compared (Birks 1985), the null hypothesis that the curves are similar in terms of swings is not rejected.

There is a major rise in human impact in the Bronze Age associated with the *Tilia*-decline. This is paralleled by an increase in richness. A short-lived phase of local *Hordeum* cultivation in the Bronze Age is clearly reflected by both CA and rarefaction. There is then a decrease in human impact (regression phase?) prior to extensive Iron Age forest-clearance. Human impact on the vegetation within the pollen-source area then increased progressively to a maximum at *ca.* 1750 A.D. This increase was paralleled by increases in richness. Effects of human impact on vegetation appear to have declined in the last 200 years, possibly as a result of the development in modern times of large, uniform and intensively cultivated fields with few weeds or ruderals.

Although strong richness gradients can cause distortion in CA results (Dargie 1986), theoretically there should be no statistical relationship between E(Sn) and CA axes (M. O. Hill personal communication), unless they are both responding to the same underlying process. In such cases their plots would be similar. The strong similarities between

Fig. 1. Stratigraphical plot of sample scores on the first correspondence analysis axis (left) and of rarefaction estimate of richness (E(Sn)) (right) for Diss Mere, England. Major pollen-stratigraphical and cultural levels are also shown. The vertical axis is depth (cm). The scale for sample scores runs from -1.0 (left) to +1.2 (right).

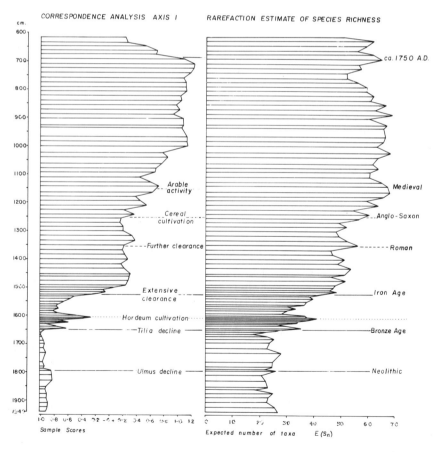

DISS MERE

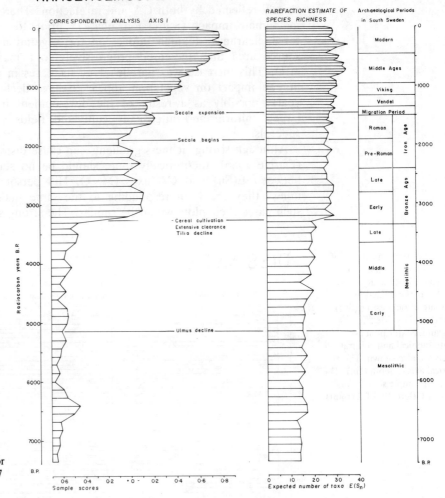

Fig. 2. Stratigraphical plot of sample scores on the first correspondence analysis axis (left) and of rarefaction estimate of richness (E(Sn)) (right) for Krageholmssjön, Sweden. Major pollen-stratigraphical horizons are shown, along with the main archaeological periods in South Sweden. The vertical axis is estimated sample age in ^{14}C-years B.P. The scale for sample scores runs from +0.7 (left) to -0.9 (right).

CA axis 1 scores and E(Sn) (Fig. 1) suggest both vegetational change and richness are responding to a common factor, namely human impact. It is necessary therefore to consider how human impact could influence richness at the landscape scale of 10^8-10^9 m^2).

The curve of E(Sn) reflects, in part, changing floristic richness, resulting from land-use changes and agricultural practices (e.g. Groenman-van Waateringe 1979). Today floristic richness is often greatest at boundaries between vegetational units (van der Maarel 1971). A diverse landscape is one with many component vegetational patches and hence many boundaries. Thus the changing patterns of pollen richness over the last 7000 years may be a record not only of biotic richness *per se* but also of the frequency of boundaries and patches within the landscape. E(Sn) may be an indirect reflection of the frequency of patches and of the mosaic structure of the landscape. Such a mosaic would be intensified by increasing human impact due to the creation of clearings, fields, tracks, coppice stands, etc. Prior to intensive human impact there would be a limited mosaic pattern of vegetation types and associated boundaries and hence a low E(Sn). During the *Homo sapiens* phase, mosaic patterns would have become more and more pronounced, thereby creating more and more boundaries and increasing E(Sn). Figure 1 suggests, on this interpretation, that maximal landscape diversity occurred between Medieval times and 1750 A.D. This interpretation explains the close parallel between the CA ('human impact') and E(Sn) curves.

A similar relationship exists at Krageholmssjön, southern Sweden (Fig. 2) studied as part of the Ystad Project by Gaillard (1984). CA, using all 66 taxa included in the pollen sum, gives a large first non-trivial axis ($\lambda_1 = 0.261$) compared with axes 2 and 3 ($\lambda_2 = 0.065$, $\lambda_3 = 0.040$). Axis 1 represents 48% of the total inertia and associates pollen types such as *Viscum*, *Hedera* and *Tilia* with low sample scores and Cerealia-type, *Centaurea cyanus* and *Secale* with large sample scores. The major onset of human impact in the Bronze Age is paralleled by an increase in richness. There is a slight gradual increase in richness as human impact increased from the Viking Age until the present day.

Fårarps Mosse and Krageholmssjön, south Sweden

In this section we present the results of using CA to compare the relative extent of human impact through time at 2 sites in the same general area. This comparative approach involves CA of data combined from 2 (or more) sites. For comparative purposes, the data should cover approximately the same time span at each site. We selected, for illustrative purposes, the Ystad area of southernmost Sweden where an intensive study of the development and functioning of the cultural landscape over the last 6000 years is currently in progress (Berglund this volume). Two contrasting sites in this area are analysed numerically here. Fårarps Mosse (Hjelmroos 1985a) is located within the coastal boulder clay area, and lies close to the Neolithic settlement at Karlsfält. Krageholmssjön (Gaillard 1984) is situated further inland on clayey sandy till in an area with Bronze Age round-barrows and an Iron Age settlement near the lake.

The data from both sites were edited to achieve consistent pollen-morphological categories and nomenclature. CA was then done using all 164 samples from the 2 sites that are 7500 B.P. or younger. All 66 pollen types were included in the calculation sum. The first non-trivial CA axis is 38.5% of the total inertia ($\lambda_1 = 0.271$) and is large compared with subsequent axes ($\lambda_2 = 0.088$, $\lambda_3 = 0.047$). Sample scores for the 2 sites are plotted against estimated sample age using provisional age estimates provided by B. E. Berglund, M.-J. Gaillard & J. Regnéll (May 1986, personal communication). Phases of high sample scores, equated with phases of high human impact, are marked by thick vertical bars (Fig. 3). The chronology used here and on Figure 2 differs from a later scheme (March 1987) used by Berglund (this volume). Both chronological schemes are necessarily tentative and will, no doubt, be revised further in light of investigations currently in progress within the Ystad project. The chronology shown in Figures 2 and 3 is provisional, but as this chapter is primarily about methodology rather than palaeoecological or archaeological interpretations, the chronology used is adequate for illustrating the methodology.

The CA results (Fig. 3) highlight differences in the intensity of human impact at the sites. There are important phases of human impact in the Early and Middle-Late Neolithic at Fårarps Mosse, whereas there are no indications for any contemporaneous impact at Krageholmssjön. Human impact of comparable magnitude occurred at both sites in the Bronze Age. Human impact appears to have been consistently less at Krageholmssjön during the Iron Age but reached a peak during the Roman Iron Age, perhaps associated with the lake settlement site. Human impact increased sharply in the Viking Period at Fårarps Mosse, in contrast to a gradual increase at Krageholmssjön. The CA results (Fig. 3) provide a useful summary and a quantitative comparison of the relative magnitude of vegetational change and presumed human impact at these contrasting sites over the last 5000 years.

There is obvious potential in extending this approach to include many sites and to use CA scores for contemporaneous samples to compare relative human impact between sites. These estimates can be mapped to provide a geographical perspective (e.g. Berglund 1985a) and correlated to construct time-space diagrams of human impact (e.g. Berglund 1969; Gräslund 1980). Results of this approach applied to several sites within the Ystad area will be presented elsewhere.

Conclusion

Within the broader context of the last glacial/interglacial cycle (Birks 1986a), the *Homo sapiens* phase of the last 5000 years is characterized by high richness and rapid rates of vegetational change. We have demonstrated here how numerical methods can be used to quantify magnitudes of vegetational change, and to estimate changing patterns of richness at the landscape scale within the *Homo sapiens* phase. Rates of vegetational change can be readily quantified by means of correspondence analysis (Jacobson & Grimm 1986). Numerical methods can contribute to our understanding of human-vegetation-environment interactions over the last 5000 years, and how human impact has

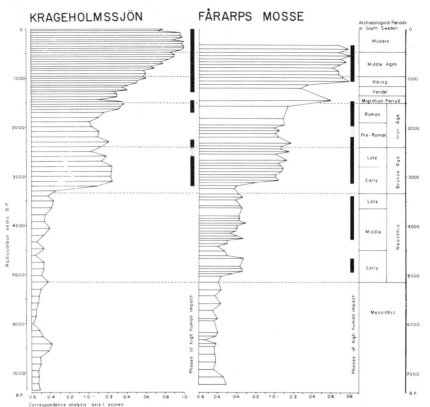

Fig. 3. Stratigraphical plot of sample scores on the first correspondence analysis axis of a combined analysis of data from Krageholmssjön (left) and Fårarps Mosse (right). The main archaeological periods in South Sweden are shown. The vertical axis is estimated sample age in ^{14}C-years B.P. The scales for sample scores run from +0.8 (left) to -1.0 (right). Major phases of human impact, suggested by the correspondence analysis scores, are marked as black bands. Uncertain phases are stippled.

influenced landscape-vegetation mosaics and richness through time. These and similar numerical methods could also have wider applications in the study of human influence on the landscape, for example in summarizing and synthesizing pollen-stratigraphical data at other spatial scales, such as the study of vegetational development within mosaic elements of the landscape using local-scale sites (e.g. Andersen this volume).

Summary

(1) The magnitude of human impact on vegetation at the landscape scale (10^8-10^9 m²) over the last 5000 years has varied spatially and temporally.

(2) Previous approaches to quantitative estimation of changing patterns of human impact have involved summing values of selected anthropogenic indicator pollen types.

(3) As human impact affects vegetational cover at the landscape scale, a multivariate approach involving entire pollen assemblages is desirable.

(4) Correspondence analysis, with its emphasis on 'shapes' rather than 'sizes' of pollen curves, its duality between samples and taxa, and its maximization of correlations between sample and taxon scores is ideal for quantifying principal directions of variation and estimating magnitudes of vegetational change from post-5000 B.P. relative pollen assemblages.

(6) Correspondence analysis of data from Diss Mere, England, gives a quantitative estimate of presumed human impact on vegetation that closely parallels changes in pollen richness. Changes in landscape mosaic following human impact may explain the close relationship between human influence and richness.

(7) Correspondence analysis is used to compare quantitatively human impact at 2 sites in the Ystad area of southern Sweden. Contrasts in human impact between sites are greatest in the Neolithic, paralleling known archaeological differences between the sites.

Acknowledgements

We are indebted to Marie-José Gaillard, Mervi Hjelmroos and Sylvia Peglar for generously providing pollen data, to Björn Berglund, Hilary Birks, Marie-José Gaillard, Mark Hill, Nils Malmer, Sylvia Peglar and Cajo ter Braak for helpful discussions or correspondence, to Björn Berglund for stimulating our interest in applying numerical methods to cultural landscape problems and for supporting this work, and to Hilary Birks and Annechen Ree for invaluable help.

Nomenclature

Plant nomenclature follows A. R. Clapham, T. G. Tutin & E. F. Warburg (1962) *Flora of the British Isles* (2nd edition), Cambridge University Press, Cambridge.

The Cultural Landscape during 6000 Years in South Sweden - An Interdisciplinary Project

Björn E. Berglund

Introduction

Research on the past cultural landscape and the interrelation between society and natural resources should preferably be performed in close cooperation between disciplines in the humanities and sciences. It is also important to find a joint aim and research hypothesis for different scholars collaborating in such an interdisciplinary project (cf. Berglund 1985a). This has been the background for the project on the long-term changes in the South Swedish cultural landscape - 'The cultural landscape during 6000 years', named after the geographical area 'The Ystad project'. This chapter describes the project and outlines our present view on the long-term changes. In this volume, chapters by Gaillard & Berglund and Olsson are part of this project, as are the abstracts by G. Regnéll and J. Regnéll. The chapters by Emanuelsson and Birks *et al.* are also closely related to this project.

The scope of the project

General aims

The general aims of the project are as follows:

(1) To describe changes in society and the landscape within a representative area of southern Sweden.

(2) To analyse the causes behind these changes and especially to emphasize the relation between land-use, vegetation and fauna, primary production and consumption on one hand, and population pressure, society structure, economy and technology on the other hand.

(3) To correlate and compare the investigation area with other areas in Sweden as well as other areas in Europe.

(4) To contribute to a scientific exchange between participating disciplines, particularly concerning research approach, methods, terminology, etc.

(5) To contribute to the management of the natural environment and ancient monuments by considering the following: (i) documentation

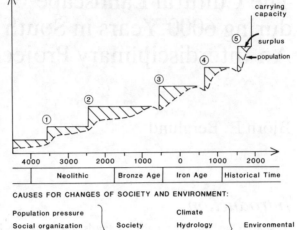

Fig. 1. Theoretical background for the cultural-landscape development following the general hypothesis of the Ystad project. The intention is to show a stepwise, increasing impact on the landscape, but there is no precision in timing and quality of these changes.

of valuable nature areas and ancient historical monuments, (ii) increasing our knowledge of ecosystem continuity and discontinuity, and (iii) giving advice for future planning, considering the accelerating changes in hydrology, erosion, nutrient balance, etc. as well as the overall fragmentation of natural ecosystems.

General hypothesis

The development of the agrarian landscape as seen in a long-term perspective is characterized by phases of expansion - increased clearance, grazing, coppicing, agriculture, etc. - often followed by stagnation/regression periods. This interpretation is based on mainly pollen-analytical studies in southern Scandinavia, including the present study area (Berglund 1969).

The expansion-regression hypothesis is further described by Berglund (1986), where references are made to other palaeoecological and archaeological studies on this problem. Causality has been discussed by archaeologists and geographers (cf. Welinder 1983a and earlier; Kristiansen 1984). In our project we have proposed a dynamic, multi-causal theory for the observed changes during pre-historic and historical time. We illustrate this like a staircase for cultural-landscape development, and the background factors are to be found in social as well as environmental factors (Fig. 1). The latter can be further classified into natural and human-induced. A preliminary classification is shown in Figure 2.

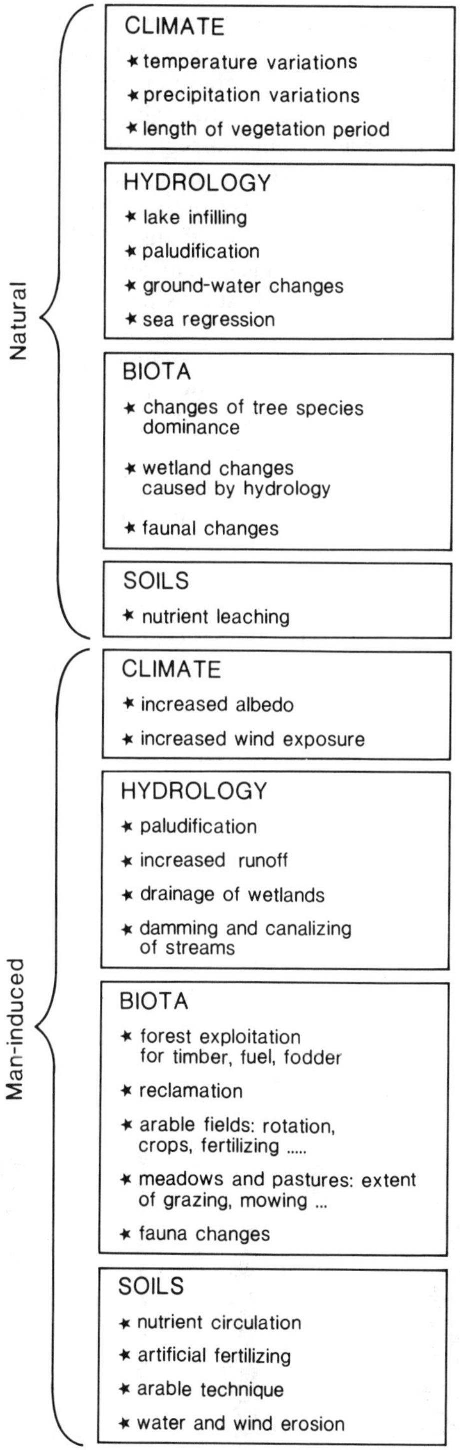

CLIMATE
* temperature variations
* precipitation variations
* length of vegetation period

HYDROLOGY
* lake infilling
* paludification
* ground-water changes
* sea regression

BIOTA
* changes of tree species dominance
* wetland changes caused by hydrology
* faunal changes

SOILS
* nutrient leaching

Natural

CLIMATE
* increased albedo
* increased wind exposure

HYDROLOGY
* paludification
* increased runoff
* drainage of wetlands
* damming and canalizing of streams

BIOTA
* forest exploitation for timber, fuel, fodder
* reclamation
* arable fields: rotation, crops, fertilizing .....
* meadows and pastures: extent of grazing, mowing ...
* fauna changes

SOILS
* nutrient circulation
* artificial fertilizing
* arable technique
* water and wind erosion

Man-induced

Fig. 2. Scheme indicating the main ecological factors behind the changes of the cultural landscape since the Neolithic.

Björn E. Berglund   244

From the study area the pollen diagram from Bjärsjöholmssjön (Fig. 3) discussed by Berglund (1969) was already available. One of the fundamental questions for the new project is the following. Is the pattern of human influence through time repeatable in new diagrams and if so, has the expansion/regression pattern a correspondence in the archaeological-historical source material? As a whole, the project has concentrated on *changes* of landscape and society. Some specific palaeoecological aims and problems can be summarized as follows:

(1) To elucidate ecological changes independent of human activity, such as long-term forest succession and climatic and hydrological changes.

(2) To establish phases of expanding human impact on the landscape, their timing and correlation between central areas (coast) and marginal areas (inland). What is the ecological implication of these expansion phases?

(3) To establish changes in society development - economy, demography, etc. - and a possible correlation between such changes and palaeoecological expansion phases.

(4) To what extent do changes in land-use depend on cultural or ecological factors (e.g. climate, hydrology, etc.)?

(5) Is there any relation between human expansion (e.g. deforestation) and palaeohydrology (e.g. water-level changes)?

(6) Is there any relation between human expansion (land-use patterns, etc.) and soil erosion?

(7) To establish the negative effects on productivity caused by increased exploitation, such as soil leaching, waterlogging, desiccation, soil erosion, etc. How have people compensated for decreased production?

Fig. 3. Simplified survey pollen diagram from the ancient lake Bjäresjöholmsjön. Changes in human impact are illustrated by the curve on the right. Phases 1-4 correspond to expansion phases (from Berglund (1969) after Nilsson's (1961) original diagram).

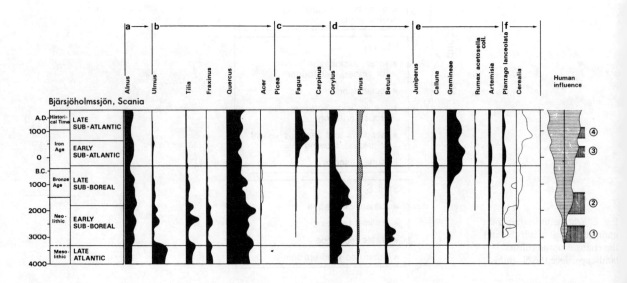

THE YSTAD PROJECT

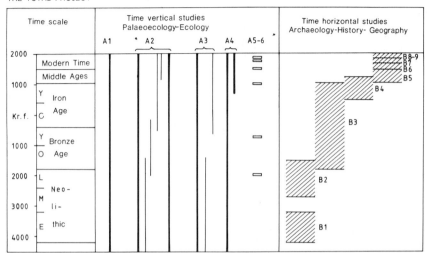

Fig. 4. Project organization with subprojects plotted against time. See Table 1 for details.

(8) What are the future ecological consequences caused by the industrialization in modern agriculture for, e.g., productivity, soil stability, hydrology, flora and fauna, etc.?

Answers to these problems will help to provide explanations for the observed expansion/regression phases illustrated in Figure 1.

Project organization

The organization of the project is based on the hypothesis of the assumed expansion phases in the cultural landscape. The focus is therefore on these phases, but research is also included on the intermediate periods. The research projects are of two kinds: (1) time-vertical studies, dealing with landscape changes in a long-term perspective, and (2) time-horizontal studies, which are multidisciplinary studies dealing with selected periods of special importance for changes in society and landscape.

The time-vertical studies are mainly palaeoecological and ecological in character, and provide the framework for the time-horizontal studies. The arrangement of the subprojects in time is shown in Figure 4. The details of the subprojects are shown in Table 1. Although one discipline is mainly responsible for each subproject, scientists from different disciplines collaborate within each. This means that there is close cooperation between researchers from 6 departments at the University of Lund during the period 1982-88, in some cases also including specialists at other universities or institutes. In addition, scientists from outside Sweden have been invited to cooperate.

Title of Subproject	Leader	Discipline	Department	Collaborator
Time vertical				
A 1 Regional changes in vegetation	Berglund	Palaeoecology	Quaternary geology	J. Regnéll Hjelmroos Göransson
A 2 Local changes in vegetation	Berglund	Palaeoecology	Quaternary geology	Kolstrup Håkansson
A 3 Palaeohydrological changes	Digerfeldt/ Gaillard	Palaeoecology	Quaternary geology	-
A 4 Soil erosion	Dearing	Palaeoecology	Quaternary geology	-
A 5 Reconstruction of former vegetation	Malmer/ Olsson	Plant ecology	Plant ecology	G. Regnéll Bengtsson
A 6 Bioproduction available to people during different periods	Olsson	Plant ecology	Plant ecology	-
Time horizontal				
B 1 Changes during the introduction of agriculture	M. Larsson	Prehistoric archaeology	Archaeology	-
B 2 Expansion during the Middle Neolithic period	L. Larsson	Prehistoric archaeology	Archaeology	-
B 3 Settlement changes during the Bronze and Iron Ages	Tesch	Prehistoric archaeology	Archaeology	Olausson Ridderspore
B 4 Village consolidation in the Late Iron Age	Callmer	Prehistoric archaeology	Archaeology	-
B 5 Settlement development, production, and social organization during the Middle Ages	Andersson	Medieval archaeology	Medieval archaeology	Billberg Sundnér
B 6 Estate management and settlement expansion since the mid-16th century	Skansjö	History	History	-
B 7 The agrarian landscape from the end of the 17th century until the enclosures during the 19th century	Persson	History	History	-
B 8 Agrarian expansion in the 19th century	Lewan	Human geography	Human geography	Möller
B 9 From production society to service society during the 20th century		Human geography	Human geography	Germundsson

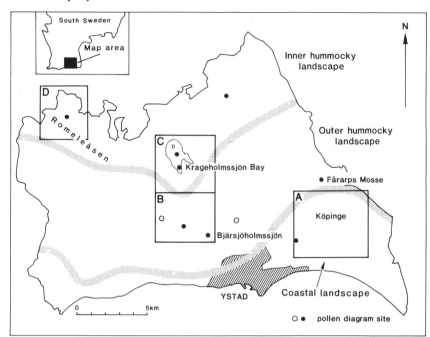

Fig. 5. Study area around the town of Ystad with 3 landscape zones. Research is concentrated in 4 key areas: A = Köpinge; B = Bjäresjö; C = Krageholm; and D = Romeleåsen. Sites with pollen diagrams are also indicated (open circles = diagram not yet completed).

The study area

After careful consideration by the representatives of the 6 disciplines, the area adjacent to the town of Ystad on the south coast of Scania was chosen as a suitable study area. Physiographically it is divisible into 3 landscape zones from the coast towards the inland (Fig. 5):

(1) a *coastal landscape* with sandy soils (below 25 m elevation), today fully exploited for agriculture and settlement,

(2) an *outer hummocky landscape* with clayey silty soils (mainly 25-75 m) today fully exploited for agriculture, and

(3) an *inner hummocky landscape* with clayey or stony soils (mainly 75-100 m) today partly forested and less suitable for agriculture.

This zonation implies a gradient from central to marginal settlement, which would also apply to prehistoric and historical times. It is representative of the present-day cultivated plain extending through southern and western Scania. With archaeological-historical and palaeoecological studies in each zone, it will be possible to make correlations and comparisons in time and space. During historical times, big estates dominated the study area, and since Medieval times, Ystad has been the commercial centre. It is therefore possible to study the relationship between this centre and the surrounding countryside during most of historical time.

◄

Table 1.

SETTLEMENT STRUCTURE AROUND 1670

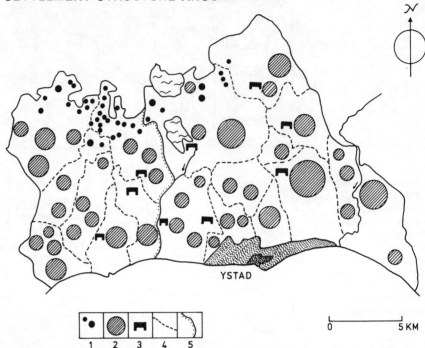

Fig. 6. Settlement structure in
the study area around 1670
(from Skansjö 1987a).
1 = isolated farm or twin farm;
2 = village with 3-45 farms;
3 = manor;
4 = parish boundary;
5 = hundred boundary.

Source material

This is of two main types.

(1) *Palaeoecological.* Although only 3 lakes still exist (Krageholms-sjön, Bjäresjön, Bussjön), ancient, infilled lakes are numerous throughout the undulating plain of the study area, but become sparser on the coastal plain. This availability of gyttja and peat deposits provides good possibilities for comparison between different parts of the area. However, ^{14}C-dates of the lake sediments are problematic, due to the high content of calcium carbonate (Gaillard & Berglund this volume). These difficulties can be largely overcome by making pollen-analytical correlations with the well-dated reference site Ageröds Mosse in central Scania (Nilsson 1964; J. Regnéll unpublished).

Natural or semi-natural vegetation is rare, mainly occurring in the inner hummocky landscape and in the south-eastern part of the coastal plain. Such areas are of importance for palaeovegetational reconstructions.

(2) *Archaeological/historical.* The area is rich in archaeological material, especially the coastal plain and the outer hummocky landscape. Ancient monuments have been well documented by RAÄ (Central Board of National Antiquities) (cf. Tesch 1983).

Written source material is also abundant, especially from Late Medieval times and onwards, and includes fiscal and demographic records, taxation registers, etc. Further, extremely informative enclosure records for the former villages and for the estates are available from the end of the 17th century. Old series of maps are followed by modern aerial photographic documentation of the landscape (Ihse this volume).

The cultural landscape differentiation since Neolithic time

Archaeological evidence

From previous documentation of ancient monuments (Tesch 1983) as well as inventories made during the project period, it is possible to trace a pattern of settlement or exploitation, as follows:

(1) Late Mesolithic settlement was concentrated at a few sites along the coastal zone.

(2) Sparse but more or less even human occupation occurred in the whole Ystad area during Early Neolithic time. Settlement expanded during the late Early Neolithic.

(3) Human occupation contracted, or at least the concentration of settlement in the coastal landscape and adjacent parts of the outer hummocky landscape declined during early Middle Neolithic time, followed by expansion inland.

(4) Settlement expanded and consolidated in the coastal and outer hummocky landscapes during Late Neolithic time and Early Bronze Age.

(5) Settlement expanded in parts of the inner hummocky landscape, possibly already in the Late Bronze Age, in any case during the Iron Age. Archaeological evidence from the upland ridge Romeleåsen is weak so far.

Historical evidence

Written sources together with an interpretation of place-name distributions indicate a more or less continuous settlement in the whole area during the Medieval Period with the exception of the upland ridge Romeleåsen. Probably this area, and a few other marginal areas, were not colonized until this time (cf. Skansjö 1987a). Taxation acts and other sources make it possible to compile a map of the settlement structure at *ca.* 1670 A.D. (Fig. 6) which probably reflects the pattern of a highly exploited cultural landscape and a less exploited (pastoral) cultural landscape in the marginal parts of the inner hummocky landscape (cf. Germundsson 1987).

Palaeoecological evidence

Pollen diagrams will be prepared from 9 sites. However, infilling of lakes and peat cuttings means that not all cover the whole Late Holocene period. The geographical differences are demonstrated here by a comparison of 3 main sites: Fårarps Mosse (Hjelmroos 1985a) and Bjäresjöholmssjön (Nilsson 1961) (two ancient lakes near the coast) and Krageholmssjön (Gaillard 1984) in the inland zone. The Bjäresjöholmssjön diagram is based on old pollen analyses (possibly with under-recorded NAP), but a new diagram is being prepared by H. Göransson. The transect of sites is shown in Figure 7. The chronology in non-calibrated ^{14}C-years is mainly based on correlation with Ageröds Mosse (Nilsson 1964). The land-use indicators in the pollen spectra are classified as follows: (a) arable land, (b) ruderal communities, (c) wet meadows, and (d) pastures including fresh meadows and pastures, dry pastures, general apophytes (*i.a.* Gramineae).

In Figure 7 I identify 6 expansion phases as defined by Berglund (1985a, 1986). However, there seem to be 3 main periods of change in the cultural landscape:

(1) Introduction of agriculture *ca.* 3000 B.C. Inland, some areas appear to have been unaffected by Early Neolithic disturbance (Birks *et al.* this volume). The compilation of all available diagrams indicates furthermore a decrease of human impact except in the coastal landscape during early Middle Neolithic, followed by a gradual spatial expansion towards the inland. The spatial changes in the area during the Neolithic have tentatively been summarized in the map sequence of Figure 8.

(2) Large expansion of pastures *ca.* 800 B.C. During this time an increased impact on the landscape in the whole area is recorded, possibly with the exception of the upland area on the ridge Romeleåsen. This interpretation means that, in particular, open or partly wooded pastures expanded during this time.

(3) Large agrarian expansion 800-1100 A.D. This change involves a more intensified land-use in the whole area, also including an increased utilization (and colonization?) of upland areas on Romeleåsen.

Landscape ecological phases

A tentative summary of the cultural landscape changes in the Ystad area is given in Figure 9. The curve is an attempt to illustrate the agrarian production in the entire area. It is derived from available pollen diagrams, based on the crude assumption that land-use indicators - complemented with information on the forest dynamics - are directly correlated with utilized plant production (crops, hay, leaf-fodder, etc.). In the Early Iron Age the diagrams seem to indicate decreased frequencies of these plants (dotted line). But if we utilize the knowledge on agrarian history in southern Sweden (e.g. Widgren 1983), we may

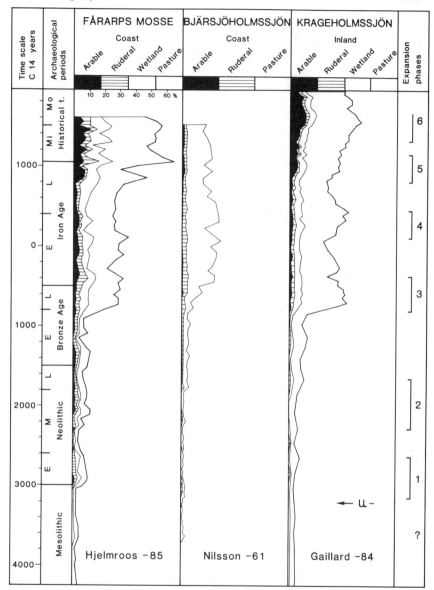

Fig. 7. Correlation of human impact diagrams from 3 sites within the study area in a transect from the coast towards the inland (cf. map in Fig. 5). Diagram construction described by Berglund and Ralska-Jasie-wiczowa (1986).
U- = *Ulmus*-decline.

assume a change towards intensified cultivation with fenced-off areas, i.e. increased production per unit area. Altogether this leads us to identify 6 different landscape phases, briefly characterized in Figure 9. Possible causes for the changes are also indicated. In many cases they are concurrent. But the nutrient strategy (= technological level) of the agrarian society at different time periods is probably of funda-mental importance. A simple model for this is described by Emanuelsson

Björn E. Berglund

252

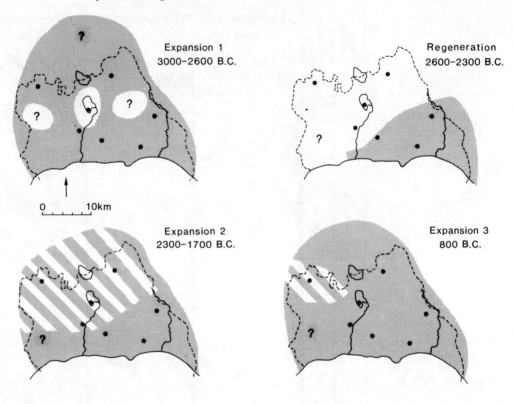

Fig. 8. A tentative
interpretation of the
pollen-analytical evidence for
agrarian settlement within the
study area from 3000-800 B.C.
(Neolithic and Bronze Age).
Shaded areas denote settled
landscape, open areas denote
landscape with no or sparse
agrarian settlement.
Chronology in non-calibrated
[14]C-years.

(this volume), and the 4 nutrient levels of his model are tentatively
indicated next to the landscape phases in Figure 9. We can therefore
summarize the landscape and society development in the following
phases A-F (chronology estimated):

Phase A. Before 3100 B.C. A virgin forest landscape with hunters
and gatherers. Possibly fine-scale cultivation occurred, but as yet we
have no definite proof of this for the study area (cf. Berglund 1985a).

Phases B & C. 3100-400 B.C. A coppiced woodland landscape with
extensive agriculture in addition to a hunting/gatherer economy. Shifting
fields and pastures characterize the open parts of the landscape.
Clearings by ring-barking and probably slash-and-burn technique.
Mainly coastal area utilized until Middle Bronze Age, when a great
expansion of pasture land affected the inland (phase C). Food production for *ca.* 20 persons km^{-2}. Leaching of mineral nutrients important.

Phase D. 400 B.C.-800 A.D. A rather open pasture landscape, with
more intensive cultivation of permanent fields within infield areas of
permanent farms or villages. Food production for *ca.* 50 persons km^{-2}.
Manuring techniques introduced at the beginning, resulting in a more
closed mineral nutrient system than before.

Phase E. 800-1900 A.D. Open landscape with expansive agriculture throughout the period, also in marginal inland/upland areas. General village formation and expansion *ca.* 800-1200. All this meant deforestation and increased erosion starting *ca.* 800-1000. Although the same mineral nutrient system as before, leaching gradually increased with erosion. Increased demand with rising population in the 19th century caused an expansion of arable fields at the expense of pastures and meadows. Food production may then have increased, to support 50-200 persons km^{-2}. Export of agrarian products has been important during Modern Time.

Phase F. 1900-today. Broad-scale, open arable field landscape with industrial agriculture. After the mineral-nutrient crisis of the 19th century, artificial fertilizers were introduced. Production rose enormously and food for *ca.* 1000 persons km^{-2} is produced, also for export. Mechanized techniques in the almost totally ploughed landscape caused increased erosion and leaching of nutrients.

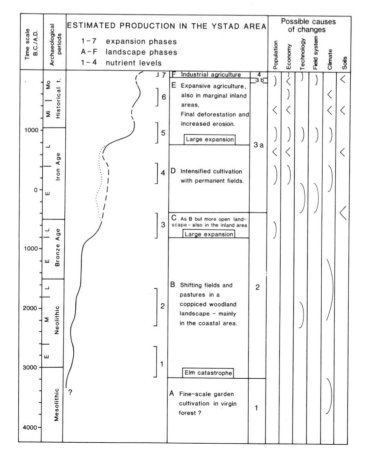

Fig. 9. Tentative scheme of long-term landscape changes in the Ystad area. The curve for 'estimated production' is based on a synthesis of available pollen diagrams. It is a relative curve which illustrates the production level for the entire area and the landscape-ecological changes. Seven expansion phases are interpreted by means of 6 landscape phases for 4 nutrient levels. Possible causes for these changes are also indicated.
< = regression or negative change;
) = expansion or positive change.

The landscape and ecosystem diversity has also changed through time. This is possible to quantify from detailed pollen diagrams. An attempt has been made by Birks *et al.* (this volume) using data from Krageholmssjön and Fårarps Mosse.

As a general conclusion at this stage of the project, we can say that climatic and hydrological factors, together with soil-nutrient status, have been of fundamental importance for agrarian production, but they have interacted in a complex way with social factors. The impact of people on the landscape has depended on the social development in the Ystad area as well as in the surrounding areas of southern Scandinavia.

Summary

(1) Six departments at the University of Lund are collaborating on a six-year interdisciplinary project 'The cultural landscape during 6000 years', devoted to the long-term changes of landscape and society in the Ystad area of southernmost Sweden.

(2) The aims, the research hypothesis, the project organization and the field area are described.

(3) The field area shows an ecological gradient from the coast with fertile, arable soils, towards the inland with less fertile soils.

(4) Following the working hypothesis of expansion/regression dynamics, the coast is a central settlement area which has expanded gradually inland. This hypothesis is supported by the cultural landscape differentiation demonstrated by archaeological, historical and palaeo-ecological material.

(5) A transect of three pollen diagrams gives evidence of six or seven expansion periods, but only three periods of major change in the landscape: (i) introduction of agriculture *ca.* 3000 B.C.; (ii) large expansion of pastures *ca.* 800 B.C.; and (iii) large agrarian expansion 800-1100 A.D.

(6) A tentative interpretation of the land-use and economy from the pollen diagrams reveals six landscape ecological phases since Late Mesolithic time. Several concurrent causes are briefly discussed, but changes in the nutrient strategy of the agrarian society are emphasized.

Acknowledgements

Verbal information from colleagues within the Ystad project is gratefully acknowledged. I would also like to thank Hilary Birks and John Birks for editorial advice and linguistic corrections. Drawings were made by Christin Andréasson and typing by Karin Price, for which I express my sincere thanks.

The Hunter-Gatherer/Agricultural Transition and the Pollen Record in the British Isles

Kevin J. Edwards

Introduction

Anthropo-palynological investigations have exercised the minds of pollen analysts ever since the appearance of Iversen's (1941) classic 'Landnam' paper. The many studies spawned by the impact of Iversen's article (which itself had related precursors - see Fægri 1981) extend beyond palynology into archaeology. The conventional view of over 40 years of research in north-west Europe can be summarized thus: at *ca*. 5100 ^{14}C-years B.P. the decline in the frequencies for elm and sometimes other arboreal pollen and the frequent concomitant appearance of cereal-type pollen, indicate the contraction of woodland (with clearance, climatic change and disease as possible causes) and the agricultural activities of Neolithic people. The possible impacts of pre-elm-decline hunter-gatherers are either ignored or minimized. It is instructive, however, to refer back to the pioneeering contributions from Scandinavia. Iversen (1941, p. 52) quoted earlier archaeological and palynological research in support of the possible contemporaneity of Mesolithic hunter-gatherer Ertebølle and Neolithic cultures in Denmark. Although Fægri (1944) did not find evidence for a contemporaneity of the Nøstvet hunter-gatherers and the Vespestad agriculturalists in western Norway, he was certainly aware of the apparent overlap in the cultural record and he discusses (p. 460) the question of agricultural adoption by indigenous hunter-gatherers as opposed to the partial or wholesale immigration of agriculturalists.

Already then, we see the uncertainty or richness of interpretations concerning the most important transition in the history of the economic life of people. Since the 1940's the palynological literature has largely ignored such issues, at least within the British Isles. Although there has been an extensive consideration of the role of Mesolithic impacts (Simmons *et al*. 1981) and Neolithic clearance activity (Smith 1981), the hunter-gatherer/agricultural transition (abbreviated to 'the Transition' hereafter) has not been discussed sufficiently in its own right by palynologists (but see Edwards 1985a and mentions by Smith

1970; Simmons 1975; Groenman-van Waateringe 1983; Edwards & Hirons 1984; Edwards & Ralston 1985). This is in contrast to the archaeological literature where the Transition has been the subject of much attention (e.g. Case 1969; Coles 1976; Ammerman & Cavalli-Sforza 1984; Bradley 1984; Dennell 1985; Zvelebil 1988).

The aims of this chapter are to examine the Transition in archaeological terms, to consider the palynological manifestations of such a transition, to examine the Transition in the context of archaeological and palynological research in Scotland and Ireland, and to raise some additional points in discussion.

The Transition and archaeology

It would be foolish to attempt to provide anything other than a rudimentary flavour of what the Transition means to prehistoric archaeologists. It has been described as "the most rigidly demarcated frontier in the whole of European prehistory" (Dennell 1983, p. 152). The technological divisions between Old (and later Middle) and New Stone Ages were also seen as chronostratigraphic markers. The diffusionist ideas of Childe (1958) led to economic bases whereby hunter-gatherer lifestyles (equated with the Mesolithic) were overcome by the spread of crop husbandry as Neolithic farmers migrated across Europe. Ambiguities and complexities in the archaeological evidence indicated that such simplified views of the Transition were inappropriate. At what point is a hunter-gatherer-fisher economy which also involves some crop growth deemed to be 'Neolithic' or 'agricultural'? Is the expenditure of energy such that hunter-gatherer activities are more efficient than those on the agricultural side of a divide (Lee & DeVore 1968; Welinder 1983b; Emanuelsson this volume)? Where do herding practices fit into a scheme? When we talk about the dispersal of agriculture are we talking of people or ideas as the agents of

Fig. 1. Radiocarbon dates of the period 6000-4000 B.P. from selected archaeological contexts in the British Isles. Dating errors of one standard deviation are shown.

Fig. 1. Radiocarbon dates of the period 6000-4000 B.P. from selected archaeological contexts in the British Isles. Dating errors of one standard deviation are shown.

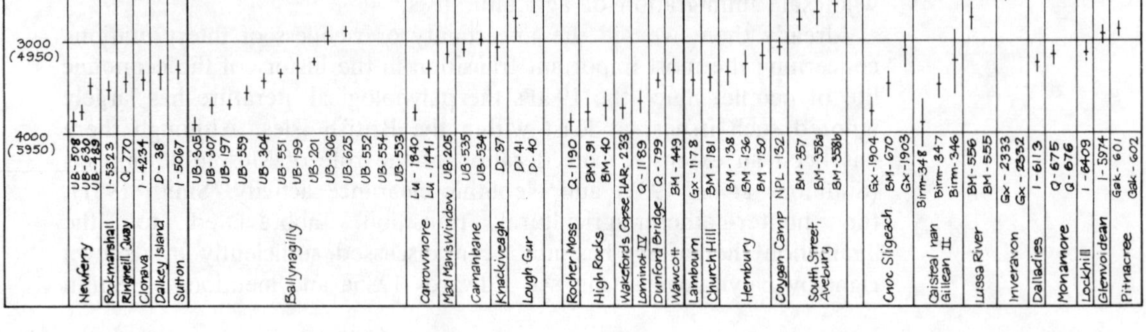

change, or both? Similarly, did the Transition involve the assimilation of agriculturalists or the displacement of hunters-gatherers? How sharp is the Transition and how does it manifest itself spatially as well as chronologically and economically (e.g. at the supra-regional or local scales)? Did precocious farming activity have its economic counterpart in the continuity of hunter-gathering lifestyles in areas where agriculture had long become dominant?

Archaeological evidence from Scotland and Ireland relevant to this discussion is not extensive. The spread of ^{14}C-dates from Britain and Ireland covering the period *ca.* 4000-6000 B.P. is displayed in Figure 1. It is evident that several sites have dates for apparent Mesolithic contexts which overlap with the dates from apparent Neolithic sites, especially in Ireland and Scotland. These overlaps (which may relate to different material contexts as much as to different cultural groups) suggest that for much of this time-band there was a co-existence of lifestyles within the area as a whole. Further analysis may show that there are some underlying spatio-economic bases for the patterns. The Scottish and Irish concentration of overlapping dates may suggest the persistence of hunter-gatherer activities in areas of land marginal for agriculture. At some sites excavation has revealed Neolithic material closely overlying that ascribed to the Mesolithic, as at Glecknabae, Isle of Bute (Bryce 1904) and Kinloch, Isle of Rhum (Wickham-Jones & Sharples 1984). It is not easy, however, to assert convincingly either continuity or survival of cultural groups. It is unfortunate that many of the interesting speculations raised in the archaeological literature (see above) are not testable using existing archaeological data.

The Transition and palynology

As far as hunter-gatherer disturbances are concerned, the palynological and related palaeoecological records provide evidence which can be interpreted as the effects of human utilization of the environment. Examples include primary and secondary peaks in *Corylus/Myrica* pollen (Smith 1970), events at the rational *Alnus* limit (Smith 1984; Chambers & Price 1985) and burning (Jacobi *et al.* 1976; Simmons *et al.* 1981; Robinson 1983) although caution regarding interpretation has been urged (Rackham 1980; Edwards 1982, 1985a; Edwards & Ralston 1985).

In terms of the Transition, the palynological evidence is largely restricted to more modest changes. These include the distinctive 'elm-decline' which is capable of a multitude of explanations singly or in combination (e.g. climate, disease, human impact, plant competition, stress; summarized in Huntley & Birks 1983). There are minor pertur-bations in arboreal and herbaceous pollen taxa before the *Ulmus*-decline which have frequently been attributed to Late Mesolithic activity (cf. Simmons 1964; Pennington 1975), but also to possible pioneer Neolithic

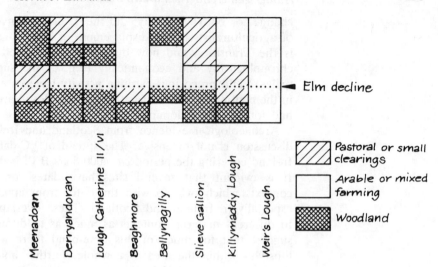

Fig. 2. Schematic diagram of Transition phase events in Co. Tyrone, Northern Ireland (after Edwards 1985b).

agriculture (cf. Smith 1981; Edwards & Hirons 1984). The frequent discovery of cereal-type pollen grains at and above the elm-decline, accompanied by pollen taxa characteristic of arable land, encourages us to infer arable activity and, by association and convention, the existence of Neolithic farming at this time.

The Transition and research in Scotland and Ireland

Syntheses of the palynological literature for Scotland (Birks 1977; Walker 1984a) and Ireland (Edwards 1985b; Watts 1985) suggest that up until the elm-decline, lowland forest cover was virtually complete (Rackham this volume). An exception to this pattern is apparent for the islands and far north of Scotland where trees were climatically restricted to locally favourable situations, though plant macrofossil (Wilkins 1984), molluscan (Burleigh *et al.* 1973) and local palynological data (Bohnke this volume) do not always confirm relative or total treelessness. The general persistence of an extensive tree cover until widespread clearance at or after elm-decline times, should potentially provide a sensitive backdrop against which anthropogenic disturbances might be expected to register. After broad-scale clearance has begun, secondary regrowth vegetation, soil deterioration and the spread of blanket peat are all likely to combine to make interpretation more difficult.

Figure 2 is a schematic presentation of Transition events for a transect of sites in mid-Ulster. The dating assignations are fairly nominal with only the elm-decline taken to be a synchronous datum (Edwards 1985b). Pastoral activity is inferred from increases in taxa such as Gramineae, *Plantago lanceolata* and *Pteridium aquilinum*, while arable agriculture is inferred from finds of cereal-type pollen or a suite of taxa which are normally associated with cultivated land in Ireland (e.g. *Artemisia*, Umbelliferae, Chenopodiaceae, Cruciferae). It should be said that the representation of land-use variations portrayed in Figure 2 differs markedly from some of the original interpretations (e.g. Beaghmore and Ballynagilly, Pilcher *et al.* 1971; and Meenadoan, Pilcher & Larmour 1982). I may have been too strict in the interpretation of pollen indicators (see further discussion in Edwards 1985b) but I do not see clearly the pattern of land-use change which Pilcher and his associates have proposed, from clearance and arable farming, to pastoral farming followed by forest regneration.

An example with interesting inferential implications concerns 2 linked inter-drumlin sites in Ireland (Hirons & Edwards 1986). At Killymaddy Lough the pre-elm-decline phase I features some fluctuations in the *Quercus* curve, increases in values for Gramineae, Cyperaceae and *Pteridium*, and the first occurrence of *Plantago major/media*-type pollen. From these indications one might infer fine-scale disturbance and, possibly, some pastoral activity. At neighbouring Weir's Lough, similar phenomena are seen but also with fluctuating *Ulmus* values and, significantly, the presence of Cerealia-type and *Urtica* pollen. The earliest of the 4 cereal-type grains in the pre-elm-decline phase I is estimated to date to 5740 B.P. The most likely inference to be made from Killymaddy Lough would have been hunter-gatherer disturbance by possible Late Mesolithic groups rather than the equally likely agricultural clearance by Early Neolithic farmers around Weir's Lough. The sites illustrate local differentiation of land-use.

I know of 13 pollen profiles from 12 sites in the British Isles which record cereal-type pollen and other indicators of disturbance prior to the elm-decline. Six of the sites are in Ireland, 3 in Scotland and 3 in England (Fig. 3). Even assuming that certain identification criteria are met (cf. Beug 1961; Andersen 1979b; Köhler & Lange 1979), it is still not possible to be absolutely certain that pollen of wild grasses are not being assigned to a Cerealia category. Of interest in this context is the recorded use of lyme-grass (*Elymus arenarius*) as a cereal crop in Iceland (Olafsson 1943; Tómasson 1973; see also Jóhansen 1985) and Greenland (Fredskild this volume), especially since *E. arenarius* pollen can fall within the morphological criteria set for Cerealia grains. The existence of possible cereal growing in pre-elm-decline times and the similarly 'early' dates for Neolithic-type monuments in Ireland at Ballynagilly and Carrowmore (Fig. 1 and Edwards & Hirons 1984) lead me to seriously question standard assumptions, such as pre-elm-decline disturbances must equal Late Mesolithic hunter-gatherer activity, or the elm-decline equates with the start of Neolithic agriculture. It would also be proposed that the earliest indications of cereal growing could be the result of farming by indigenous hunter-gatherer communities who had acquired the methods and materials of cultivation from

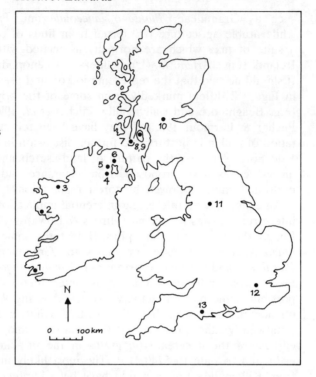

Fig. 3. Sites in the British Isles from which pre-elm-decline cereal-type pollen grains have been reported (based on data summarized in Edwards & Hirons 1984; Edwards *et al.* 1986; Edwards 1988).
1 = Cashelkeelty;
2 = Dolan;
3 = Lough Doo;
4 = Weir's Lough;
5 = Ballynagilly;
6 = Newferry;
7 = Aros Moss;
8,9 = Machrie Moor;
10 = North Mains;
11 = Soyland Moor;
12 = West Heath Spa;
13 = Rimsmoor.

contacts with farming groups either in the British Isles or mainland Europe.

The few reports of early cereal-type pollen until fairly recently may be due to its unexpectedness in pre-elm-decline deposits (and palynologists therefore exercise caution and allocate such finds to their Gramineae category), or it may be a result of the slim chance of discovery. Cereal pollen is produced in low amounts, most cereals are self-pollinating and their large pollen grains travel short distances. If cereal cultivation in pre-elm-decline times was more common than thought hitherto, it is reasonable to develop methods to detect the relatively rare grains. In seeking to optimize detection, several methods have been proposed (Edwards *et al.* 1986). These may be summarized as (1) the selective sieving of samples in an attempt to filter out large grains, (2) differential settling in a liquid column (cf. Flenley 1981), (3) rapid scanning, and (4) favourable coring-site location. The last two have proved useful and will now be described briefly.

Assuming that a core from close to the edge of a bog is more likely to contain the pollen derived from nearby field crops (and, of course, the large pollen of species such as *Agropyron repens* or *Glyceria fluitans*) than a central bog profile (Edwards 1982), a core was extracted from the northern edge of Aros Moss on the Kintyre Peninsula of south-west Scotland (McIntosh 1986). Analyses of a core from the bog centre (Nichols 1967) recorded no cereal-type grains until well above the elm-decline. Contiguous samples from just above the elm-decline to well below it in the new core were scanned at low magnification in a search for cereal-type grains. Prior counts across sample transects of individual microscope slides enable an estimate to be made of total pollen (or exotics) per slide. Using this strategy on any number of replicate slides enables the analyst to 'count' the equivalent of many thousands of pollen in a short time in the search for cereal grains. At Aros Moss, pollen concentrations meant that an estimated 5000 grains could be scanned in *ca.* 30 minutes. Cereal type pollen (identified on the criteria of Andersen 1979b) were found at and immediately above the elm-decline, and 3 possible grains below it. Similar research at Machrie Moor on the Isle of Arran, close to Kintyre, has produced comparable results (McIntosh 1986).

The Transition and the elm-decline

If the cereal-type grains really do derive from Cerealia (and acknowledging the identification difficulties yet noting the evidence already discussed for the archaeological records of Neolithic-type materials in

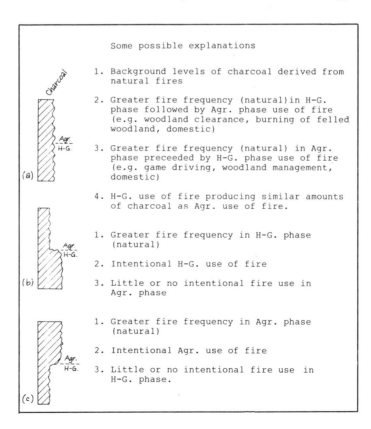

Fig. 4. Model patterns of charcoal representation at the hunter-gatherer (H-G.) to agricultural (Agr.) transition.

contexts dating to before 'elm-decline' times), then what is the status of the elm-decline? Its relative synchroneity and its ubiquity across north-west Europe might best suggest a primary catastrophic cause, of which disease and climate would be the best contenders - it is difficult to imagine people everywhere in Europe doing the same thing for the first time at the same time! This topic will not be discussed in detail here, but it is worth noting fossil records of the elm bark beetle (*Scolytus scolytus* F.) just before the elm-decline at West Heath Spa near London together with early finds of cereal-type pollen made recently by Girling & Greig (1985). *S. scolytus* is the main carrier of *Ceratocystis ulmi* (Buis.) Moreau, the fungus causing Dutch elm disease, and it typically inhabits clearings, hedges, isolated copses or isolated trees rather than dense forest. The *Scolytus* finds occur together with a rise in the representation of dung-dependent species of beetle. This raises the possibility that animal husbandry practices of primarily hunter-gatherer or farming peoples could have provided optimal conditions for the spread of disease, by, for instance, damaging elm trees by foliage or bark stripping and both weakening trees and providing clearings in woodland. In her review of pre-elm-decline events in pollen diagrams from Ireland, Groenman-van Waateringe (1983) favoured a disease explanation for the *Ulmus*-decline (cf. Rackham this volume).

The Transition and charcoal stratigraphy

Microscopic charcoal in pollen preparations has been widely used to infer fire history (reviewed in Patterson *et al.* 1987). The widespread use of fire by hunter-gatherer groups, and its hypothesized use in prehistory as part of a hunting and browse-creating regime, is well attested (Mellars 1976; Simmons *et al.* 1981). Therefore the relationship between pollen and charcoal records during the Transition is of special interest (Fig. 4). For example, if the quantitities of charcoal during inferred hunter-gatherer phases in the palaeoecological record are much higher than in agricultural phases, one possible explanation would be a change towards a decreased use of fire. There are, in fact, many hypotheses which could explain charcoal patterns and pollen-charcoal relationships, and some of these are discussed elsewhere (Edwards & Ralston 1985; Patterson *et al.* 1987; Edwards 1988). Curves for microscopic charcoal concentrations across the elm-decline sections of pollen diagrams from Aros Moss and Machrie Moor (McIntosh 1987) conform with the pattern in Figure 4b. At both sites charcoal nearly disappears at about the elm-decline level, which might suggest a fire-decrease model to explain events at these sites. If charcoal concentrations increased after the elm-decline (cf. Fig. 4c) it is possible that they indicate continued hunter-gathering practices into the period conventionally assigned to the Neolithic or perhaps fresh hunting activity (see below). If this was the situation, then the charcoal rise could denote fire associated with domestic activity rather than the burning of woodland. The multiplicity of possible explanations of such patterns will be returned to in the Discussion.

The Transition and combined archaeological and palynological evidence

Attention was drawn above to archaeological evidence for the persistence of hunter-gatherer activities into the period traditionally allocated to the Neolithic. In the context of research in the British Isles at least, this would suggest that post-elm-decline clearances or disturbances may not necessarily be due to agricultural activities. This does not, of course, preclude hunter-gatherer activities on the part of communities which follow predominantly agricultural lifestyles. In an example published by Edwards & Ralston (1985), the possibility of a continuity of hunter-gatherer activities in an upland area with contemporaneous farming in adjacent lowlands is examined in the light of archaeological and palaeoecological data. The availability of co-spatial archaeological and palynological evidence was also an element in the strategy employed in the search for pre-elm-decline pollen sites discussed above.

Ideally the juxtaposition of archaeological and palaeoecological evidence is what we require but seldom manage to find. Archaeological sites may not show an unbroken continuity from Mesolithic to Neolithic levels. Where they might, our confidence in the microfossil records is perhaps reduced because the polleniferous site is most likely to be mineral soil and is therefore subject to the difficulties inherent in pedological studies (Dimbleby 1985). An example of this is shown in Figure 5. Dam I is a pitchstone-flaking site on the Isle of Arran,

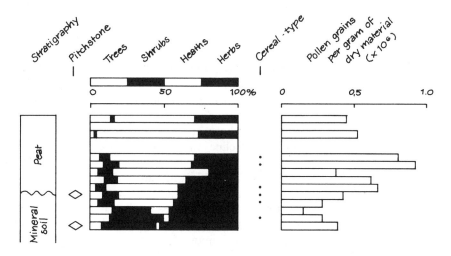

Fig. 5. Summary pollen data from the Dam I site, Isle of Arran.

Scotland. The pitchstone lithics from the site bear the hallmarks of Mesolithic-type manufacture. The stratigraphy at the point where the pollen profile was obtained consists of hill peat grading into a weakly-developed podsol profile. Cereal-type pollen is present throughout the pollen profile including the levels where pitchstone is found at the base and apparent top of the soil. Although various permutations of cultural/land-use activity could be cited in explanation, the propensity of pollen to move downwards through a mineral-soil profile *via* downwash or bioturbation (Aaby 1983) and the decrease in pollen concentrations with depth (Fig. 5, suggesting problems of contemporaneity or destruction, though a volumetric or organic matter calculation base could alter the pattern) make inference difficult.

Discussion

Some archaeologists have long suggested that the transition between hunter-gatherer and agricultural activities in prehistory was not a simple sharp temporal change. However, within the British Isles both palynologists and archaeologists have largely slipped into a complacent acceptance of the elm-decline boundary as a demarcation between hunter-gatherer (= Mesolithic) and agricultural (= Neolithic) lifestyles. My more limited reading of the Continental European literature suggests that the situation is not perhaps very much different in north-west Europe. A possible way forward is in joint research or discussions between palynologists and archaeologists.

In terms of the Transition, pollen analysis has shown that pre-elm-decline cereal-type pollen, especially in conjunction with weed taxa typical of arable activity, indicates the probability of pioneer agriculture and thus validates archaeological speculation. This provides a further challenge to the archaeologist to find early agricultural settlements (which may be difficult if we envisage relatively nomadic pioneering immigrants), and to explain the processes of agricultural adoption. It is perhaps worthy of note that "aside from the palynological evidence, there is surprisingly little evidence for the economic basis of the Irish neolithic" (Woodman 1985, p. 260). The clear signs of vegetational disturbance after the elm-decline in many pollen diagrams, especially in the absence of arable-type pollen indicators, may well indicate hunter-gathering, but we do not have unequivocal archaeological evidence. There are ^{14}C-dates from Mesolithic contexts extending into post-elm-decline times and natural faunal and floral produce would continue to please the palates of settled agriculturalists.

Major problems continue to exist with anthropo-palynological inference and the lack of objectivity (cf. Edwards 1979, 1982; Maguire *et al.* 1983). The difficulties implied in this rather obvious pronouncement, however, are clearly shown in Behre (1986) where the wide ecological amplitude of many 'cultural indicators' entreats us to be wary of over-precision in our inferences. It is possible that we cannot go much further in our interpretations than a set of broad categories based partly on Edwards (1988):

(1) mixed farming, with the presence of Cerealia, Gramineae and other herbaceous taxa frequent on pastoral land;

(2) posssible cereal cultivation as part of a mixed farming system (weed taxa common on arable land are present, but Cerealia are absent); and

(3) non-specific disturbances (frequently with pastoral-type indicators). Category (3) could arise from a great variety of vegetational changes including those due to natural grazing (Buckland & Edwards 1984), animal husbandry, wind-throw of trees, clearings made for settlement, clearings made for crop growing but where the indicators of arable activity did not reach the site of pollen deposition, or even changes in lake levels or bog-surface drying.

Inferential problems are also met with microscopic (and macroscopic) charcoal data where numerous possible causes can be evoked to explain the patterns. The difficulties can be briefly illustrated by the 3 simple model patterns (Figs. 4a, b, c). These models assume that the spatial distribution of charcoal sources and charcoal sedimentation are fairly constant, and that the transition between hunter-gatherer and agricultural activities is locally sharp (temporally and in terms of predominant lifestyles). They exclude consideration of seasonality of occupation and population size. Given these restrictions and if the range of possible explanations is accepted, it is apparent that researchers must also consider the complexities arising from the above factors and such additional ones as the upland or lowland locations of activity, different vegetational communities (as sources of charcoal), and hunter-gatherer and agricultural groups in operation at the same time.

Progress out of such impasses may not be possible from archaeology or palaeoecology because the evidence is simply unavailable or unsuitable. Potentially informative lines of enquiry include more surface pollen and charcoal sampling programmes (e.g. Vuorela 1973; Hicks this volume) orientated towards anthropogenic problems, a greater consideration of ethnographic parallels for land-use practices, and perhaps simulation modelling of both (Hulthén & Welinder 1981; Keene 1981). On the subject of the elm-decline, I wish to say nothing other than that it should probably assume less importance as a cultural indicator of definitive land-use significance, but that it continues to represent a useful dating horizon.

Summary

(1) The economic change from hunter-gatherer (usually equated with 'Mesolithic') activity to agricultural (cf. 'Neolithic') activity represents the most important transition in human history. It was of

great significance for the nature of food procurement, land-use and settlement.

(2) The Transition has been insufficiently discussed by palynologists (who generally give it token consideration as they report inferred events at the elm-decline), but it has been the subject of intense debate by archaeologists.

(3) The Transition is examined in archaeological and palynological terms, primarily in the context of research in Scotland and Ireland.

(4) The objectivity surrounding the use of cultural pollen indicators and the interpretation of microscopic charcoal and fire history are also discussed.

Acknowledgements

I wish to thank K. R. Hirons, C. J. McIntosh and P. J. Newell, my research colleagues at Birmingham, for their assistance and comments on the text. The financial support of the Science and Engineering Research Council for work at Aros Moss and Machrie Moor is gratefully acknowledged. The word-processing aid of R. G. Ford has been invaluable.

Nomenclature

Plant nomenclature follows A. R. Clapham, T. G. Tutin & E. F. Warburg (1962) *Flora of the British Isles* (2nd edition), Cambridge University Press, Cambridge.

Post-glacial Landscape Evolution in Connemara, Western Ireland with Particular Reference to Woodland History

Michael O'Connell, Karen Molloy and Máire Bowler

Introduction

Connemara, in west County Galway, is one of the most botanically interesting regions of Ireland on account of its Lusitanian ericaceous element of *Daboecia cantabrica*, *Erica erigena*, *E. mackaiana* and *E. ciliaris* and its North American element, most notably *Eriocaulon aquaticum* (Webb & Scannell 1983). In an attempt to document the history of these elements, Jessen (1949) studied pollen and macrofossils at sites near Roundstone (Fig. 1). He found pollen of *E. aquaticum* in zone VIIa and macrofossils of *E. mackaiana*, the identification of which, however, has since been questioned (Godwin 1975). The general trends in woodland history elucidated by Jessen have been confirmed by Teunissen & Teunissen-van Oorschot (1980) at Dolan, west of Roundstone (Fig. 1). The lack of ¹⁴C-dating here is a serious drawback, particularly in a region where no well-dated diagrams are available for comparison. Furthermore, these studies, being from a single peninsula, cannot provide an adequate basis for the reconstruction of the vegetation history of a region as extensive and diverse as Connemara.

Our research aims at filling this lacuna and, in particular, documenting the impact of prehistoric human activity on woodlands and for landscape development in general. Although our investigations are incomplete, we present the major results obtained so far.

The region and the sites studied

Description of the region

Connemara, which lies between 9⁰4' and 10⁰10' W and 53⁰12' and 53⁰36' N, has never been a political or administrative unit but it forms a geographical unit comprising mainly western Co. Galway (Fig. 1) which is well characterized by extensive blanket-bog, very limited

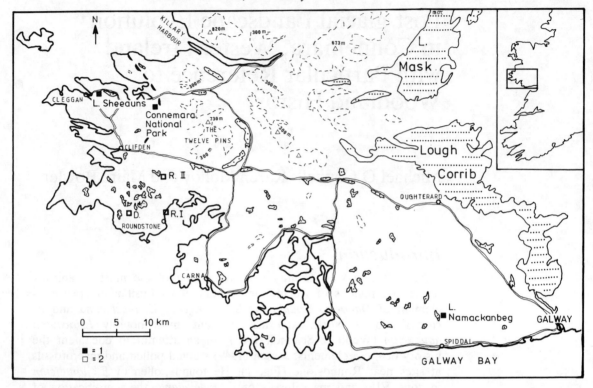

Fig. 1. Map of Connemara, Co. Galway.
1 = sites investigated in the present study;
2 = published diagrams; RI and RII are Roundstone I and Roundstone II (Jessen 1949), D is Dolan (Teunissen & Teunissen-van Oorschot 1980). Megaliths, standing stones and pre-bog field boundaries are concentrated to the west of the thick broken line.

woodland, many lakes and a present-day settlement pattern which is mainly coastal.

The geology of the region is complex (Leake *et al*. 1981; Tanner & Shackleton 1979). Most of it consists of siliceous rocks (granite, schists, gneiss, quartzite), yielding acid soils. Some of the schists, such as those classified as Lakes Marble Formation at Lough Sheeauns, have a considerable calcareous content (Tanner & Shackleton 1979) and their distribution has a marked influence on present-day vegetation cover and soil fertility (see below). Considerable depths of drift occur in western Connemara, particularly in the Kylemore-Cleggan-Clifden area (Orme 1967). In south-eastern Connemara, and especially about Spiddal, the Midlandian (Weichselian) glaciation resulted in a landscape strewn with erratics resting directly on granitic bedrock (Whittow 1974).

The climate is markedly oceanic with mild winters (January temperature 6°C), cool summers (July temperature 15°C) and frequent though, for the most part, not excessively heavy rainfall. In the coastal areas precipitation is normally less than 1300 mm yr⁻¹ while in the mountains it may reach 2500 mm (Webb & Scannell 1983). Strong south-westerly or westerly winds are frequent, particularly at the west coast.

Connemara National Park, Letterfrack, north-west Connemara (profile FRKII)

The site, at *ca.* 60 m elevation, is in a valley running north-west - south-east at the western side of Diamond Hill, an outlier of the Twelve Pins (grid reference L 71 57; Fig. 1). Peat cutting in the valley floor, which is *ca.* 120 m across, has exposed an extensive layer of pine stumps. At the sampling point, this pine-stump layer rested on 2.4 m of peat and, in turn, was covered by a 1.2 m peat bank left by the turf cutters. Marginal farming land is present *ca.* 1 km to the north-west; otherwise the landscape is dominated by blanket-bog, which frequently contains pine stumps, normally resting on less than 1 m of peat, or by heath where the peat has been cut away. The patchy drift cover and the exposed bedrock, consisting of alternating bands of quartzites, calcareous schists and some marble (Tanner & Shackleton 1979), probably yielded reasonably fertile soils prior to blanket-bog growth.

The profile FRKII was taken 40 m from the eastern edge of the valley bog as a monolith in 5 sections, the lowest, 0.2 m long, occupying a small hollow in the bedrock. The profile is considered to reflect mainly regional vegetation development, particularly after widespread initiation of blanket-bog had taken place.

Lough Namackanbeg, Spiddal, south-east Connemara (profile NMKI)

Lough Namackanbeg is a small kidney-shaped lake at 90 m elevation, situated 4.6 km north of Spiddal and 50 km from the Connemara National Park (grid reference M 132 269; Fig. 1). The gently sloping sides of the small closed basin were once covered by blanket bog. The lake is now a *schwingmoor*, a surface scraw having replaced open water in this century. Bog to the west has been cut away and the area now supports marginal farming. Pine stumps, usually resting on a thin layer of peat, are frequent nearby and throughout the expanses of bog north of Spiddal. To the south-west, a relict *Quercus petraea* Atlantic woodland (34 ha) survives in the Boliska river valley (Kirby & O'Connell 1982). Apart from wooded lake-islands, this is one of the few semi-natural woodlands in Connemara.

The site was chosen to document local vegetation development in a part of Connemara which contrasts sharply with the other areas studied. Here, the edaphic conditions associated with granite bedrock largely devoid of drift (see above), appear particularly unfavourable for higher plants. There are no archaeological records from the Neolithic or Bronze Ages apart from a recently discovered, as yet undated, bog trackway (P. Gosling personal communication). The nearest field monuments to the west are megaliths at Cashel (Cooney

Fig. 2. Percentage pollen diagram based on an AP sum from the Connemara National Park, Letterfrack, north-west Connemara. Charcoal particles (>37 μm) are calculated on a total terrestrial pollen (T.T.P.) sum plus charcoal particles. + = presence outside the count; a dot = a low percentage level (usually <0.2%) where that might not be obvious because of the scale selected. The horizontal scale (continued) ▶

1987) and Maam (M. Gibbons personal communication), over 35 km away.

The core NMKI was taken using a Livingstone sampler from the centre of the deeper arm of the former lake, where the basin is *ca.* 100 m wide. The uppermost 0.5 m, i.e. the surface scraw, proved impossible to sample satisfactorily.

Lough Sheeauns, Cleggan, north-west Connemara (profile SHEIII)

Lough Sheeauns is a small, steep-sided, approximately circular lake, 120 m greatest diameter, situated at *ca.* 18 m elevation, 2.5 km east

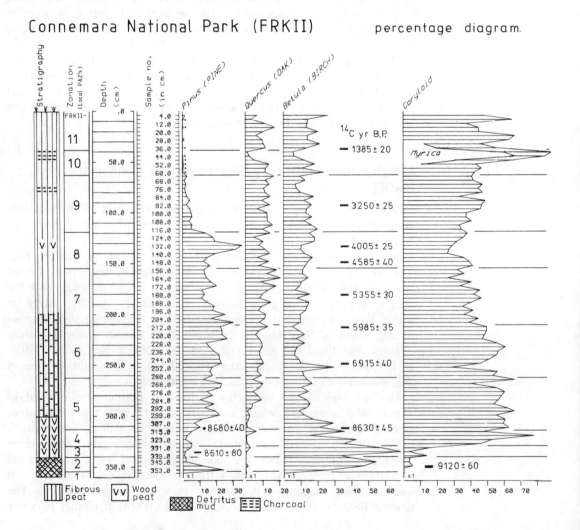

Connemara National Park (FRKII) percentage diagram.

of Cleggan and 9 km west of the Connemara National Park site (grid reference L 625 582; Fig. 1). It lies in a long, narrow (*ca.* 700 m wide), relatively fertile valley, where the bedrock, classified as Lakes Marble Formation (see above), is probably responsible for the favourable edaphic conditions. Drainage is by a small stream into the nearby Ballynakill Lough. Both this and the small inflowing stream are probably artificial. Bog is present in the valley floor between the lakes and, on higher ground to the south-east there is extensive blanket-bog with numerous pine and occasional oak stumps.

The lake is in an area particularly rich in megaliths and pre-bog enclosures (de Valera & Ó Nualláin 1972; M. Gibbons personal communication; Fig. 1). It was sampled to study in detail prehistoric human impact on the landscape. The 5 m long core, SHEIII, was

values are given at the base of each curve and the magnification, e.g. x10, is also indicated to highlight the different scales employed. The composite summary diagram is based on a T.T.P. sum from which bog/heath taxa are excluded.

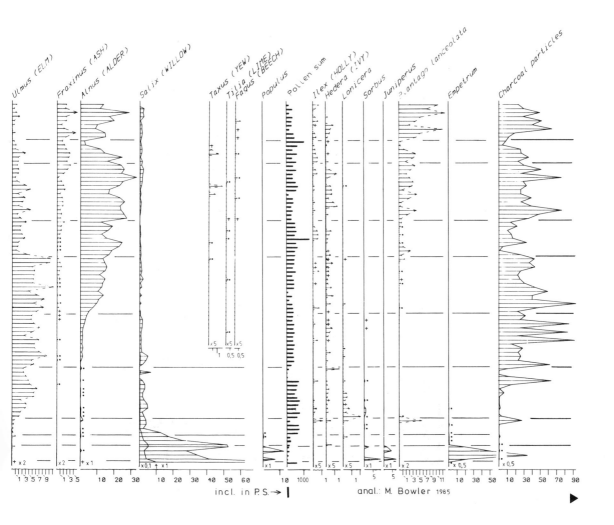

taken with a Livingstone corer, from the centre, under *ca*. 5.5 m water. The uppermost *ca*. 2 m was not sampled.

Vegetation history

In the main pollen diagrams (Figs. 2, 3, 4), curves for individual taxa are based on an arboreal pollen sum. Distortion resulting from high local pollen producers such as *Molinia*, which is an important component of blanket-bog vegetation, is thereby avoided and comparison with published isopollen maps (e.g. Huntley & Birks 1983) is facilitated. Composite summary diagrams are included to show the relative importance of woodland, grassland and bog/heath communities. In

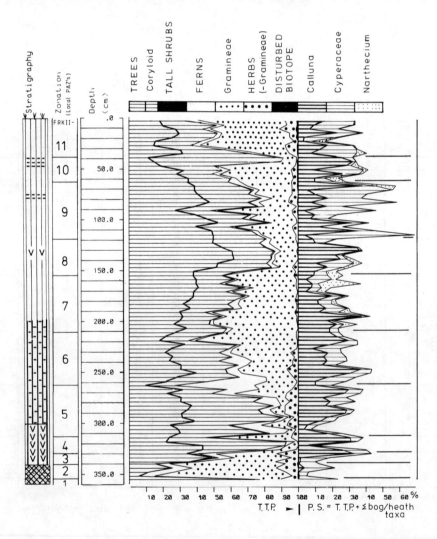

Fig. 2. (continued)

each diagram the *Plantago lanceolata* curve is also shown, this being regarded, in this data-set, as the most reliable indicator of extent of open areas without bog or heath development. In FRKII, *Corylus* and *Myrica* are combined in the pollen taxon Coryloid which is included in the pollen sum. In the other profiles these taxa are separated and *Myrica* is regarded as a bog/heath species. Local pollen assemblage zones (p.a.z.) are delimited using changes in the percentage represent-ation of all taxa based on a total terrestrial pollen sum.

A linear time-depth relationship between dated levels is assumed for the purposes of estimating ages for intermediate levels. In Figure 5 the ages assigned to the spectra from the three profiles are based on these dates except at the top and bottom of the profiles where, in some instances, the ^{14}C-dates are unreliable. Here ages based mainly on pollen-analytical evidence are used. All dates are quoted in uncor-rected ^{14}C-years B.P., unless otherwise indicated.

Profile FRKII from the Connemara National Park

Pollen assemblage zones FRKII-1, 2 and 3: Late-/post-glacial transition

The base of the diagram (Fig. 2) indicates an early post-glacial suc-cession with *Salix* being followed by *Empetrum*, *Betula* and finally *Corylus*. *Empetrum* was probably the dominant dwarf-shrub in FRKII-2, with values between 19-95%. Similarly high levels were recorded by Jessen (1949) in the Roundstone I, 6 and Roundstone II, 11 diagrams. *Juniperus* was sparse, as is usual in parts of Ireland with base-poor soils (Watts 1977). *Populus* and *Sorbus* are also represented. All the Connemara diagrams show that *Salix* species were much more frequent in western Connemara than at sites elsewhere in Ireland during this time (e.g. Craig 1978; Singh 1970; Watts 1977). High Gramineae, *Rumex* and *Filipendula* values in FRKII-2 and 3 suggest ungrazed grassland and tall-herb vegetation (Fig. 2, composite diagram).

The dates 9120 ± 60 and 8610 ± 80 B.P. appear too young, the former by at least 500 years (cf. Watts 1977), though a date of 9015 ± 335 B.P. has been reported by Pilcher & Larmour (1982) for an early post-glacial pollen assemblage from an upland site in Co. Tyrone.

Pollen assemblage zones FRKII-4 and 5: Early woodland development

The maximum expansion of *Corylus* recorded in FRKII-4 is followed in FRKII-5 by the expansion of *Pinus*, *Quercus* and *Ulmus*, with some *Sorbus* present early on. The high *Betula* values probably arise mainly from *Betula* growing on the mire surface, its wood being frequent at these levels (Fig. 2).

Peat and *Betula* wood from the base of FRKII-5 yielded ^{14}C-dates of 8630 ± 45 B.P. and 8680 ± 40 B.P., respectively. Unfortunately, the diagram does not permit pinpointing with certainty the arrival of the main arboreal taxa. However, *Pinus* and *Corylus* were probably present at the end of FRKII-3, (*ca.* 9200 B.P.). During FRKII-4, (*ca.* 9100-8700 B.P.), *Corylus* (presumably) assumed a dominance in the landscape equivalent to that recorded on more base-rich soils in central Ireland (cf. O'Connell 1980). At the base of FRKII-5, *Pinus* began to expand rapidly, followed quickly by a smaller expansion of *Ulmus* and later (*ca.* 8100 B.P., 292 cm) by *Quercus*. The resulting woodlands were dominated by *Pinus* with lesser amounts of *Quercus* and *Ulmus*. *Corylus* was a major component as an undershrub and possibly also formed pure stands. The early presence (before 8700 B.P.) of *Ilex, Hedera* and *Lonicera*, all indicators of climatic amelioration, is noteworthy.

Pollen assemblage zones FRKII-6 and 7: Development of Atlantic woodlands

In these zones, the invasion and expansion of *Alnus* and the accompanying adjustments of the canopy-forming species are recorded. Initial low values of *Alnus* at 252 cm (7025 B.P.) increase slowly to 3.7% at the top of FRKII-6 (6000 B.P.). Chambers & Price (1985) suggest that low values, such as these, may reflect local presence rather than long-distance transport. At L. Sheeauns, 9 km to the west, *Alnus* arrives at 6700 B.P. and expands rapidly. Alder may have been present earlier in the Letterfrack area but did not expand until shortly after 6000 B.P. The expansion was probably favoured by a change to wetter conditions, indicated by the simultaneous expansion of *Narthecium* and *Assulina*. During FRKII-7, the continued expansion of *Alnus* is paralleled by a decline in *Pinus* and *Corylus*. The NAP curves suggest that the canopy became more closed as *Quercus* and to a lesser extent *Ulmus* increased at the expense of *Pinus* and *Corylus*. *Fraxinus*, probably present from 6000 B.P., remained a minor component.

Pollen assemblage zone FRKII-8: The elm-decline and the establishment and extinction of pine on the bog surface

Immediately prior to the elm-decline, *Ulmus* approximately doubled its earlier pollen representation to 11%, so that the decline at 152 cm to 4.3 % and 4 cm higher to 1.0%, is sharply defined. At the elm-decline, Coryloid, *Betula* and *Alnus* curves rise while *Quercus* falls. That open areas existed prior to the elm-decline is suggested by the

Plantago lanceolata curve which begins at 156 cm with a value of 1.7% (see below).

By linear interpolation, the *P. lanceolata* curve begins at 4780 B.P. and the elm-decline (152 cm) commences at 4680 B.P. This is rather late but it should be noted that, across the elm-decline, the Gramineae curve decreases dramatically and *Calluna* begins to rise, indicating changes in the bog vegetation (see composite diagram). This may influence directly the ^{14}C-dates (cf. Olsson 1986) and, most probably, also affected peat-accumulation rates.

Between 140-120 cm, at the level of the pine-stump layer in the profile, a *Pinus* peak (36%) reflects the local establishment of pine on the bog. This bog-woodland phase lasted for almost 500 years centred on 4005 ± 25 B.P. McNally & Doyle (1984) have shown that *Pinus* established itself on 3 raised bogs in counties Offaly and Kildare between 4000 and 3500 B.P. After surviving for 500 years at one of the sites, pine died out, not as a result of a cataclysmic event, but through failure to produce viable seedlings due to increased wetness on the bog surface. They suggest that climatic changes resulted in dry and then wet habitat conditions, which facilitated establishment and later hindered regeneration. The bog taxa in FRKII indicate similar changes but the possibility exists that these are the consequence of the initiation and decline of bog woodland rather than the primary determining factor. Investigations in Scotland by Birks (1975) and Dubois & Ferguson (1985) suggest that a climatic increase in wetness predated 4000 B.P. Significantly, diagrams from the Roundstone area (Jessen 1949; Teunissen & Teunissen-van Oorschot 1980) record even more pronounced *Pinus* peaks, which, although not ^{14}C-dated, occur in a similar stratigraphic position *vis-a-vis* the elm-decline. It appears likely that the invasion of *Pinus* on to bog surfaces in western Connemara, in Clare Island, 30 km to the north (Coxon 1987) and in parts at least of north Co. Mayo (cf. date of 4400 B.P. in Caulfield 1978; Moore 1979) was broadly contemporaneous and was mediated by climatic shifts in the Subboreal.

Pollen assemblage zones FRKII-9, 10 and 11: Decline and extinction of Pinus and Ulmus and the expansion of Fraxinus

As FRKII-9 opens at 3700 B.P., the first substantial decline in *Pinus* is registered. For the next 1500 years it played a steadily declining role before finally becoming extinct at *ca.* 200 B.C. At Roundstone, the *Pinus* curve follows a similar and, it is assumed, broadly synchronous course (Jessen 1949; Teunissen & Teunissen-van Oorschot 1980). Other features in the FRKII profile at this time are the consolidation of the *Quercus* curve and expansion of *Alnus* and *P. lanceolata*. This suggests a significant shift in woodland composition accompanied by a decline in total woodland area. Towards the top of FRKII-9 (*ca.* 3000 B.P.), *Ulmus* declined to very low levels. Shortly afterwards, the curve for *Fraxinus* rises and no longer parallels that of *Ulmus*, as a result possibly of a further decline in total woodland, rather than any significant expansion of ash (see below).

In FRKII-10, the dramatic expansion of the Coryloid curve reflects a local expansion of *Myrica*, possibly on to the bog surface (Fig. 2). *M. gale* had probably expanded much earlier (3500 B.P.) and is possibly entirely responsible for the high Coryloid values in the upper part of the diagram.

At the base of FRKII-11 (565 A.D.), *Ulmus* became extinct, possibly because of increased farming in the early Christian Period (cf. *P. lanceolata* curve). *Pinus* and *Fagus* do not expand at the top of the diagram, so the record may not extend beyond the 18th century.

Fig. 3. Percentage pollen diagram based on an AP sum from Lough Namackanbeg, Spiddal, south-east Connemara (see also legend to Fig. 2, but note that the T.T.P. sum used in the summary diagram includes bog/heath taxa).

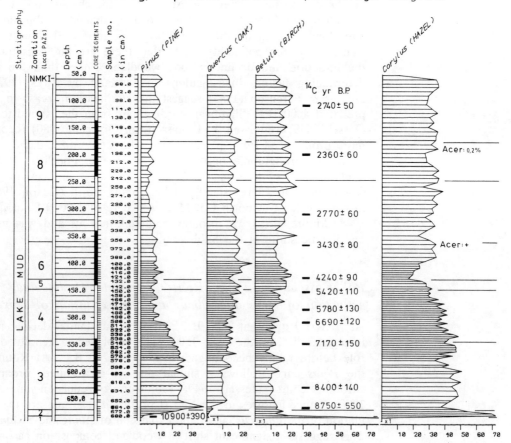

Profile NMKI from Lough Namackanbeg, Spiddal

Pollen assemblage zones NMKI-1, 2 and 3: Opening of the post-glacial and early woodland development

The base of NMKI (Fig. 3) is characterized by rapidly changing curves, with very high *Betula* and moderate *Salix* values (NMKI-1) being replaced by *Corylus* and *Pinus* (NMKI-2). *Juniperus* does not exceed 1% and *Empetrum* is not recorded. This indicates that open *Betula* scrub with willows was replaced by *Corylus* scrub in which *Pinus* was important. *Ulmus* had not yet arrived in NMKI-2, and *Quercus* was rare. In NMKI-1 herbaceous vegetation, including Gramineae, *Rumex* and *Filipendula*, was important but declined rapidly in NMKI-2. A noticeable silt/clay component at the base of the profile indicates that soil erosion was taking place in the catchment.

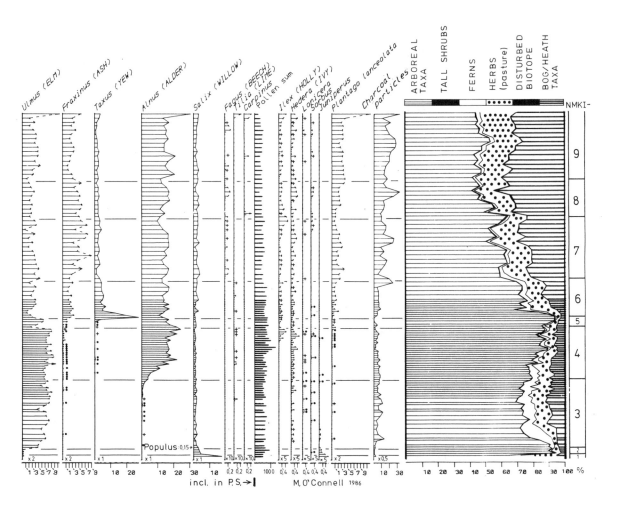

As in the FRKII-2 profile, there are uncertainties regarding chronology. The date of 10 900 ± 390 B.P. (681-684 cm) appears to be too
old. If the date of 8750 ± 550 B.P. (665-668 cm; the only sample not
acid and alkali pretreated) is correct, then the base of the diagram
may be somewhat older than 9000 B.P. This appears acceptable,
especially since p.a.z. NMKI-1 is broadly comparable with the top of
FRKII-3 (Fig. 2).

At the base of NMKI-3 (*ca.* 8800 B.P.), *Pinus* achieves a maximum
of 40%, but this value is then depressed by the rapid increase in
Quercus and also *Ulmus*. *Corylus*, at 45%, is lower than in FRKII-5
(60% including *Myrica*, which was probably insignificant). The percentages of *Pinus* and *Quercus* are correspondingly higher, but high
Pteridium and *Calluna* values, and consistent presence of *Melampyrum*
suggest a rather open, acidic woodland. These differences are probably
a reflection of poorer soils and granitic bedrock.

Pollen assemblage zone NMKI-4: Development of Atlantic woodlands

The expansion of *Alnus* to 3.9% and a steady decline in *Pinus* define
the lower boundary of NMKI-4 (542 cm; 7100 B.P.). *Alnus* is more
or less consistently present from 618 cm (8250 B.P.) and crosses the
1% threshold at 550 cm (7200 B.P.). The latter is interpreted as
indicating local establishment, the date being earlier than that recorded
in the FRKII and SHEIII profiles. Maximum expansion is achieved
by 6700 B.P. (510 cm), while in the FRKII profile the corresponding
feature is as late as 5600 B.P. (Fig. 5). Immediately above this peak
in *Alnus, Pinus* representation is halved, *Quercus* expands and *Ulmus*
levels are consolidated. In the upper part of NMKI-4, not only did
Pinus become less frequent in the landscape than *Quercus* and *Ulmus*,
but the woodland structure became more closed, as indicated by a
substantial decline in *Pteridium* and lower *Calluna* and Gramineae
percentages. As in the Connemara National Park profile, *Corylus*
declined and *Betula* assumed greater importance. Finally, *Ulmus*
expanded and, at *ca.* 5600 B.P., *Ilex* entered and *P. lanceolata* formed
a more or less unbroken curve. The latter features suggest an opening
up of the woodland structure (see below).

Pollen assemblage zones NMKI-5 and 6: The elm-decline and the expansion of Taxus

As in the FRKII profile, the elm-decline is well defined, *Ulmus*
representation declining from 8.9% (450 cm) to 2.3% (446 cm) at
5150 B.P. The rise in *P. lanceolata* indicates a substantial opening up
of the canopy, which probably favoured *Quercus* pollen production

(Aaby this volume) and the expansion of *Betula*. In contrast to FRKII and also to the Dolan and Roundstone profiles, a distinct woodland recovery phase follows in which *Ulmus* achieves 6.8%, *Fraxinus* expands (earlier than in FRKII) and *Taxus* invades. The ^{14}C-dates suggest a significant decline in sedimentation-rate over the elm-decline interval. The opposite might be expected to follow a major disturbance in the terrestrial environment (cf. Edwards 1985a), such as a landnam phase. Here, however, it seems that human activity is not the decisive factor determining the behaviour of the *Ulmus* curve at this time (see below).

The *Taxus* rise marks the lower boundary of NMKI-6 (428 cm; 4350 B.P.). By 416 cm (4100 B.P.), it had declined again to 5%, which coincides with a blip in the *P. lanceolata* curve, a sharp rise in Gramineae and a major expansion in *Pteridium* from 0.95 through 5.2 to 11.9%. Obviously, *Taxus* became a major component in the woodlands and, after expansion, was subjected to widescale clearance. That *Taxus* flourished in this area is unexpected in view of the base-poor soils. However, *T. baccata* is frequent in woodlands on lake islands in western Connemara on acid bedrock (Webb & Glanville 1962; Connolly 1930). Our evidence confirms Webb & Glanville's view that it should be regarded as a relict species in Connemara.

Pollen assemblage zones NMKI-7, 8 and 9: The decline of woodland and the expansion of bog and heath

At the lower boundary of NMKI-7 (3200 B.P.), *P. lanceolata*, Gramineae and especially *Calluna*, *Myrica* and *Narthecium* expand dramatically and the curve for *Erica tetralix* begins. The trend towards bog and heath, which began after the *Taxus* maximum (4100 B.P.), accelerated, most probably in the context of woodland clearance. Later (*ca.* 2800 B.P.), further bog expansion took place at the expense of *Pinus* and *Ulmus*, while the *Fraxinus* population expanded at least relative to other trees. At the top of the zone, there was some woodland recovery (cf. *Betula*, *Ulmus* and, especially, *Ilex*) as a result of reduced farming, indicated by lower *P. lanceolata* values.

The uppermost two ^{14}C-dates are too old (Fig. 3) probably due to inwash of organic material which increases upwards in the profile. The spectra immediately above the date of 2740 ± 50 B.P. contain *Solanum tuberosum* pollen, which suggests that these relate to the late-18th/19th centuries when the potato was first widely sown as a crop (Drury 1984). NMKI-8 is defined by a significant expansion of the cereal curve, steady *Urtica* values and the beginning of a broken *Artemisia* curve, suggesting that the base of the zone dates to *ca.* 300 A.D., the beginning of the Christian Period (cf. Mitchell 1986). At Knocknacarra, on the western outskirts of Galway city, a horizontal mill dated to 1335 ± 45 B.P. (Pearson & Pilcher 1975) indicates that arable farming was of considerable importance in that locality in the early Christian Period. Around the lake, it appears that increased farming activity led to soil erosion - in this instance peat - with the consequence that older non-contemporaneous pollen of taxa which were frequent earlier, e.g. *Pinus* and *Taxus*, were incorporated in the upper sediment, thus giving continuous curves for these taxa to the top of the diagram. Widescale extinction of *Pinus* had generally taken place by this time in Ireland (Bradshaw & Browne 1987), but there remains some doubt as to whether the species actually became extinct in Connemara. Wade (1802, p. 122) reports it to be "scattered in a few places in Cunnamara, of a diminutive stunted growth, although apparently very old". To-day, the situation is complicated as a result of colonization of isolated areas, such as lake islands, by seeds arising from pine plantations

Fig. 4. Percentage pollen diagram based on an AP sum from Lough Sheeauns, Cleggan, north-west Connemara (see also legends to Figs. 2 and 3).

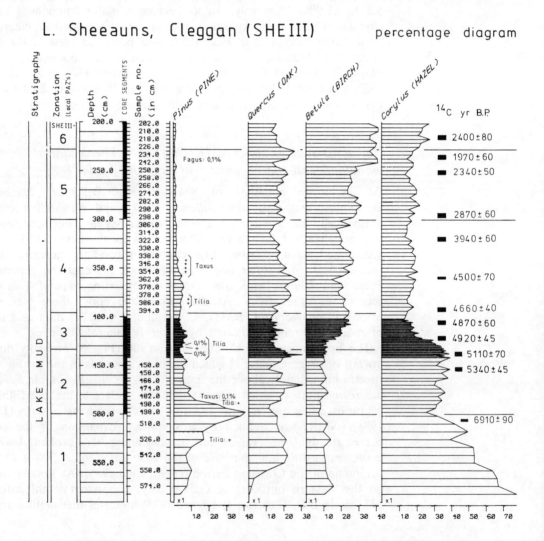

dating to the last century or earlier.

The occasional occurrence of *Tilia* and *Fagus* pollen in this and the other profiles are probably the result of long-distance transport. These trees are not native in Ireland (Mitchell 1986).

Profile SHEIII from Lough Sheeauns, Cleggan

Pollen assemblage zones SHEIII-1 and 2 (lower part): Early post-glacial woodland development

At the base of SHEIII-1 (Fig. 4), *Corylus* has its maximum value (78%), *Pinus* representation is low and *Quercus* and *Ulmus* are less than 0.6%. The lowermost date (9270 ± 100 B.P.) suggests that the

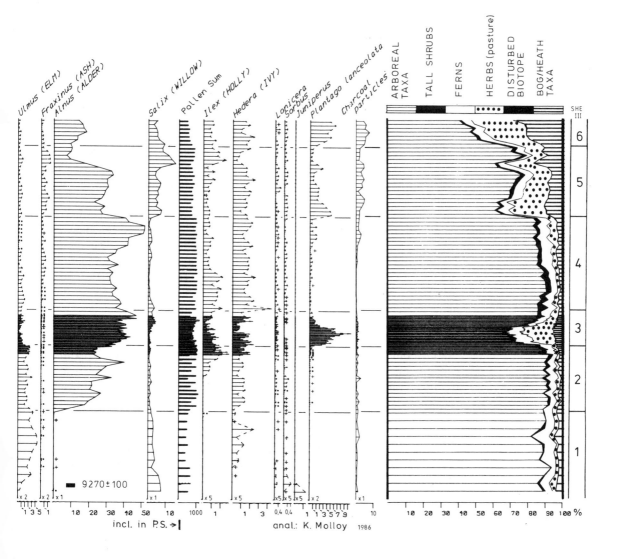

early Boreal *Corylus* maximum is recorded at the base of the diagram. As at other sites, *Pinus* is present before *Quercus* and *Ulmus*, but its major expansion is delayed, and then it is of short duration. *Quercus* dominated the Boreal and Atlantic woodlands, and surprisingly in view of the favourable soils, *Ulmus* was less frequent than elsewhere in the region. *Corylus* declined steadily throughout SHEIII-1, as full-canopy woodland developed. Other features in the Boreal woodlands are the importance of *Sorbus* (probably *S. aucuparia*) and the early records of *Lonicera* (9200 B.P.), which pre-dates *Hedera*. The invasion of *Alnus* at 6700 B.P. is preceded by the *Pinus* rise and its sharp expansion is matched by the dramatic *Pinus* decline. Similar features appear in several Irish lake diagrams and seem to reflect local changes in vegetation around the sampling site, possibly connected with a lowering of water-table (O'Connell *et al.* 1987).

Pollen assemblage zone SHEIII-2 (upper part), 3 and 4: The elm-decline, landnam and secondary woodland

The detailed changes around the elm-decline have been recorded by continuous sampling (Molloy & O'Connell 1987). The main features are as follows. The *Ilex* curve starts late at *ca.* 5400 B.P., and is followed by steady levels of *Ulmus* and the initiation of a low *P. lanceolata* curve. The elm-decline spans 2 spectra and is not accompanied by a rise in NAP (base of SHEIII-3). *Quercus* and *Corylus* decline subsequently and NAP shows a dramatic expansion reflecting changes brought about by landnam (see below).

After the phase of woodland destruction associated with landnam, estimated to have lasted *ca.* 200 years, the woodlands regenerated with *Quercus* as the main canopy tree and a little *Ulmus*. The steady *Pinus* curve suggests that pine was growing outside the main pollen-source area. In the secondary woodlands (top of SHEIII-3 and SHEIII-4), *Corylus* appears to have been replaced by *Betula* as the main understorey shrub, *Quercus* was the dominant tree, *Ilex* was important (until 4500 B.P.) and *Alnus* expanded, probably along the wet valley floor to the east of the lake.

At about 4500 B.P. *Quercus, Ulmus* and *Ilex* decline, *Fraxinus* expands and the plus values for *Taxus* (Fig. 5) (grains present in low frequency may have been overlooked in the routine counting) indicate its possible presence. Throughout most of SHEIII-4, anthropogenic indicator and bog taxa are very poorly represented (small rise begins at 326 cm, *ca.* 4000 B.P.), suggesting an almost completely wooded landscape. Human activity appears to have been minimal but there may be a filtering effect by local *Alnus*, resulting in reduced NAP representation.

Pollen assemblage zones SHEIII-5 and 6: Woodland clearance and bog/heath expansion

The base of SHEIII-5 is marked by a sharp decline in *Alnus* accompanied by an expansion of NAP taxa, especially Gramineae, *P. lanceolata*, *Filipendula*, cereal-type and *P. major/media*, and bog/heath taxa. A substantial reduction of *Alnus* occurred and bog/heath expanded in the context of renewed farming. The intensity of farming may, however, be exaggerated in the pollen record as a result of the percentage method of calculation and the reduction in the filtering fringe of *Alnus*. This farming activity between 2900 and 2200 B.P. occurs in the Late Bronze Age and Early Iron Age. Then a distinct lull set in which facilitated regeneration of *Betula* and *Ilex*, and then *Quercus*. The decline in *Alnus* and the expansions of bog taxa and *Salix* appear to signify the replacement of alder carr by bog, thus favouring the transport of *Salix* pollen to the lake. At the base of SHEIII-6 (1900 B.P.), a major woodland-clearance phase commenced which may have been associated with renewal of farming in the Christian Period. As at L. Namackanbeg, this became sufficiently intense to result in substantial inwash of organics, giving a reversal in the uppermost ^{14}C-date.

Discussion

Woodland history and regional differentiation

Our results show an unexpected degree of regional variation and of time transgression in the courses of the AP curves, especially *Alnus* (Fig. 5), but there are also some unifying features. The importance of *Pinus sylvestris* (Jessen 1949) is confirmed and this probably applies to much of Connemara. It was the first canopy species to invade at *ca.* 9200 B.P., continued until *ca.* 3500 B.P. to be an important tree in the landscape, albeit with a good deal of spatial variation, and probably became extinct over most of the region between 200 B.C.-100 A.D. This parallels its history in the Burren, a karst limestone region to the south (Crabtree 1982; Watts 1984). Its success in these very different habitats is undoubtedly related to its ability to tolerate wet acid and dry alkaline conditions which are largely outside the ecological amplitude of the major, more competitive tree species (cf. Fig. 53, Ellenberg 1978).

Quercus (probably *Q. petraea*) was important everywhere. At L. Sheeauns, with more favourable soils, it was dominant. Surprisingly, *Corylus* and especially *Ulmus* and later *Fraxinus* played a much reduced role there compared with the other sites investigated.

After the elm-decline, there was a dramatic expansion of *Taxus* at L. Namackanbeg dating to 4300 B.P. Investigations in progress record a similar expansion at L. Begglenmore, 0.5 km distant (O'Connell unpublished), which suggests that the species was of considerable importance in the Spiddal area at this time. Yet, elsewhere in western Connemara, only a hint of such a feature is recorded at L. Sheeauns. In the Burren, Watts (1984) has shown that *Taxus* expanded shortly after the elm-decline and again in the first millennium B.C. while O'Connell *et al.* (1987) (also O'Connell 1986) showed that *Taxus* expanded in north-east Mayo at 1800 B.P. Its expansion during the later post-glacial in western Ireland was obviously favoured by woodland disturbance.

Fig. 5. Selected AP curves from each of the 3 profiles expressed as a percentage of total terrestrial pollen and plotted against a time scale derived from the ¹⁴C-dates and the pollen-analytical evidence (base and top of profiles; see text). The establishment (FRKII only) and expansion of *Alnus*, and the elm-decline are indicated on the respective curves.

Early prehistoric human impact

The high levels of charcoal in profiles FRKII and NMKI suggest that fire occurred throughout the earlier post-glacial and may have been important in giving *Pinus* the competitive advantage over *Quercus* at sites where the former remained dominant (cf. Bradshaw & Browne 1987; Moore 1987). It is unlikely that the fires were the result of human activity. Herbaceous pollen taxa, including so-called anthropogenic indicator taxa, are well represented in the relevant levels (see especially the *P. lanceolata* peak in FRKII-4 at 8600 B.P.), but these may also be interpreted as arising from natural openings in the

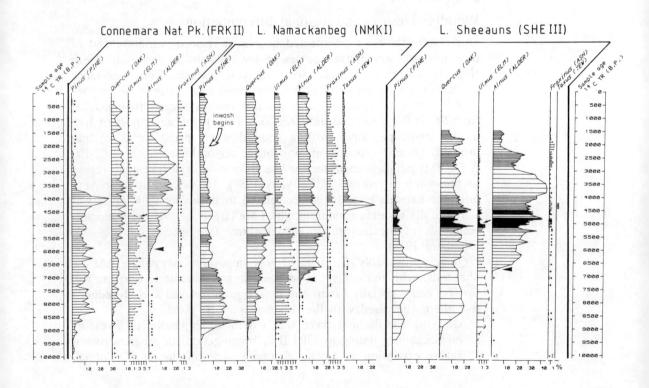

woodland cover (O'Connell 1987). A substantial Mesolithic population interfering with the woodland cover appears unlikely, in view of the scanty archaeological evidence relating to this period in western Ireland generally and Connemara in particular (Woodman 1985; J. Higgins personal communication). However, this may reflect as much lack of survey as absence of artifacts.

The elm-decline is distinct in all 3 diagrams. Other features are the consistent presence of *P. lanceolata* some centuries prior to the elm-decline, and the initiation and expansion of the *Ilex* curve in the lake diagrams. The entry of these two species indicates opening up of the woodland cover (cf. Hirons & Edwards 1986) but uncertainty remains as to whether this is ascribable to human disturbance. Only in the uppermost spectra of SHEIII-2 have cereal-type pollen, clearly of cereal origin, been recorded, indicating that arable farming was occurring here in the decades immediately preceeding the elm-decline (Molloy & O'Connell 1987).

In the SHEIII profile, vegetational changes associated with the elm-decline and subsequent landnam suggest disease rather than anthropogenic activity as the primary cause of the decline (Molloy & O'Connell 1987; Rackham this volume; Edwards this volume). In the FRKII profile, the pollen record is more regional, and, furthermore, it is complicated at this time by changes on the bog surface which include invasion by *Pinus*. At L. Namackanbeg, on the other hand, the post-elm-decline recovery of *Ulmus* is particularly pronounced. The local archaeological evidence indicates that, in this period, people were very few. This would have facilitated a more or less complete recovery of *Ulmus* from the proposed disease-induced collapse.

Late- and post-Neolithic landscape evolution

The initiation and expansion of blanket bog is a key development in the creation of the present landscape. At L. Namackanbeg, bog initiation is signalled by increases in the *Calluna* and Cyperaceae curves and the start of a *Narthecium* curve immediately following the decline of *Taxus* at 4100 B.P. The first major expansion of bog probably dates to 3400 B.P., when the above curves and also *Myrica* expand considerably and the *Rhynchospora* and *Erica tetralix* curves start (NMKI-6/7 boundary). Midway between these events in the pollen diagram, a more or less continuous cereal curve begins and charcoal representation increases at the top of NMKI-6 (Fig. 3). This suggests that, from the Early/Middle Bronze Age (*ca.* 3500 B.P.), human impact was considerable and that fire was important in effecting the transition from woodland to bog here.

In the FRKII profile, indications of widespread blanket-bog growth are obscured by the predominance of local bog pollen. Investigations of 2 short soil/peat profiles taken nearby, where peat accumulation is not merely a consequence of a local depression in the topography, show that blanket bog was initiated at *ca.* 3000 B.P. in the context of woodland clearance (O'Connell unpublished). This evidence points to human activity rather than climatic change leading directly to bog growth.

The final general demise of woodland in the early centuries A.D., can be ascribed to the upsurge in farming activity at the beginning of the Christian Period, recorded in many parts of Ireland (Mitchell 1986). The next major impact probably begins in the 17th century, when population expanded following land confiscations elsewhere in Ireland. Woodland remnants were cleared, bog and heath were used for summer pasturage and in parts reclaimed, tillage expanded as evidenced by the numerous small mills in the study area (P. Duffy personal communication), and nucleated settlements or *clachans* became a feature of the landscape here, as in other parts of western Ireland (Graham 1970; MacAodha 1965). The landscape still bears witness to the severe population pressures that built up in the first half of the 19th century, in the form of cultivation ridges often extending into the most inhospitable environs (cf. Dodgshon this volume). Nature, however, is again regaining the upper hand, as blanket-bog growth slowly obliterates the physical evidence of recent human incursion into its domain.

Summary

(1) Radiocarbon-dated pollen diagrams, spanning most of the post-glacial, from two lakes and a bog in Connemara, Co. Galway are presented.

(2) The course of woodland development is outlined. *Pinus sylvestris* was important from *ca.* 9200 B.P., maintained a dominant role in the earlier post-glacial and remained an important woodland species in western Connemara until as late as *ca.* 3500 B.P. Widespread extinction occurred between 200 B.C. and 100 A.D. At Lough Sheeauns, *Quercus* dominated on the more favourable soils, and *Pinus* and *Ulmus* were minor components. At L. Namackanbeg, a major expansion of *Taxus* is recorded at 4300 B.P.

(3) Fire was important throughout most of the post-glacial. The earlier changes in woodland composition, including those associated with the elm-decline, are ascribed to factors other than human impact. The main period of blanket-bog growth took place between 4000 and 3000 B.P., in the context of extensive woodland clearance by fire.

(4) All diagrams show an upsurge in human activity in the later Bronze Age, i.e. post-3500 B.P. A lull at the end of the Iron Age is followed by widespread woodland clearance at the beginning of the Christian Period (*ca.* 300 A.D.).

Acknowledgements

We thank P. Cooke, P. O'Rafferty and N. Kirby and members of the staff of the Connemara National Park for assistance in the field and Clare Walsh for typing the manuscript. The ^{14}C-determinations were carried out by W. G. Mook, Rijksuniversiteit, Groningen. The authors are particularly indebted to John Birks and Hilary Birks for their constructive criticism of the draft manuscript. Generous financial support from I.C.I. (Ireland) Ltd and the Office of Public Works, Dublin is gratefully acknowledged.

Nomenclature

Plant nomenclature follows D. A. Webb & M. J. P. Scannell (1983) *Flora of Connemara and the Burren,* Royal Dublin Society, Dublin and Cambridge University Press, Cambridge.

Palynological Evidence of Prehistoric Anthropogenic Forest Changes on the Swiss Plateau

Studies in the Late Quaternary of Lobsigensee no 16

Brigitta Ammann

Introduction

Palynology has two main ways of tracing early human influence on vegetation: first, by searching for indicator species, such as apophytic and anthropochorous species (Behre 1981); and second, by evaluating changes in the forests, i.e. changes in their composition and in their proportions to non-arboreal vegetation (Iversen 1941). Here, I concentrate on the second approach and use indicator species only for defining the lower and upper limits of the period concerned: the prehistoric, i.e. pre-Roman, periods with a distinct human impact on vegetation. The first Cerealia and accompanying weeds mark the beginning of the Neolithic revolution (Firbas 1937); pollen of *Juglans* and *Castanea* mark the Roman colonization.

The site and forest-clearance phases at Lobsigensee

Between the Alpine arch and its side-branch of the Jura mountains, lies the trough of the Molasse, filled with Tertiary sandstones and marls, also called the Molasse Plateau. Figure 1 shows the situation of Lobsigensee and the 3 localities compared with it. Lobsigensee is a reference site of the Swiss Plateau in IGCP 158b (Lang 1983, 1985; Ralska-Jasiewiczowa 1986) and therefore many studies are published or are in progress (e.g. Ammann *et al.* 1983, 1985). This chapter concentrates on the forest development between the second of the four steps of the elm-decline and the Roman colonization (i.e. between *ca.* 6500 B.P. and 2000 B.P. conventional ^{14}C-years, Figs. 2, 3). As illustrated in Figure 2 there are 5 pre-Roman and one post-Roman *Fagus*-peaks. Instead of refining the sampling intervals as detailed as Tolonen (1981) did, several cores in the basin were checked. Figure 4 shows the consistency of the *Fagus* peaks in 4 pollen diagrams across the lake basin.

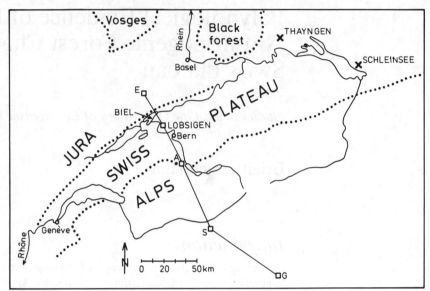

Fig. 1. Map of the Northern
Alpine Foreland between the
Jura mountains and the Alps.
E to G is the cross section
of the Swiss IGCP 158b project
(Lang 1985).
E = Les Embreux, Lobsigen;
A = Amsoldingen;
S = Simplon;
G = Lago di Ganna;
BIEL = Bielersee.

Neither *Quercus* nor any other genus of the mixed-oak-forest respond to these fluctuations of *Fagus* (Fig. 2), but *Alnus, Betula* and *Corylus* do. Peaks of *Alnus, Betula* and *Corylus* are synchronous with the *Fagus*-declines after the *Fagus*-peaks. The three genera *Alnus, Betula* and *Corylus* are here called the ABC-group. After *Fagus*-peak F1 we observe ABC-peak α; after F2a it is ABC-β; after F2b it is ABC-γ (lacking *Alnus*); after F2c it is ABC-δ (lacking *Corylus*); after F3 it is ABC-ϵ (lacking *Alnus*); after F4 it is ABC-ξ (lacking *Corylus*); after F5a it is ABC-ϑ. In addition, the ABC-peaks η and ι are to be found in *Fagus*-rises rather than *Fagus*-declines.

These alternations between *Fagus*-peaks and ABC-peaks are reminiscent of the work by Iversen (1941) who found ABC-peaks synchronous with declines of mixed-oak-forest trees. He discussed the ecological and successional mechanism behind them: declines of Quercetum mixtum reflect deforestations in the main forest community of an area, the ABC-shrubs are the heliophilous pioneers on abandoned areas flowering at an early age (see also the "trivialisation of the forests" described by Fægri 1944).

The earliest description of such a succession might be by Albertus Magnus (1193-1280) in his "de vegetabilibus libri VII": "Videbimus enim in omnibus silvis, quod, praecisis arboribus, quae quercus vocantur vel fagi, recrescunt arbores, quae tremulae dicuntur, et arbores, quae vocantum miricae*..." with the comment by Jessen (1867): *sunt *Populus tremula* et *Betula alba* Lin. But Albertus Magnus concluded that out of the rotting stumps of *Quercus* and *Fagus* the poplars and birches sprout, an example of "transmutationis unius plantae in aliam". I hope that my interpretations of correct observations are no worse than his.

Two comments offered by U. Willerding (personal communication) add to the interpretation by Iversen (1941). First, the impoverishment of soils was probably much faster in the Danish areas not covered by tills of the last glaciation than in the Northern Alpine Foreland. Second, the fact that many walls of Neolithic houses seem to have been woven out of twigs raises the possibility that pioneer shrubs were cultivated and coppiced for that purpose (see Rackham this volume). It should be born in mind that natural short-term successions exist, as described by Müller (1962) for the early part of the Younger Atlantic (before the expansion of *Fagus*) and by Andersen (1975) for the Eemian interglacial.

The question arises if the antagonism between *Fagus*-peaks and ABC-peaks is a result of percentage calculations and is therefore an artefact. Concentration- and influx-diagrams may answer this, even though it is known that each method brings along its own problems, connected with sedimentological changes for concentration- and, with errors from ^{14}C-dating, for influx-diagrams. Figure 2 presents the concentration curves for *Fagus* and ABC-shrubs. The peaks obviously remain at the same levels as in the percentage diagram (displaced by one sample in F1 and F5). In the concentrations there are some peaks that are missing in the percentages, i.e. β, γ and η for *Alnus* and δ and ξ for *Corylus*. The *Betula*- and *Corylus*-peak registered as percentages in the local pollen-assemblage zone L19 disappears in the concentration curves. This local L19 zone was formed during the Neolithic settlement of the Cortaillod-culture on the shore of Lobsigensee. Its sediment is a distinctly laminated gyttja rich in carbonates, signs of an early human

Fig. 2. Comparison of *Fagus*-peaks 1-6 and *Alnus*-, *Betula*- and *Corylus*-peaks alpha to kappa in percentages, concentrations and influx. Local pollen assemblage zone L19 corresponds to the Neolithic Cortaillod culture (deforestation 2) and L23 to the Roman occupation (deforestation 6). U2 to U4 indicate steps of the Elm-decline.

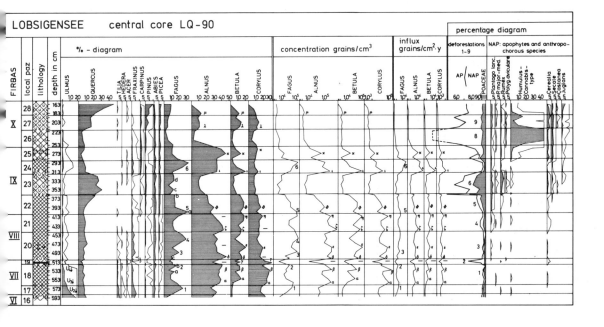

eutrophication (Ammann 1986). The gyttja seems to be diluted by carbonates and thus all pollen concentrations decrease.

The influx values were calculated on the basis of 7 ^{14}C-dates (Oeschger *et al.* 1985) (Fig. 3). The influx peaks correspond well with the concentration peaks. From the consistency of the peaks in percentages, concentrations and influx, I conclude that the antagonism between *Fagus*-peaks and ABC-peaks is not an artefact but reflects real vegetational changes (see also Pennington 1975).

Comparison of Lobsigensee with other sites on the Molasse plateau

I will now compare the results from Lobsigensee with results from 3 other sites on the Molasse Plateau.

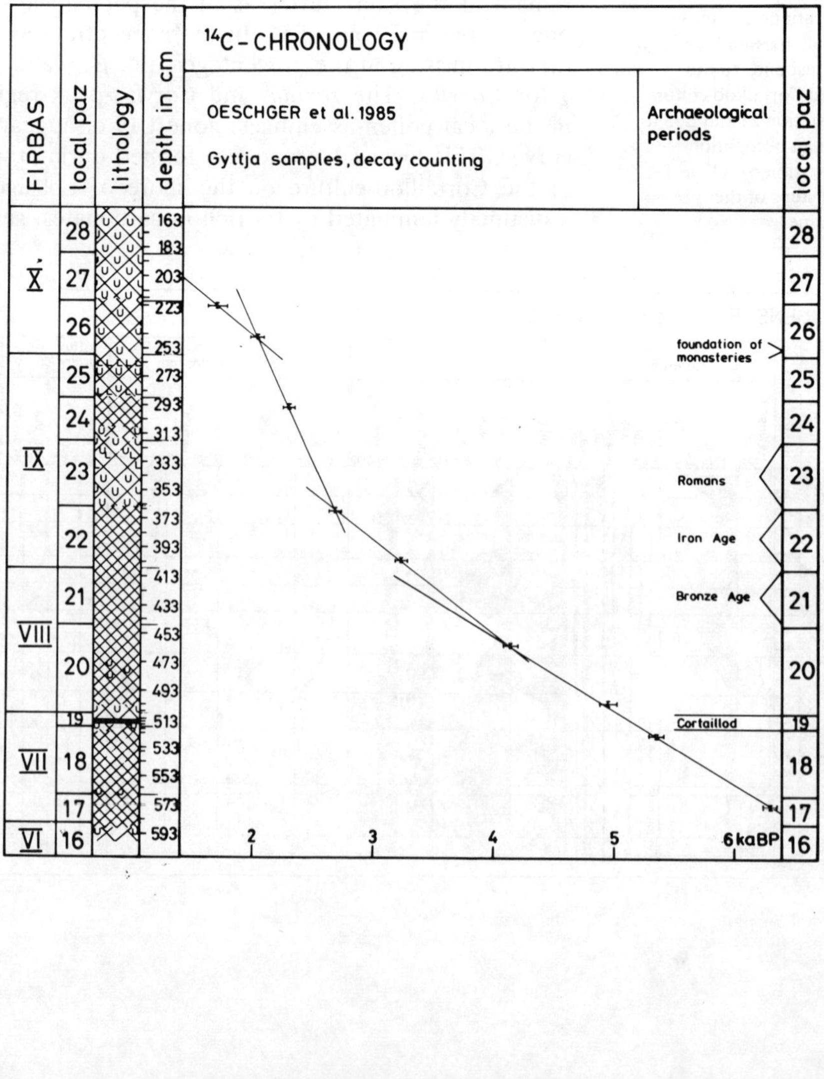

Fig. 3. Radiocarbon ages of core LQ-90 given in conventional ^{14}C-years. The oldest date of 6300 B.P. corresponds to the first *Fagus*-peak. The 2 dates below and above the Cortaillod layer bracket the age of 5100 B.P., which is in good agreement with other sites. The depth-age relationship shows a major change at the beginning of local zone L23, i.e. with the Roman colonization: increased sedimentation rate together with a higher proportion of clay point to enhanced erosional inwash, most probably due to extensive deforestation. Steeper curves since 4000 B.P. can also be due to erosive input of older carbon (either organic carbon or carbonate).

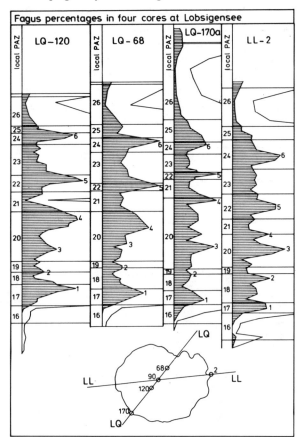

Fig. 4. Percentage curves of *Fagus* in 4 cores from Lobsigensee comparable to core LQ-90 (Fig. 2). The depth scales are identical. The 6 *Fagus*-peaks are found in all diagrams. The subdivisons of *Fagus*-peaks 2 and 5 are not always found; this may partly be due to coarser sampling intervals.

Thayngen-Egelsee

Figure 5 illustrates a pollen diagram which is part of a larger investigation at Thayngen-Egelsee near Schaffhausen (130 km north-east of Lobsigensee) by Troels-Smith (1981). There is a good correlation of the 5 pre-Roman *Fagus*-peaks with the ones from Lobsigensee; only peak F2 and its relationship to the Neolithic culture of Pfyn (corresponding to Cortaillod) is not very clear. Troels-Smith (1981) proposes two arguments for not regarding the *Fagus*-declines as real. First, wood analysis at the Thayngen-Weiher settlement showed that less than 1% of the wood chips or charcoal were *Fagus* - thus very little *Fagus* seems to have been cut by Neolithic people in this area. Second, the 2 pollen-influx values for *Fagus*, one before and one after a *Fagus*-decline, were of similar order of magnitude. Troels-Smith (1981) then calculated how much larger the pollen production of *Quercus* (and of other trees in the mixed-oak-forest) must have been in order to produce the *Fagus*-declines observed in the percentage diagram. Woodland clearing is considered to be the main reason for the increased pollen production of *Quercus*: *Quercus* in a dense forest canopy flowers relatively poorly, whereas single, free-standing *Quercus* specimens (after forest clearances) flower much more abundantly.

There are two contrasts between Thayngen-Egelsee and Lobsigensee: (1) at the latter site the *Fagus*-declines are not fictitious but real

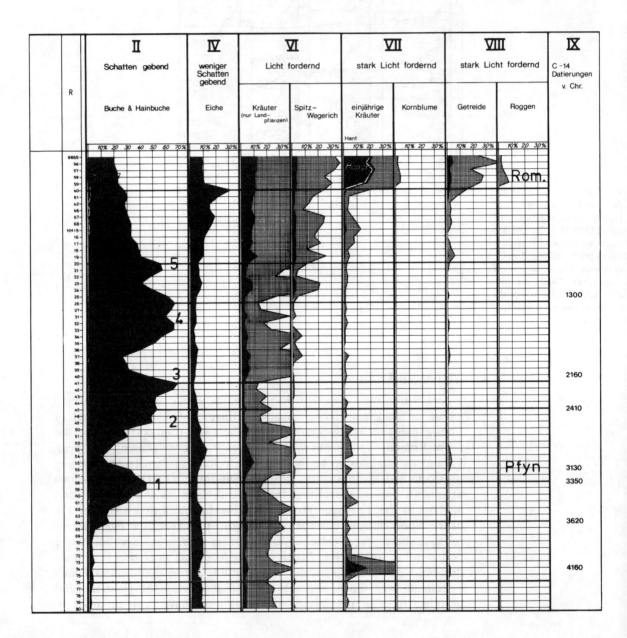

(consistent in percentages, concentrations, influx), and (2) periods of *Fagus*-minima do not show *Quercus*-maxima.

An explanation for this second feature comes from comparing modern potential vegetation maps (although we cannot assume that the forest communities around 5000 B.P. were the same as today since, for example, *Carpinus* had not expanded yet). Around Lobsigen there would mainly be *Fagus* forests with a few *Quercus* (Hegg 1980: mainly Galio odorati-Fagetum; on soils rich in carbonates Carici-Fagetum, and on poor soils Luzulo-Fagetum). Around Thayngen, however, more thermophilous forest communities previously grouped into Querco-Carpinetum would occur (Ellenberg & Klötzli 1972: Galio odorati-Fagetum typicum, Pulmonario-Fagetum typicum, Pulmonario-Fagetum melittetosum, Coronillo coronatae-Quercetum). If *Quercus* was used by Neolithic people for constructing lake dwellings, only very few *Quercus* were left around Lobsigen, just producing a 'background pollen rain'. Around Thayngen, on the other hand, many *Quercus* may have been left which then may have flowered more abundantly than before partial deforestation (see Aaby 1986b, this volume). The intriguing result from wood analysis in Thayngen-Weiher with so little *Fagus* wood is different for Neolithic sites in central and western Switzerland where Schweingruber (1976) showed that *Fagus* was used extensively for different tools (large numbers of chips were found) and for burning.

Schleinsee

The second site to be compared with Lobsigensee is Schleinsee just east of Lake Constance. Müller (1962) described the following sequence which is consistently observed at Schleinsee: peak of *Fagus* - decline of *Fagus* together with an increase of NAP - peaks of *Alnus*, *Betula* and *Corylus* - peak of *Fraxinus* with an increase of *Fagus* - *Fagus*-peak, etc. (see Fig. 6). Müller showed that these fluctuations reflect the influence of early people on the vegetation by deforestation. Even when the process of clearance was not necessarily carried out in the same manner each time, the reaction of the fast-growing and early-flowering shrubs ('ABC') was very similar. (In addition Müller gives an example of a pre-Neolithic change in forest composition ascribed to succession during the early part of the Younger Atlantic).

In Figure 6 a correlation between the 5 pre-Roman *Fagus*-peaks at Lobsigensee with the ones from Schleinsee is proposed. In general the agreement is good, although the the sites are 190 km apart, and human impact cannot be expected to have been exactly synchronous. The subdivison of peak number 5 is more pronounced and peak number 4 is also subdivided at Schleinsee. The correlation with archaeological periods is good, for example the Roman colonization begins between peak 5b and 5c. It may appear that at Lobsigensee the first phase

◄

Fig. 5. Pollen diagram from Thayngen-Egelsee showing some selected taxa.
II: Schatten gebend = shade-demanding plants; Buche = *Fagus*; Hainbuche = *Carpinus*; 1-5 = tentative correlation with *Fagus*-peaks at Lobsigensee.
IV: weniger Schatten gebend = less shade-demanding plants; Eiche = *Quercus*.
VI: Licht fordernd = light-demanding plants; Kräuter = herbs (terrestrial plants only); Spitz-Wegerich = *Plantago lanceolata*.
VII: stark Licht fordernd = strongly light-demanding plants; einjährige Kräuter = annual herbs; Kornblume = *Centaurea cyanus*.
VIII: Getreide = Cerealia; Roggen = *Secale*.
Pfyn = level of the Neolithic culture of Pfyn, equivalent to Cortaillod; Rom = level of the Roman colonization; (after Troels-Smith 1981, reproduced with permission of *Archäologie der Schweiz*).

of deforestation is rather weak and late compared to the first *Fagus*-decline. This latter, however, corresponds with the first records of Cerealia. This means that in Figure 2 deforestation no. 1 reflects the Neolithic landnam in the area, deforestation no. 2, on the other hand, the local settlement at the lake shore. The main difference from Schleinsee is that at Lobsigensee *Fraxinus* peaks during the increase of *Fagus*.

In the area of Lake Constance the sequence of *Fagus*-peaks was also found by Rösch (1983, 1985) and Liese-Kleiber (1985).

Bielersee

The third comparison is between the small closed basin of Lobsigensee with the large (42 km²) basin of Bielersee with through-flowing rivers (distance between these sites is 13 km). In a pollen diagram for the corresponding period from Heidenweg/Bielersee (Fig. 7 and Ammann-Moser 1975) there are only minor fluctuations in *Fagus* and ABC-shrubs. The peaks 1 and 6 are comparable, but the others are not. Moreover, the amplitudes are different: the fluctuations vary between 5 and 35% at Lobsigensee but only between 8 and 22% at Heidenweg/Bielersee.

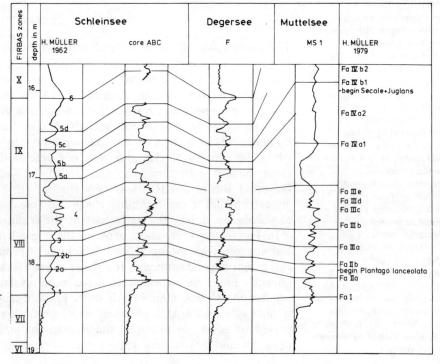

Fig. 6. *Fagus* percentages (of AP+NAP=100%) in 4 cores from 3 small lakes east of Lake Constance, two from Schleinsee, one from Degersee and one from Muttelsee. The numbers of *Fagus*-peaks 1 to 6 by the left curve give the correlation with Lobsigensee. The numbers FAI to FAIVb2 to the right are the denominations given by Müller (1979) (from Müller in Merkt, Müller and Streif (1979) with permission from Helmut Müller).

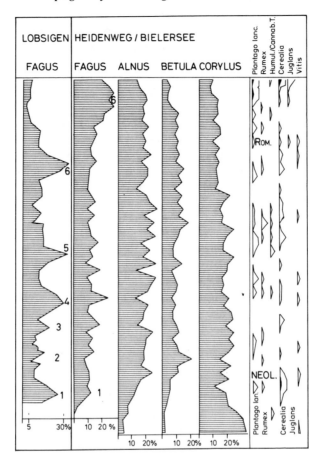

Fig. 7. Comparison of *Fagus* percentages from two neighbouring lakes: the small (2 ha) closed basin of Lobsigensee registered marked fluctuations between 5% and 35%, the large (42 km²) basin of Bielersee with inflowing streams only shows changes between 8% and 22%. Only the first and the post-Roman *Fagus*-peaks are comparable. Neither depth nor time scale are identical.

Conclusions

It turns out that among the 3 comparisons, the correlations between the two closest sites are the poorest. In other words, one can go as far as nearly 200 km when comparing small closed basins and the correlations remain good. Basin size seems to play an important role in recording short-term local events (Oldfield 1970; Jacobson & Bradshaw 1981; Behre & Kučan 1986). It controls, together with the hydrological regime (Pennington 1979), the size of the pollen-source area (Prentice 1985). A large basin like Bielersee represents an example of a meso-scale site (Birks 1986a, Fig. 1.1.b) registering regional vegetation (pollen-source area 100 km² or more). Because of the relatively minor effects of pollen filtration (Tauber 1967), a large basin is well suited to trace early indicator taxa (Edwards 1982) especially small, light grains such as *Castanea* (Ammann-Moser 1975). A small lake or pond like Lobsigensee would be a basin intermediate between type 1 and type 2 in Figure 1.1.b of Birks (1986a). The *Fagus*-declines described for Lobsigensee are about 100-350 years apart from each other. Thus, single events of deforestation are not registered, but rather pulses of human activity. This composite nature of such pulses is mentioned by many authors (e.g. Troels-Smith 1953; Turner 1970; Edwards 1979).

The single clearance or single field-abandonment would most probably be registered only in an 'Andersen-sized' hollow (Andersen 1973, 1978, this volume; Aaby 1986b, this volume) and with high-resolution analysis (Moore 1980). Possible reasons for local and regional cyclicity are still very much open to debate (e.g. Lambert *et al.* 1983; Edwards 1979).

By linking the archaeological information of an area (as summarized by von Känel *et al.* 1980 for the Bernese Seeland) with a network of pollen diagrams from different basin sizes and with numerical dispersal models (e.g. Prentice 1985), one could hopefully obtain some answers to archaeological and botanical questions such as how did population density and settlement continuity change during prehistoric periods and how did successions and vegetation dynamics function after disturbances.

Summary

(1) Pollen diagrams from Lobsigensee (Swiss Plateau) show five distinct pre-Roman *Fagus*-peaks between 6300 B.P. and 2000 B.P. They can be correlated with peaks at other sites in the Northern Alpine Foreland, as far as 190 km away.

(2) The *Fagus*-declines coincide with phases of deforestation. Forest regeneration involves colonization by *Alnus*, *Betula* and *Corylus*.

(3) The classical pattern of forest clearance and regeneration is confirmed by concentration and influx diagrams.

(4) The resolution of short-term local changes in vegetation is good in cores from small, closed basins, but poor in the large Bielersee, due to differences in basin size and pollen-source area.

Acknowledgements

My cordial thanks to all who helped during this project: H. J. B. Birks, G. Lang, A. Lotter, M. Moser, H. Müller, H. Murray, C. Scherrer, J. Troels-Smith, E. Venanzoni, U. Willerding.

Nomenclature

Plant nomenclature follows A. Binz & Ch. Heitz (1986) *Schulund Exkursionsflora der Schweiz*, Schwabe Verlag, Basel.

The History of the Landscape around Auerberg, Southern Bavaria - A Pollen Analytical Study

Hansjörg Küster

Introduction

Auerberg (1055 m elevation) is an isolated hill 20 km north of the Bavarian Alps and 5 km west of the Lech river (Fig. 1). A Roman town on top of the hill has been recently excavated and dated to between the second and the early fifth decade A.D. (Ulbert 1975). About 45 A.D. the Via Claudia, a Roman road following the Lech, was improved, and became more important when the town was abandoned. Later, during the Dark Ages and Medieval times it was still an important line of communication. Churches, cloisters, villages and farmsteads have existed along its route since early Middle Ages.

Nothing else was known about the history of the area before the present study commenced. Paul and Ruoff's (1927, 1932) pollen diagrams are good for their time, but no cereal or cultural indicator pollen were determined, and they had no ^{14}C-dates.

The present study set out to answer the following questions:
- What was the overall pattern of forest history in the region?
- Were there any prehistoric settlements in the region?
- Did the Romans build the Auerberg settlement in a previously cultivated or in a virgin landscape?
- What was the Roman impact on natural woodlands?
- Were there other Roman settlements?
- Was there continuous settlement during the Migration Period or was there a gap in settlement?

To answer these questions cores were taken from three bogs each situated *ca.* 3 km distant from Auerberg: Langegger Filz in the south, Geltnachmoor in the west, and the bog at Haslacher See near the Via Claudia, to the north-east of the hill (Figs. 1, 2).

Methods

Pollen samples counted at 5 cm intervals give a general impression of the Holocene vegetational history. During the settlement phases,

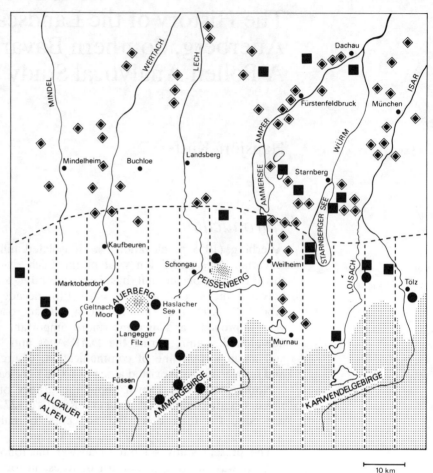

Fig. 1. Map of southern
Bavaria. Pollen diagrams
showing a dominance of
Quercus during Atlantic times
are marked by squares, those
showing a dominance of *Ulmus*
by large dots. Rhomboids
mark burial mounds of the
Hallstatt C/D-period (Early
Iron Age; after Kossack 1959).
The area of present-day
montane forests is shaded;
further north natural beech
forests are common (after
Hornstein 1951).

samples were counted at every cm. Peat-accumulation rates are such
that 1 cm represents, on average, 30 years; thus one pollen spectrum
corresponds in time to a single human generation. For the 3 profiles
ca. 100 ¹⁴C-determinations are available which provide a detailed
chronology.

Cereal pollen was separated from that of wild grasses, particularly
Bromus, using the criteria of Grohne (1957) and Beug (1961).
Examination of the modern pollen rain in the Auerberg region shows
that dispersal of cereal pollen does not extend more than a few km
and, indeed, in places only a few 100 m. The presence of cereal pollen
in the bog profiles therefore provides good evidence for cereal cultivation nearby.

Preliminary results are reported in Küster (1984, 1986).

Forest history

Selected pollen curves from Haslacher See (Fig. 2) are shown in Figure 3.

About 8500 years ago, *Corylus* and *Ulmus* invaded the prevailing *Pinus* forests, soon followed by *Tilia* and *Picea*. The forest of the Atlantic period was dominated by *Ulmus*. The first *Ulmus*-decline is very early (6500 B.P.) and is accompanied by a rise in *Abies* and *Fagus*, and at 6390 B.P. (pollen assemblage zone (p.a.z.) 7/8 boundary) Cerealia pollen is first recorded. *Abies* and *Picea* become abundant, but in p.a.z. 10 the sudden increase in *Fagus* suggests that there may be a hiatus, perhaps associated with a stratigraphic change from cyperaceous to *Sphagnum* peat. Subsequently, *Fagus* is more important than *Abies*, and *Quercus* slowly increases in abundance.

Large reductions in *Abies* and *Fagus*, associated with peaks in cereals, cultural indicators, NAP and Poaceae, in p.a.z. 13 and 14 probably reflect Roman influence on the landscape.

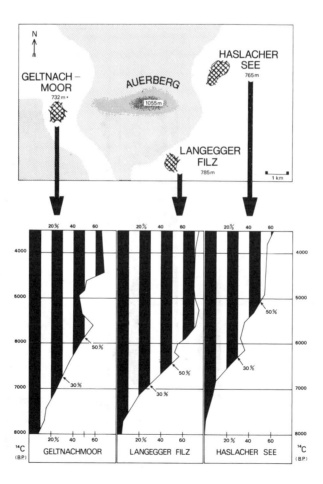

Fig. 2. Simplified pollen curves of montane forest-trees (*Picea, Abies, Fagus*) from 3 bogs near Auerberg.

The forest, dominated by *Fagus, Picea, Quercus, Carpinus* and with some *Corylus*, regenerated through a *Betula* phase (p.a.z. 15). However, low (prehistoric) levels of cultural indicators suggest that there was a continuous, if substantially reduced, human activity during the Migration Period.

Substantial and permanent deforestation occurred during Medieval times. Remnants of montane forest were dominated by *Picea, Fagus* and *Abies*. Today, *Picea* forests are most common, and *Pinus* has expanded on bog surfaces.

Local variations in forest history around Auerberg

Fig. 3. Simplified pollen diagram from Haslacher See 3 km north-east of Auerberg. Tree pollen as %AP.

Simplified pollen curves from the three bogs studied (Fig. 2) show that the invasion of montane forests began at *ca*. 8000 B.P. at all

HASLACHER SEE

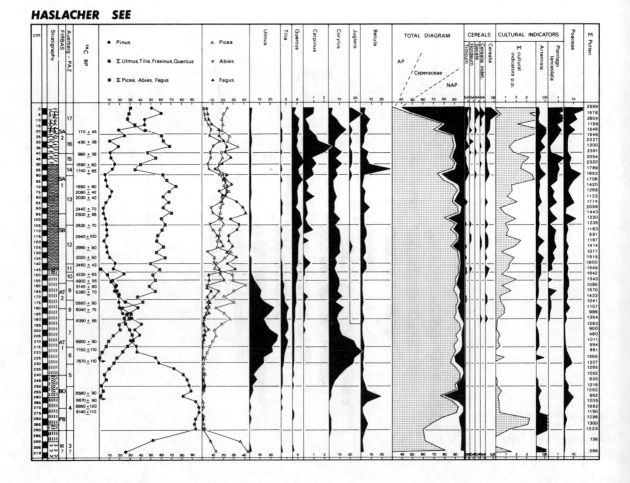

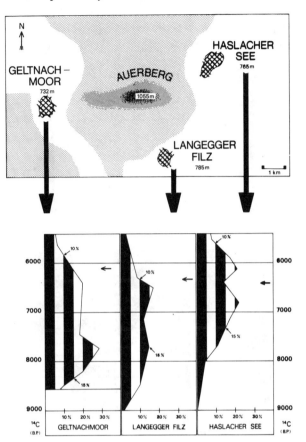

Fig. 4. Simplified pollen curves of *Ulmus* from 3 bogs near Auerberg. The oldest records of cereal pollen are indicated by arrows.

three sites, but the speed of colonization was different. At 7000 B.P. the montane species comprised nearly 30% AP at Langegger Filz and Geltnachmoor, while at Haslacher See 10% was not reached. The speed of colonization was highest at Langegger Filz, 50% being reached by 6500 B.P., about 500 years earlier than at Geltnachmoor and more than 1000 years earlier than at Haslacher See.

The main causes of these differences may be the different distances from the northern edge of the Alps. However, Haslacher See is only 5 km further north than Langegger Filz. The overall lower percentages and later dominance at Haslacher See may be due to local ecological factors restricting the extent of montane forest.

The arrival and expansion of *Ulmus* follow a similar pattern (Fig. 4), being later at Haslacher See. The elm-decline occurs at a very early date, as elsewhere in southern Germany and the Alps. It is latest at Haslacher See. It coincides with the first cultivation of cereals in the region, indicated in Figure 4 by arrows.

The opposite behaviour of the pollen curves of *Ulmus* and of *Abies*

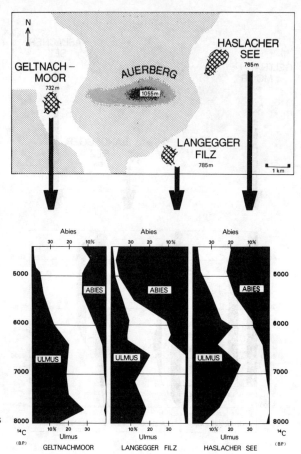

Fig. 5. Simplified pollen curves of *Ulmus* and *Abies* from 3 bogs near Auerberg.

is illustrated in Figure 5. Between 7000 and 5000 B.P. elm-dominated forests were replaced by montane forests characteristic of the region today.

The forest pattern over a larger area is shown in Figure 1. It can be seen that the present montane forest only occupies the area dominated by *Ulmus* in Atlantic times, whereas the former *Quercus*-dominated forests to the north were largely replaced by *Fagus*. The influence of this vegetational boundary in the past is reflected by the distribution of Iron Age burial mounds of the Hallstatt C/D period. They lie mostly north of the boundary with some exceptions near Weilheim and Murnay, where penetration into the montane forests occurred (Kossack 1959). However, the boundary did not prohibit people from colonizing the southern areas during the period in question or earlier in prehistory. The pollen diagrams suggest that the Auerberg region was settled since Neolithic times, even though no burial mounds were erected.

The history of human settlement

As mentioned above, continuous sampling of the profiles was carried out from the point where cereal pollen was first recorded in order to obtain a detailed picture of settlement history. The results obtained are summarized in Figure 6 which shows the general outline of settlement history in the Auerberg region.

Pre-Roman

From 6500 B.P. onwards, cereal and cultural indicator pollen are present. This suggests a surprisingly early beginning for agricultural activity in a region were there is little archaeological evidence for prehistoric settlement recorded until now. Records of very early cereal

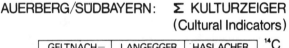

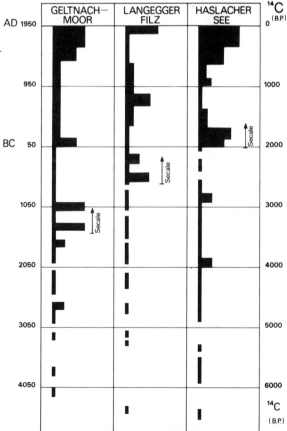

Fig. 6. Schematic fluctuations in percentage representation of cultural indicator pollen during the last 6500 years in the 3 pollen profiles.

pollen in this region (Kossack & Schmeidl 1974/75) are not, however, infrequent, so that the early records presented here are not isolated examples. During the Neolithic there are only scattered records of cultural indicators near Auerberg. From 5000 B.P. onwards there is a more or less continuous curve of cultural indicators. But there are peaks in this curve at different times, probably reflecting local settlements near the bogs. Around 4500 B.P. (Late Neolithic), a maximum of cultural indicators occurs in the Geltnachmoor profile. At *ca*. 4000 B.P. a settlement is recorded which was probably situated near Haslacher See. During the Bronze Age (3500-3000 B.P.), maxima of cultural indicators occur in the Geltnachmoor diagram. Two of them are extremely distinct, giving the impression that settlements were probably situated in or at the edge of the bog (settlements of this type are well known from areas in Upper Suebia, *ca*. 100 km to the west of Auerberg; Schlichtherle 1985). After a maximum of cultural indicators at Haslacher See shortly after 3000 B.P., cultivation obviously shifted to the south around Langegger Filz. From scattered excavation sites (Maier 1985) and local tales (Rump 1977) we can assume that people lived here during the Iron Age and worked in the very important iron mines of the northern border of the Alps. The pollen diagram shows that *Fagus* was preferentially cut in this region, probably to provide charcoal for iron-smelting.

Roman

Shortly after 0 A.D. a very distinct settlement phase is recorded in the profiles of Geltnachmoor and Haslacher See. It is the first to be recorded at two of the three sites, and, therefore, it is of greater significance in terms of landscape history than the earlier ones. It is contemporaneous with the establishment of the Roman town on top of Auerberg. It is obvious that the Romans did not come into a virgin landscape, but it is surprising that they did not colonize the part that was most densely settled in the immediately preceeding period around Langegger Filz. During the Roman Age the cutting of *Fagus* continued near Langegger Filz, indicating that the 'Iron Age people' continued to live there during the Roman occupation. However, the percentages of cultural indicators declined.

The Roman impact on the landscape is best recorded in the profile of Haslacher See, between Auerberg and Via Claudia. Here *Abies* was cut down, probably as a source of timber. This marked decline of *Abies* (see above) does not occur in the Langegger Filz diagram, where there is no record of Roman impact.

The duration of Roman impact on the landscape extends beyond that of the ancient town of Auerberg. Roman influence is recorded for several hundred years, especially at Haslacher See, suggesting that other Roman settlements were founded after Auerberg had been abandoned.

Post-Roman

During the Migration Period, the percentages of cultural indicators decline, indicating a continuous but much reduced level of settlement. During the Early Middle Ages a settlement was established near Langegger Filz but was abandoned soon afterwards. Intensive Medieval agricultural activity is recorded around Haslacher See. At Geltnachmoor, a similar level of settlement activity commenced somewhat later. Intensive cultivation has begun only very recently in the vicinity of Langegger Filz.

In general, Figure 6 shows that farming began rather early around Auerberg; during prehistoric times the typical pattern of shifting cultivation is recorded, and settlements - or only fields - were founded and abandoned 50 or 100 years later. The Roman invasion is a very distinct turning point in the history of the cultural landscape when settlement phases became more important features of landscape history in general. Once again shifting settlements are recorded for the Migration Period. During early Middle Ages more permanent settlement became the norm.

In Figure 6 first records of *Secale* pollen are marked. During the Bronze and Iron Ages they seem to be very early, but *Secale* is recorded in macrofossil analyses from comparably early dates in southern Bavaria; there are some Bronze Age records and greater amounts of *Secale* grain are recorded from Iron Age settlements (Küster unpublished). Scattered pollen and small numbers of macrofossils from the Bronze Age suggest that *Secale* grew in the fields only as a weed and was probably not cultivated as a crop.

Conclusions

As in most areas, the present-day landscape is the product of several factors including natural succession and migration of plants, climate and fluctuations in climate, wild animals, geology and hydrology and - last but not least - the human factor. Human impact on the landscape has become increasingly important during the last millenia. It is clear that there is no distinct transition from the so-called natural to the so-called cultural landscape. Instead, there is a gradual progression from one transitional to another transitional state.

The history of the human environment can be fairly precisely reconstructed by detailed pollen analysis, even in a landscape where archaeological excavations are lacking. Pollen-analytical records of settlement phases should stimulate archaeologists to undertake intensive field research, which has not been undertaken to-date in many parts of southern Bavaria.

Summary

(1) Pollen analyses at three sites around Auerberg (southern Bavaria) show that *Ulmus*-dominated forests were succeeded in late Atlantic time by montane forests characterized by *Abies, Picea* and *Fagus*. Local variations can be detected.

(2) Although local archaeological evidence is lacking, cultural indicator pollen occurs surprisingly early at *ca.* 6500 B.P., coinciding with an *Ulmus*-decline.

(3) There was a shifting settlement pattern through the Neolithic and Bronze Age. An Iron Age settlement was established south of Auerberg.

(4) The first widespread human impact on the landscape was during the Roman settlement on Auerberg. It did not affect the nearby Iron Age settlement.

(5) After the small shifting settlements of the Migration Period, the pattern of more permanent Medieval settlement was similar to the Roman pattern.

Acknowledgements

The above research has been supported by a grant of the Deutsche Forschungsgemeinschaft. I am very grateful to U. Körber-Grohne, G. Kossack and G. Ulbert for advice and many helpful discussions, to H. Willkomm for carrying out the many ^{14}C-determinations, to M. O'Connell for improvements to the English text and to S. Sutt for the diagrams.

Nomenclature

Plant nomenclature follows E. Ehrendorfer (1973) *Liste der Gefässpflanzen Mitteleuropas*, Stuttgart.

Heathland History in Western Jutland, Denmark

Bent Vad Odgaard

Introduction

The geomorphology of western Jutland is dominated by two elements: (1) sandy, fluvioglacial plains - 'heath plains' - of Weichselian age, and (2) flat hills of mostly sandy till of Saalian age. Because these till areas lie as isolated elevated islands in a 'sea' of fluvioglacial plains they are termed 'hill islands'.

Until the end of the 19th century this landscape was covered by vast heathlands. Small areas of scrub with low, crooked and twisted oaks were found scattered on the hill islands. Today most of the heathland has been reclaimed for intensive agriculture or conifer plantation but the oak-scrubs persist. They have traditionally been regarded as relics of the natural vegetation of western Jutland (Gram *et al*. 1944).

The history of West Jutland's vegetation was investigated by Jonassen (1950). Based on several regional pollen diagrams he demonstrated that western Jutland - heath plains as well as hill islands - had been covered by forest, and that the Holocene vegetational development followed the same general scheme as in eastern Denmark. Although the possibilities for dating were then very limited, Jonassen (1950) interpreted a *Calluna* increase in the upper part of his diagrams as a heath expansion at the transition from the Subboreal to the Subatlantic. He explained this expansion in the following way. During the Subboreal and early Subatlantic he assumed agriculture to be very primitive with brief cultivation of the infertile soils. He imagined that after abandonment of the fields, *Calluna* and trees would invade the area and compete for the new habitat. He supposed that the climate was favourable for tree growth in the Subboreal, and hence forest dominated the fields in this period, while in the Subatlantic, after an assumed climatic deterioration, *Calluna* was favoured and spread on the abandoned fields. Thus Jonassen interpreted the heath expansion as the combined effect of primitive, temporary cultivation and climatic deterioration.

The vegetational development of western Jutland is now being reinvestigated by means of ^{14}C-dated regional and local pollen diagrams, interpreted on the basis of present heathland ecology (e.g. Gimingham

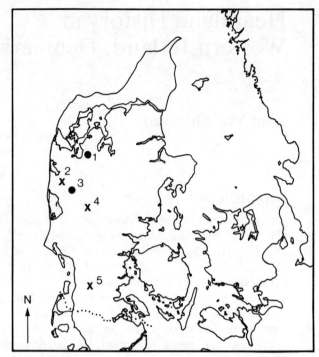

Fig. 1. Sites in western Jutland where studies of vegetational history are being carried out. Dots = sites of regional pollen diagrams; X = local pollen diagrams; 1 = Skånsø; 2 = Grøntoft; 3 = Solsø; 4 = Harreskov and Skarrild; 5 = Hønning.

1972; Böcher & Jørgensen 1972; Hansen 1976; Christensen 1981). The regional sites are lakes with non-calcareous sediments and the local sites are recent or fossil soil sections. The sites chosen so far are shown in Figure 1. This chapter presents briefly the preliminary results of this investigation.

The age of the heathlands

The ^{14}C-dated pollen diagram from Lake Solsø (Fig. 2) reflects the broad-scale Holocene vegetational development. Through the first 5 millenia there is a development from the open Preboreal *Betula-Pinus* forest, through the more dense Boreal *Corylus* forest to the richer Atlantic communities of *Tilia, Corylus, Quercus, Ulmus, Alnus* and, later, *Fraxinus*. The differences from the well-known eastern Danish development (e.g. Aaby in Andersen *et al*. 1983) are (1) that the forest of western Jutland was more open with *Calluna*, grasses and much *Betula*, and (2) that the demanding trees *Tilia* and *Ulmus* were less common.

Soon after the *Ulmus*-decline, *Plantago lanceolata* appears and the *Calluna*-curve increases dramatically, indicating heathland expansion. Soon the first cereal pollen are also found and already at about 2000 B.C. the landscape was probably very open with large heathland areas.

A new *Calluna* expansion starts at about the Birth of Christ and continues for more than 1000 years.

In Denmark today, *Calluna*-heaths need burning, grazing or other sorts of management to prevent them from turning into *Empetrum*-dominated heath or grassland and eventually into scrub and forest (Böcher & Jørgensen 1972; Christensen 1981). It is only in very oceanic climates, as on the Faroe Islands, that heaths with *Calluna* as an important element may be the climax vegetation (Jóhansen 1985). In order to explain the heathland expansion seen in the Solsø diagram by climatic changes, a very dramatic deterioration would have to be assumed, an event for which there is no evidence in the time period concerned. A climatic oscillation at the Subboreal/Subatlantic transition is recorded in peat layers as one of a series of oscillations (Aaby 1976), but this does not cause any heath expansion at Solsø. In other pollen diagrams from western Jutland where heathland expansions can be dated with some accuracy the expansions are not synchronous: middle Subboreal at Filsø (Jonassen 1957), early Subatlantic at Tinglev Sø (Andersen 1954), and middle/late Subatlantic at Flyndersø (Jonassen 1950) and Bølling Sø (Iversen 1941).

The metachroneity of the heathland expansion and the fact that this expansion always occurs after the first appearance of cultural indicators in the pollen diagrams suggest human impact as the main reason for the *Calluna*-increase.

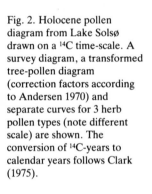

Fig. 2. Holocene pollen diagram from Lake Solsø drawn on a ¹⁴C time-scale. A survey diagram, a transformed tree-pollen diagram (correction factors according to Andersen 1970) and separate curves for 3 herb pollen types (note different scale) are shown. The conversion of ¹⁴C-years to calendar years follows Clark (1975).

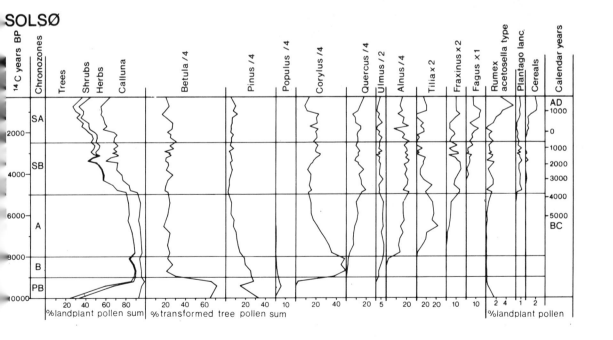

HARRESKOV

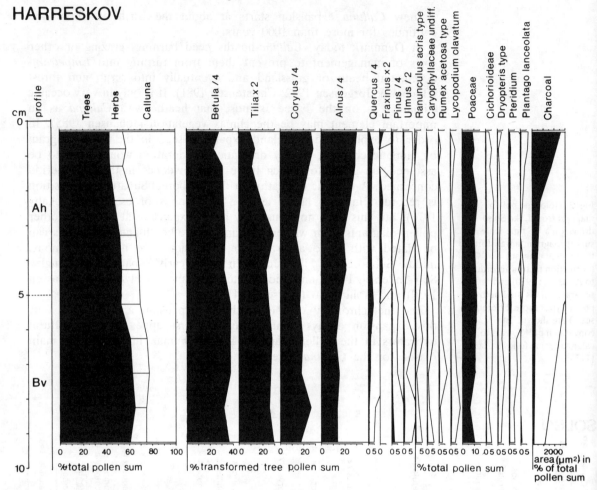

Fig. 3. Pollen diagram from
the soil section at Harreskov,
buried at 2600 B.C. (calendar
years). The diagram shows a
survey diagram, a transformed
tree-pollen diagram, curves for
major herb-pollen components
and a curve for charcoal dust.

The origin of heathlands

Regional pollen diagrams like the one from Solsø date heath expansion,
but local diagrams provide an opportunity of testing Jonassen's (1950)
hypothesis of *Calluna*-expansion on abandoned fields. Two pollen
diagrams from soil sections preserved beneath burial mounds dated to
2600 B.C. (calendar years, according to archaeological and ^{14}C-dates)
show a vegetational development from forest to heath. But whereas one
- Harreskov - is situated on elevated dry ground, the Skarrild mound
has been built on slightly moist ground.

Pollen diagrams from mineral soils cannot be interpreted as straight-
forwardly as can pollen diagrams from lakes and bogs, since the
original stratification has been altered by bioturbation, especially in
the lower parts. The pollen assemblages are thus somewhat mixed but

the major trends in the original pollen stratification can still be detected and evaluated (Andersen 1979a; Aaby 1983).

The soil profile underneath the burial mound of Harreskov is an oligotrophic brown-earth developed on medium-grained fluvio-glacial sand. The lower part of the pollen diagram (Fig. 3) is dominated by tree pollen, but *Calluna* is present in considerable amounts. *Plantago lanceolata* and other cultural indicators are virtually absent and charcoal dust is infrequent. The transformed tree-pollen diagram reflects an open forest of *Betula*, *Tilia* and *Corylus* with *Calluna* and grasses on the forest floor. This forest type is probably the virgin forest on dry soil in the late Atlantic or early Subboreal of western Jutland. In the upper part of the diagram *Plantago lanceolata*, *Calluna* and charcoal increase. Among the trees *Tilia* decreases while *Alnus* values rise, probably a result of extra-local transport. This part of the diagram reflects a heath with scattered tree growth or perhaps a very open forest. The vegetational changes recorded by this local pollen diagram are obviously the result of cultural impact, most likely grazing and burning.

The Skarrild soil section from the same period is a humus podzol developed on late-glacial eolian sand. It is situated a few km from Harreskov but on more moist ground close to a bog area.

The pollen diagram is divided into three stages (Fig. 4). In stage A tree pollen dominates together with grasses while *Calluna* and *Plantago lanceolata* are absent. Diversity is low and only small amounts of charcoal are recorded. The transformed tree-pollen diagram reflects

Fig. 4. Pollen diagram from the soil section at Skarrild, buried at 2600 B.C. (calendar years). The diagram shows a survey diagram, a transformed tree-pollen diagram, selected shrub and herb pollen-types, and curves for diversity and charcoal dust.

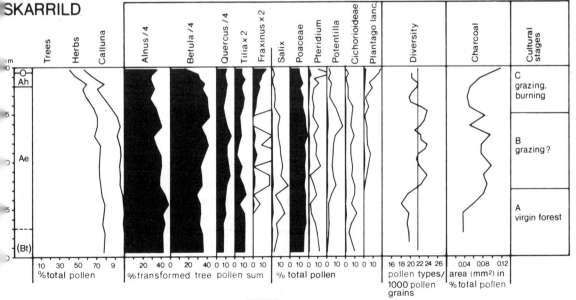

an open forest of *Alnus* and *Betula* with *Quercus*, *Tilia* and *Salix* as subordinate elements. This must be seen as the virgin forest on slightly moist ground. In the next stage, B, *Plantago lanceolata* appears and diversity as well as charcoal frequency increase. Among the trees *Tilia* decreases while pollen of *Fraxinus* is recorded. Furthermore a rise in *Calluna* frequences is observed. This is a phase of some but still low cultural impact on the vegetation, probably with some grazing and cutting in the forest. Stage C is characterized by a dramatic increase in *Calluna* and also *Plantago* and charcoal values rise. This is a phase of forest destruction and heath expansion. Based on pollen concentrations and assuming a constant but cautiously very high pollen deposition of 50 000 pollen grains cm^{-2} yr^{-1}, minimum ages for the two upper phases can be calculated (cf. Aaby 1983). This gives minimum durations of *ca*. 400 years for stage B and 200 years for stage C. Since no *Empetrum* pollen has been found and the grass curve decreases through stage C, the heath must have been kept in a *Calluna*-dominated stage. By modern analogy (Böcher & Jørgensen 1972; Christensen 1981) it is known that this is only possible by some sort of heath management, in this case probably grazing and burning.

These soil pollen-diagrams give examples of how, on dry soil, early farming changed the virgin *Betula-Tilia-Corylus-Calluna* forest to a more open *Calluna*-dominated vegetation while on slightly moist soil the primeval *Alnus-Betula*-Poaceae forest was similarly turned into a *Calluna*-heath. In both cases the change was apparently brought about by grazing, burning and cutting while there is no indication of tilling of any of the soils. Thus there is, so far, no support for Jonassen's (1950) hypothesis that heathlands originated from *Calluna* spreading on abandoned fields. Macrofossils of cereals (Rostholm 1982) document crop cultivation in western Jutland through the last 5000 years but it has apparently been of minor importance compared to animal husbandry. The open forest of western Jutland was probably very attractive for cattle breeding and was furthermore easily changed into *Calluna*-heaths which, as a grazing resource, are superior to grass-dominated pasture in their ability to provide winter grazing.

The history of oak-scrubs

The oak-scrubs are coppice woods that have been grazed and cut down regularly through recent centuries (Worsøe 1980). Oksbjerg (1964) suggested that this practice had a selective effect on the flora, favouring trees capable of withstanding regular cutting. A pollen diagram from a soil section in an oak-scrub illustrates the influence of human impact on the vegetational development.

The pollen diagram from the oak-scrub at Hønning (Fig. 5) is from a recent podzol profile with a 10 cm humus cover developed on sandy till. The scrub is dominated by *Quercus robur* and *Q. petraea*, and

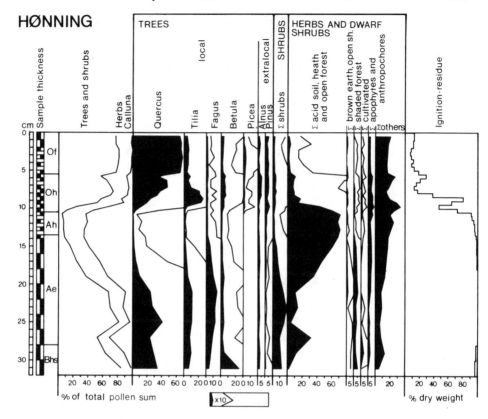

Fig. 5. Pollen diagram from a recent soil section in the Hønning oak-scrub. The diagram shows a survey diagram, untransformed curves for trees, sum curves for shrubs and ecological groups and ignition-residue values.

has scattered individuals of *Populus tremula*, *Fagus sylvatica* and *Tilia cordata*. The sampling site is situated underneath a *Tilia* tree. Only one level in the pollen diagram can be dated, namely the start of the *Picea* curve. According to historical information, *Picea* was introduced locally in 1836 A.D. so allowing for a non-flowering juvenile period of about 30 years, the *Picea* curve starts at *ca.* 1865.

The diagram reflects a development from a mixed forest of *Quercus*, *Tilia*, *Fagus* and *Corylus* - the latter is included in the shrub curve - through a heath stage and again to a forest now totally dominated by *Quercus*. There is a short interval in the lower forest stage with a peak in *Secale* - included in the curve 'Σ cultivated' - followed by a *Calluna* peak. This can be interpreted as a brief cultivation of *Secale* in a field within the forest. After abandonment *Calluna* spread over the field and eventually trees dominated. Later, the forest was destroyed, probably by cutting and grazing. After grazing ceased at *ca.* 1850, trees spread on to the heath and an almost pure oak-scrub was formed. Parallel to this vegetational development there is a soil development from a brown-earth in the early forest stage through podzolization in the heath stage to a podzol in the *Quercus*-forest stage. This soil pollen diagram gives an example of how heavy grazing and cutting cause soil degradation and at the same time an impoverishment of the flora. It also highlights the role of *Calluna* as an accelerator of podzolization.

The *Quercus*-dominance at Hønning is young - hardly more than 100 years. At other sites, however, historical sources document oak-scrubs further back in time (Worsøe 1980).

Heathland - a grazing resource

The results presented here are limited in time and space and do not allow a synthesis of heathland and forest history of western Jutland. However, these preliminary results indicate that the dominance of heathlands through the last millenia must be understood in the light of the importance of this vegetation type as a grazing resource.

Summary

(1) The predominantly sandy soils of western Jutland were covered by vast *Calluna*-dominated heathland until extensive reclamations and afforestations in the late 19th century. Scattered oak-scrubs occur, and are traditionally assumed to be relics of the natural forest vegetation.

(2) The history of heathland and oak-scrubs is being studied with regional pollen diagrams from lakes and local pollen diagrams from recent and fossil soil profiles.

(3) The pollen stratigraphy from Lake Solsø shows essentially the broad-scale vegetational development of the Holocene well known from eastern Denmark. It differs in the more open forest type of the early and middle Holocene, and the dominant role of *Calluna*-heaths through-out the last 5000 years. Cultural indicators in the pollen diagram suggest that expansion and maintenance of the heath were due to human impact.

(4) Two pollen diagrams from soil profiles preserved beneath burial mounds (2600 B.C.) reflect a development from open forest types to a *Calluna*-dominated vegetation. These local pollen diagrams indicate that grazing, cutting and frequent human-induced fires were the main causes of forest destruction and heathland maintenance, whereas short-term tilling with long intervening fallow periods was probably of limited importance.

(5) A soil pollen diagram from inside an oak-scrub emphasizes the influence of grazing and cutting on soil degradation, and the formation of oak-dominated scrub.

Nomenclature

Plant nomenclature follows *Flora Europaea* except for Poaceae (= Gramineae).

Waterland - Environmental and Economic Changes in a Dutch Bog Area, 1000 A.D. to 2000 A.D.

Jurjen M. Bos, Bas van Geel
and Jan Peter Pals

Introduction

Waterland, the former raised-bog area just north of Amsterdam (Fig. 1), is part of Western Europe's belt of wetlands. It is a famous bird sanctuary, although its agricultural aspect is dominant. It presents the visitor with a cliché view of Holland: flat green polders, surrounded by massive dykes and divided by many glittering ditches. From this description it is immediately apparent that there is actually no 'nature' to be found; the landscape of Waterland is entirely man-made. The appearance of tranquil eternity is misleading.

Most of the landscape changes were directly or indirectly induced by people. Those changes had profound effects on economic and social life in the area, in turn causing new changes in the landscape (Fig. 2). Recent archaeological research (Bos 1985a, 1985b, 1986) has revealed that this resulted in relocation of the settlements. Hundreds of deserted Medieval and post-Medieval houses are scattered around the area, the waterlogging providing good conditions for the preservation of both archaeological and palaeoecological evidence. The potential for the study of the turbulent relationship between Waterland and the Waterlanders or between environment and occupants, is enormous. This was only very recently realized, and research programmes are now being formulated. The Dutch Ministry of Agriculture has asked the University of Amsterdam to map all relevant sites and to report on ways to preserve them and to harmonize agricultural and archaeological interests. These topics fall beyond the scope of this chapter. Here we are concerned with the vicissitudes of people in a changing cultural landscape.

The natural landscape

At the end of the Weichselian, sea-level rose as a result of melting inland-ice. During the Holocene, sandy beach ridges and dunes were formed in the western Netherlands. Clayey and peaty sediments were

Fig. 1. Position of Waterland
within The Netherlands.

deposited in the coastal plain, between the coastal ridges and the
Pleistocene deposits of central and eastern Netherlands. Under the
influence of several large rivers, the saline conditions changed to a
freshwater environment. At many sites the eutrophic environment of
marshes, fens and carrs gradually became more oligotrophic. When
the vegetation became dependent solely on rain water for its nutrient
supply, raised bogs developed (Pons & van Oosten 1974).

Palaeobotanical research (Polak 1929) showed the presence of submerged raised-bog complexes in the province of Noord Holland.
Recently, their development and palaeoecology have been studied in
detail (Bakker & van Smeerdijk 1982; Pals *et al.* 1983; Witte & van
Geel 1985). At the Ilperveld (Bakker & van Smeerdijk 1982), situated
in the former raised-bog complex of Waterland, peat growth started
at *ca.* 4600 B.P. From *ca.* 4100 B.P. onwards, Ericaceae, *Sphagnum*
and other oligotrophic taxa became dominant. During peat formation,
especially during the raised-bog phase, which lasted more than 3000
years, the wet, open landscape was unattractive for human occupation.

In the 10th century, the natural boundaries of Waterland were strips
of oligotrophic peat in the north, west and south, and Almere lake in
the east. The area (Fig. 3) was drained by the Waterlandse Die river
and its tributaries, running into Almere (de Cock 1975). The occurrence of many toponyms with the suffix *-woud* (forest) or *-broek*
(carr) suggests that Waterland was once largely covered by forest (de
Cock 1975). Palynological evidence does not corroborate this assumption. The *-woud* and *-brouk* toponyms may refer to small strips of
woodland that were restricted to the river banks and were a striking
element in the landscape. Since most transport was by water, the
impression of a forested area would have been evoked.

There is no archaeological or palaeoecological evidence for human presence in Waterland before the 10th century. Earlier temporary Iron Age occupation of bogs is known only from areas nearer to the inhabitable coastal area (Brandt 1983). The first permanent bog reclamations in the western Netherlands occur in Carolingian times (Besteman 1974), but the great *hausse* began at the end of the 10th century. This is also the time of the reclamation of Waterland.

The impact of human intrusion

The exact circumstances which led to broad-scale reclamation activities are not yet known. However, changes in economy, technology, demography and social structure of the older settlements outside the bog area certainly interacted with the effects of climatic variation that made the 10th century a period of drought (Heidinga 1984). Maybe the formation of the Younger Dunes along the coast, which meant the loss of a considerable amount of arable land, was a stimulus for people to leave the dunes and invade the bog areas. A more important consequence of the drought, however, may have been the postulated end of bog growth. The upper peat is supposed to have dried out, thus making the area more accessible. Oxidation of the surface peat increased available nutrients, and the original raised-bog plants gave way to plants requiring meso- to eutrophic conditions (van Geel *et al*. 1983). With the insight and skills developed in the Carolingian reclamations, it was possible to construct drainage systems that allowed permanent exploitation and occupation.

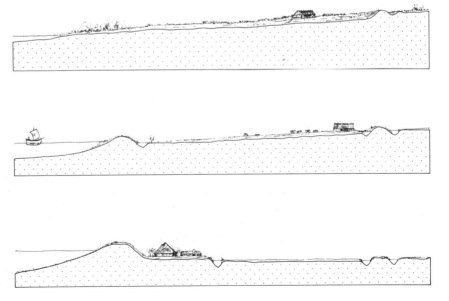

Fig. 2. The 3 main stages of landscape development in Waterland under human influence: (top to bottom) Arable land, shortly after the reclamation (11th-12th century);
Extensively used pasture land enclosed by dykes (late 13th-16th century);
Present-day intensively exploited pasture land, which has sunk below sea-level since the introduction of mechanical drainage in the early 17th century.
Drawing: B. Donker.

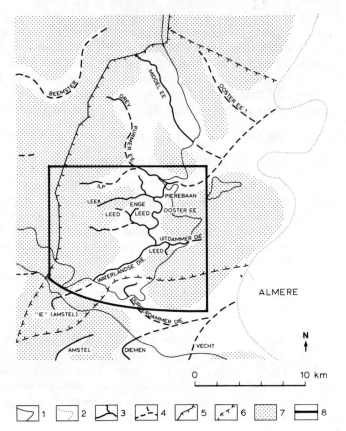

Fig. 3. Reconstruction of the natural landscape, adapted by H. A. Heidinga from de Cock (1975).
1 = present coast line;
2 = reconstructed Medieval coast line;
3 = natural rivulet;
4 = id., supposed;
5 = border of Waterland;
6 = id., supposed;
7 = oligotrophic peat;
8 = approximate outline of Fig. 4.

In a span of probably only a few generations most of Waterland was transformed into agricultural land. Drainage was so successful that the higher parts could be used as arable, mainly for rye cultivation (van Geel *et al.* 1983; van Geel 1984; Pals 1984). At this point a process was set into motion that proved to be irreversible and with consequences that are still felt even today (Fig. 2).

If water is removed from peat, the volume is strongly reduced, and this reduction is increased by oxidation of the plant remains. The surface starts to sink to a level nearer to the new water-table. Further drainage becomes imperative. The landscape sinks even further, and so on. It must have seemed as if the environment had become hostile. Although Waterland was not 'wrested' from the water like the later 'true' polders, it is certainly the case that a long struggle ensued.

The subsidence allowed the surface water to attack the land successfully. Large parts of reclaimed land along the eastern coast of Waterland were washed away in the formation of the Zuiderzee (de Cock 1975). Dykes were constructed, but these were frequently breached. The one advantage of the floods was the deposition of a fertile layer of clay (Besteman & Guiran 1983). This advantage disappeared, however, as the water of the encroaching Zuiderzee became increasingly brackish, as shown by the occurrence of salt-marsh plants such as *Triglochin maritima* L. and *Spergula marina* (L.) Griseb./*S. media* (L.) C. Presl. in the centre of the bog area (van Geel *et al.* 1983). When this became a real problem, due to the continuing subsidence, the larger rivers, which all debouched into the Zuiderzee, were dammed, the dams being provided with an outlet sluice that could be opened at low tide. In this way the dyke ring was completely closed in the second half of the 13th century (Bos 1988). This did not solve the problem completely, as the dykes proved to be rather vulnerable. At the end of the 13th century there was an increasing tendency to build houses on *terpen* (artificial mounds); palaeoecological evidence points to problems with the loose subsoil as well as continuing problems with surface water (van Geel *et al.* 1983). The most extreme example excavated so far is the town *terp* of Monnickendam, where an infill of almost 5 m was achieved in one phase (Bos 1988). Floods became a lasting aspect of the history of Waterland. The present landscape has many scars that bear witness to that: larger and smaller lakes immediately behind the dykes where these were breached; and straight stretches of dyke, where land had to be given up after a flood and a new 'safe' dyke had to be constructed, etc. The last great flood occurred in 1916.

Not only the Zuiderzee caused floods and loss of land. The natural rivulets, swept up by the wind, increased in size to form lakes, the largest of which also had to be enclosed by dykes.

Re-orientation of the economy

The relative rise of the water-table during the first 3 centuries of occupation forced the people of Waterland to reorganize their economy. Crop growing became increasingly difficult after *ca.* 1300 A.D., as the available technology failed to provide adequate drainage. Waterland became a uniform pasture area (Fig. 2). Its inhabitants turned to new forms of subsistence and cereals had to be imported.

In the late 13th century towns such as Amsterdam and Monnickendam developed, where agriculture was secondary to craft, shipping and trade. This development was encouraged by the integration of Waterland into the powerful county of Holland in 1282, when the self-styled lords of Waterland, the Persijn family, sold their rights to Count Floris V, who was a man of truly international stature. During his reign, Holland became part of the newly developing Western European economy. The expansion of Holland in the second half of the 13th century was both territorial and economic. The people of Waterland needed to import certain necessities such as corn, and they found new means of making a living using the opportunities presented by a growing demand and supply system, in which money became the lubricant.

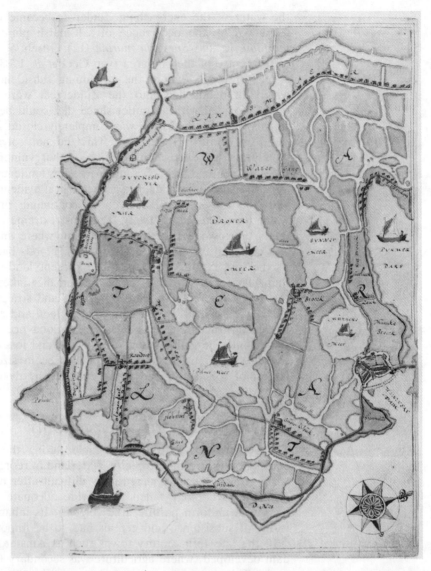

Fig. 4. Waterland in 1588 (North is 90 degrees to the right in relation to Fig. 3). Note the still elongated structure of the settlements. Map by B. de Vyll at the Rijksarchief Noord-Holland, Haarlem.

The new traders and seamen were independent of the nobility and the clergy. The new towns did not develop in the shadow of manorial or ecclesiastical residences, but mostly on the sites of the dams in the main rivers, which were readily accessible by ship, both from within and from outside.

Waterland in the 14th century is poorly documented, though we may assume that it was a period of transition and adaptation. Archaeological evidence suggests an initial decline in the population of the countryside; this was only partly mitigated by the increase in population of the urban centres.

Dairy farming is considerably less labour-intensive than crop growing. However, Waterland became no more than marginal land, even to the less demanding dairy farmer. Historical sources from the 15th and 16th century are eloquent on this point (although admittedly they are from tax declarations): "We can only use our lands from May to the middle of September, and then we have to stable our animals" (1544); "We can hardly get the hay properly off the fields once every three years" (1553); "Yes, we do keep cattle, but we can't make a living out of that" (1494); "Our lands hardly yield enough to pay the land taxes alone" (1556); "We have no way of making a living on our lands, except that some of our wives may keep one or two cows" (Bos 1985a).

It is therefore not surprising that the change from crop growing was not only to dairy farming. The people of Waterland switched mainly to craft, shipping and trade, without actually moving to the towns: "We make our living on the seas, sailing east and west, and some of us also engage in trade" (1514); "We are all fishermen and sea-faring people" (1543). In the first half of the 16th century almost half of all Dutch skippers sailing to the Baltic came from Waterland. All supporting craftsmen were to be found in the villages; the home-industry was principally spinning and sewing of clothes. Although the towns certainly profited by these activities of country people, competition was feared. Amsterdam issued dozens of regulations to curb rural activities. Such regulations have been preserved from the 15th century onwards (Bos 1985a).

Fig. 5. In the lower left corner 2 house *terpen* can be recognized from the ditch pattern. As the *terpen* are still distinguishable as small mounds, the present farmer adapted the drain system to the difference in height.

The archaeological survey has discovered hundreds of houses which were deserted in the period of transition, from *ca.* 1250 to 1350 (Bos 1985a, 1985b). These desertions were not due to the decline in population alone; it is now apparent that almost all settlements were relocated. The former villages were all on the fields, forming long strips parallel to both the rivulets and the small dykes at the end of the reclaimed village-area. Now settlement shifted towards dykes, roads and waterways. This shift may not be due to drainage problems, but be related to the re-orientation of the economy; the focus shifted from fields to connections with the outside world (Bos 1985a, 1986). Not only was this more practical, it was also an expression of a changing attitude to the environment and to life in general.

The end of the 'urban' life in rural Waterland was quick. It was brought about, both directly and indirectly, by the war with Spain. The damage from the storm surges of 1570 was not completely repaired when the troubles began. In 1572 the Protestant Dutch raided parts of Waterland, but soon Waterland declared itself for the Prince of Orange. Amsterdam delayed till 1578. Waterland became the scene of many skirmishes. Villages were pillaged and burned. The dykes were breached in many places, but the disintegration had gone so far that these were not repaired. Many people fled from their homes. What was even worse, their ships were plundered, burned or confiscated throughout the country; their capital was destroyed. When the situation stabilized, it was Amsterdam that profited. It had a huge influx of capital, brought by the religious refugees from the southern Netherlands, and it had not suffered such severe losses as Waterland.

Two things happened. Amsterdam practically acquired a trade monopoly. In the process, it became a profitable market for agricultural products, as its population doubled in the following decades. The merchants of Amsterdam sought ways to invest their abundant capital. A new re-orientation of the economy of Waterland followed.

Return of the importance of agriculture

Land subsidence had not stopped since the end of the corn-growing period and the area had now almost lost its agricultural value. Helped by capital from the towns, the countryside was thoroughly reorganized. Dykes were repaired and reinforced. The larger lakes were drained, fishing and sailing water was transformed into pasture land. An important new feature was the introduction of windmills to pump out the water. These were also used to lower the water-table in the ditches of the old land, which resulted in better pastures and hay meadows, but also accelerated subsidence. From this time (*ca.* 1600) to the present day, Waterland has sunk at least a further 2 m. Only at this stage was the present 'typically Dutch' landscape formed (Fig. 2).

Fig. 6. Loam floors seal off the palaeoecological evidence from 5 successive phases of occupation, dating from the late 10th to the 12th century. When the first layer of infill (with the first farm house on it) had sunk into the peat, a second layer was added, the house was elevated or rebuilt and a new floor was spread. This process was repeated 4 times. Four of the 5 floor levels can be seen in this section.

Milk was brought to Amsterdam once or even twice a day; milk and cheese production became the major sources of income for most people in Waterland. The economy of the villages of Waterland completely changed. They no longer said: "Oh yes, that is true, our wives may have one or two cows" (see above); now it had become: "Our people occupy themselves almost exclusively with dairy farming", "The villages are solely concerned with the production of milk, which they bring to the market at Amsterdam" (1750) (Bos 1985a).

The shift from craft, sea-faring and trade to dairy farming was the second major change in the economy of Waterland, and for the second time some of the settlements were relocated. Some smaller hamlets disappeared completely, as did many scattered farms. The elongated structure of the villages (Fig. 4) was, in several instances, given up in favour of nuclei near the church (Bos 1983, 1986).

It is curious to see that at a time of growing importance of the land, the focus of the settlements turned inward. The background of this development is as yet unknown. Was it a matter of greater security to live closer together, or was it a matter of an 'urbanization' of thought? Much interdisciplinary research remains to be done. Another new element in the landscape was the half-land, half-water area in western Waterland, where peat was cut in large quantities for sale in towns. The peat cutting was, so to say, the last possible use of the land. These areas are now, for the most part, nature reserves. Peat cutting enriched the wildlife, but obviously it decreased the archaeological value, as many sites have been cut away. In some places we see a regeneration of peat growth. At others, humans have created a totally new landscape, as in the recreation area Twiske, where trees and hills abound as if the architect suffered from a *horror vacui*!

The housing developments of Amsterdam are also new. The multi-storied apartment blocks present a sharp contrast to the adjacent polder landscape of the remaining part of Waterland.

Today, dairy farming is still important in Waterland. Problems with the irreversible process of lowering the water-table, however, have not ended. During the present reallotment programme the water-table will be lowered by, on average, a further 60 cm. It can be predicted, looking at the past, that one day yet another phase in the landscape development can be expected: one day the continuous lowering of the water-table must come to an end.

Relics from the past

It will be clear that the most important historical monument of Waterland is the landscape itself. It comprises both the natural elements, such as the remaining original rivulets, and the cultural landscape, made through, and in spite of, human activities.

This landscape is furnished with many small monuments of the past. Many of the Medieval sites can be discovered by systematic field-walking (Bos 1985b), inquiries into field names (Bos 1983) and other such methods. Others are recognizable as features in the landscape itself. House *terpen* may sometimes be easily distinguished, although the general tendency of the peat is towards a level surface. *Terpen* that have sunk can be recognized by the way they seem to have diverted the ditches (Fig. 5). Each anomaly in the ditch pattern may result from the ditch being made around or along a former yard.

The remains of these former settlements form a very important outdoor archive. Their archaeological potential is self-evident, but their palaeoecological potential is no less great; each time a loam floor was spread or a layer of infill was added (Fig. 6), information about vegetation and landscape was sealed. Continuous waterlogging guarantees good preservation of organic material. The contrast between the original raised-bog vegetation and the successive later vegetation types is considerable. Many hundreds of house sites have already been mapped and described by means of cores. A programme for palaeo-ecological research of samples from a representative number of these sites has been established.

The landscape of Waterland not only bears witness to its own history. It is also an important source of information for the study of the rise of Holland as a commercial nation. It is therefore fortunate that the importance of this cultural landscape has been realized just in time to preserve it for the future.

Summary

(1) Integrated archaeological and palaeoecological investigations of Waterland, a former raised-bog area in the western Netherlands, have concentrated on natural and human-induced changes in the landscape (from raised bog to damp pasture area) and related subsequent shifts in economy and settlement pattern.

(2) Active growth of the raised bogs ceased in the 10th century, when the bog-surface dried out during frequent droughts.

(3) The bogs were colonized and crops grown until the end of the 13th century, when subsidence caused by drainage lead to flooding.

(4) From 1300-1600 the area was used as pasture, until continued subsidence lead to declining productivity. Local people took up crafts and trades such as sea-faring and fishing, and moved their settlements from fields to dykes and waterways.

(5) After 1600, war and technical innovations resulted in further drainage to permit profitable dairy farming, and the villages and landscape were again remodelled.

(6) The land surface is now below sea level, and subsidence problems will continue until all the peat has disappeared.

Acknowledgement

We wish to express our gratitude to Mrs C. van Driel-Murray, who kindly corrected the English text.

Development of the Late-Prehistoric Cultural Landscape in Upland Ardudwy, North-west Wales

F. M. Chambers, R. S. Kelly and S.-M. Price

Introduction

The county of Gwynedd in North Wales is unique in its abundance of stone-built hut-circle settlements (*Cytiau'r Gwyddelod*). Academic opinion has regarded these as Iron Age and Romano-British farmsteads, until one was proved to have a Bronze Age origin (Smith 1985) and another a post-Roman demise (R. B. White personal communication). The prevailing view of the origin of their farming economies is summarized by Briggs (1985, p. 285): "A conventional interpretation of prehistoric farming is of a mixed economy during the Neolithic, which was replaced during Beaker times and the Early Bronze Age, at least in the uplands, by a more pastoral regime. This pastoral presence is considered to have persisted into Iron Age times, when it is believed to have been characterized by herding societies based upon hill-forts, and utilizing, to a greater or lesser degree, the hut-groups of the type common to North Wales".

Up to the late 1970's, palaeoecological assessments of prehistoric human influence upon the landscapes of North and mid-Wales had been hindered by a lack of ^{14}C-dates, e.g. Moore & Chater (1969), Moore (1973), Taylor (1973), Walker & Taylor (1976); hence, speculation on the environmental setting of and the systems of land-use associated with these hut settlements was largely based on extrapolation back in time from documented or suspected Medieval practice. Thus, in historical geography, Jones (1961, 1963, 1979) interpreted 'early' settlement evidence in the light of Welsh Law governing tenurial organization; prehistorians interpreted the function of some of the hut settlements in the context of assumed Medieval transhumance practice (Hemp & Gresham 1944); whereas others have suggested a prehistoric origin for transhumance in Wales (cf. Allen 1979). Recently, Briggs (1985) criticized the use of Medieval parallels because farming practices may have been more complex than formerly appreciated.

In parts of North Wales, the hut-circles and associated 'field' walls

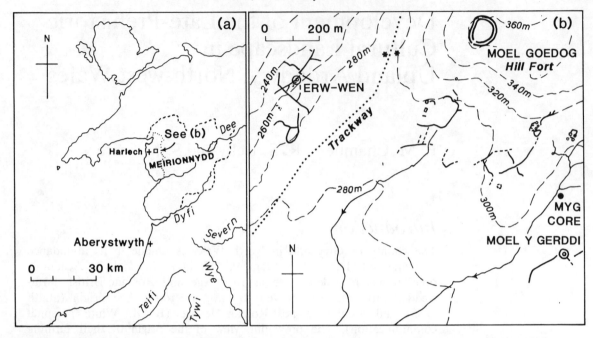

Fig. 1. (a) Map of part of North Wales showing location of study area. The eastern boundary of the District of Ardudwy is indicated by the dotted line within the former County of Meirionnydd. (b) The study area showing location of valley mire core, circular enclosures Erw-wen and Moel y Gerddi, and Moel Goedog hill-fort. Stars show location of Bronze Age complex-ring cairns. Ancient walls, hut-circle and enclosed homesteads are shown below Moel Goedog. An additional single hut lies on the 320 m contour line, south-west of the hill-fort.

remain in an essentially pastoral landscape. In the late 1970's, C. Smith recognized the need for the conservation of these 'areas of ancient landscape' and as a result, Gwynedd County Council and the Conservation and Land Division II of the Welsh Office (now *Cadw*: Welsh Heritage) commissioned a survey of antiquities in the District of Ardudwy (Fig. 1a). Detailed archaeological field work was carried out there by R. S. K. from 1979-1981. A subsequent unavoidable threat of upland pasture improvement in an area of ancient landscape, identified north-east of Harlech, prompted the rescue excavation of 2 late-prehistoric hut-circle sites, and by integrating the excavations with a programme of palaeoenvironmental work, permitted a reconstruction of the vegetational history of the local area and an evaluation of the environmental setting of the 2 sites.

Presented here are the results of pollen analyses conducted to ascertain local environmental history. These results are interpreted in the light of archaeological fieldwork and excavations in the locality and discussed in the context of the development of the late-prehistoric cultural landscape in Ardudwy. The validity of a 'transhumance' land-use model in late prehistory is then assessed.

Study area

The study area (Fig. 1b) is 4 km north-east of Harlech between 250 and 360 m elevation and today consists largely of rough pasture (*ffridd*) for sheep and store cattle. Their access and movement are restricted by 2 m high stone walls which were planned in estate offices in the 19th century and which bear no relation to the natural topography or to the remains of earlier boundaries and structures, dating mostly from prehistoric times, which comprise the 'ancient landscape' features in Figure 1b. These include two hut settlements - Erw-wen and Moel y Gerddi - which were excavated in 1980-1981 by R. S. K. An area of valley bog (*sensu* Tansley 1939) along the major stream to the west of Moel y Gerddi provided a sediment core for pollen analysis and the reconstruction of local environmental history.

Hut settlement types

Of the various morphological classifications of hut settlements in Gwynedd (e.g. Hemp & Gresham 1944; RCAHM 1964; Smith 1974), the classification adopted for Meirionnydd by Bowen & Gresham (1967) was retained for the Ardudwy survey by Kelly (1982a). Hut-circle settlements in the Ardudwy study area (Fig. 1b) include: single hut-circles - one located on the 320 m contour south-west of Moel Goedog; concentric circles and circular enclosures - exemplified by Erw-wen and Moel y Gerddi; and enclosed homesteads - two examples south-east of Moel Goedog.

Methods

Field and laboratory sampling

For palynological work, a vertical peat section (MYG) was taken in 20 cm x 20 cm monolith boxes from the deepest area of peat in the Moel y Gerddi valley mire (Fig. 1b). In the laboratory, the section was sub-sampled contiguously in 0.5 cm vertical increments for pollen analysis. Samples of 150-300 g wet weight were subsequently taken for ^{14}C-dating.

Soil samples taken during the course of the archaeological excavations from old ground surfaces (o.g.s.) were also analysed for their pollen content. The archaeological methods and detailed accounts of the excavations will be published elsewhere by R. S. K.

Pollen analysis

Lycopodium spore tablets were added to each 0.5 g sample before preparation, to permit calculation of pollen concentrations (Stockmarr 1971). Samples were prepared after Barber (1976), with silicone oil as mounting medium. A pollen sum of 400-500 total land pollen (TLP) was counted by S.-M. P. in each sample. Spores and aquatic pollen

Phase	Duration (B.P.)	Features
j	*ca.* 1645 - present	High Gramineae, NAP
i	*ca.* 3590 - *ca.* 1645	Lower AP; high Cyperaceae
h	*ca.* 4625 - *ca.* 3590	Rising AP; high *Alnus*
g	*ca.* 5140 - *ca.* 4625	Lower *Alnus* higher NAP; presence of *P. lanceolata*
f	*ca.* 5640 - *ca.* 5140	High *Alnus*
e	*ca.* 6105 - *ca.* 5640	High pollen concentration; high AP
d	*ca.* 7270 - *ca.* 6105	High *Alnus*, higher Coryloid; low NAP
c	*ca.* 7485 - *ca.* 7270	High Gramineae, Cyperaceae; lower AP
b	*ca.* 8465 - *ca.* 7485	High *Alnus*; lower NAP
a	*ca.* 8700 - *ca.* 8465	High Cyperaceae, Gramineae; low *Alnus*, Coryloid

Table 1. Estimated duration and distinguishing features of phases in Figures 2 and 3 (comparisons are made with earlier phases).

types were excluded from the pollen sum in the relative diagram (Fig. 2). Pollen-influx data are presented in Figure 3. Cerealia pollen were not convincingly separated by established criteria and so two Gramineae curves are presented, separated on a size-criterion alone.

The percentage and influx pollen diagrams were divided into phases *a* to *j*. Summary AP pollen diagrams from the MYG core were used by Chambers & Price (1985) to illustrate the early appearance of *Alnus*.

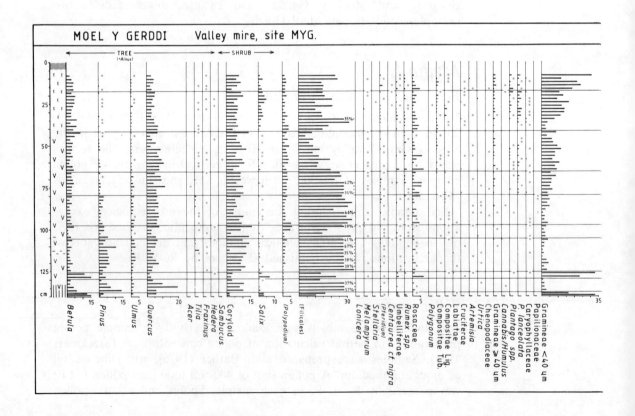

Radiocarbon dating

Samples were pretreated and dated at the Cardiff Radiocarbon Dating Laboratory. Dates were obtained for the fine particulate fraction (<240 μm), as this is considered to give the most reliable estimate of ^{14}C-age for peat horizons, and are quoted in conventional ^{14}C-years B.P. (Dresser 1985). The ages of phase boundaries in Table 1 were interpolated from peat-accumulation rates between adjacent pairs of dates and are acknowledged as approximations by the prefix *circa* (*ca.*).

All other ^{14}C-dates cited below are on samples from archaeological contexts in the study area and are reported in Dresser (1985).

Pollen diagrams

The sampling point was 180 m north of the Moel y Gerddi excavation (Fig. 1b, National Grid Reference SH 6164 3190). From the base, the stratigraphy was recorded in the laboratory as fen and alder carr peat with some silt lenses, succeeded by *Sphagnum*-sedge peat, and finally by *Sphagnum*-rich valley-bog peat.

In the pollen diagrams, NAP types have been arranged, left to

Fig. 2. Percentage pollen diagram from valley mire site MYG. Taxa in parentheses were excluded from the pollen sum of total land pollen (TLP) and from the AP/NAP summary, but are expressed as % total pollen and spores. All other taxa are expressed as % TLP. An open circle denotes one grain. The relative abundance of charcoal dust in the pollen slides is indicated.

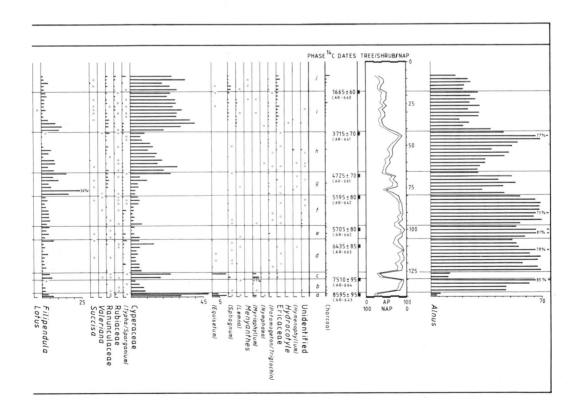

Table 2. Summary of percentage pollen data in pollen sums
of 1000 TLP from selected
o.g.s. samples beneath Erw-
wen (EW) and Moel y Gerddi
(MG).

Sample name	EW5006	EW5007	MG5005	MG5022	MG5026
Tree	4	2	5	1	2
Shrub	15	26	11	6	6
Gramineae	30	19	40	29	56
Cerealia	2	1	2	1	2
Ericaceae	-	1	9	7	7
Plantago spp.	39	40	20	46	21
Other NAP	11	11	13	12	6

right, into woodland, woodland margin, ruderal, arable, grassland, fen, aquatic and bog habitat groups. These groups are not delimited because there is inevitable overlap and there are some taxa with several affinities. AP types are arranged in conventional order, except that in Figure 2 the *Alnus* curve appears on the far right of the diagram to permit easy identification of horizons in which *Alnus* has clear local over-representation. The disturbing effect of *Alnus* on this percentage-based diagram (cf. Janssen 1959) can be appreciated by reference to

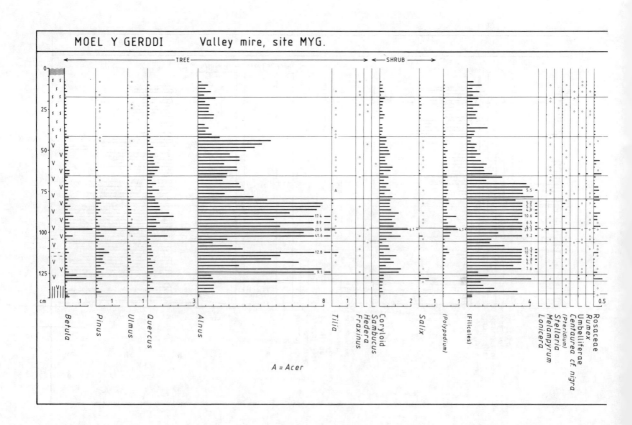

the pollen influx data (Fig. 3). The approximate duration and main distinguishing features of each phase are summarized in Table 1.

Soil pollen (o.g.s.) data are summarized in Table 2.

Interpretation

Ecological history of the valley mire

In Figure 2, the high pollen percentages of Cyperaceae and Gramineae, accompanied by *Filipendula* and *Myriophyllum*, and by *Equisetum* spores, together indicate a damp, open fen environment in phase *a* (from *ca.* 8700 B.P.). The increasing *Alnus* pollen influx and abundant Filicales spores indicate that alder carr became established in the mire by the start of phase *b* (*ca.* 8465 B.P.).

Reduced AP/NAP ratios in phase *c* indicate a more open local environment; the influx of *Salix* and *Betula* pollen here suggests a successional response to disturbance. The abundance of *Myriophyllum* pollen, with records of *Nymphaea* and substantial Cyperaceae values, confirm a very wet local environment during phase *c*. Whilst such a

Fig. 3. Pollen influx diagram in thousands of grains cm⁻² yr⁻¹. Taxa in parentheses were excluded from the percentage AP/NAP summary curve. An open circle denotes one grain only of the taxon recorded in the pollen count for that horizon.

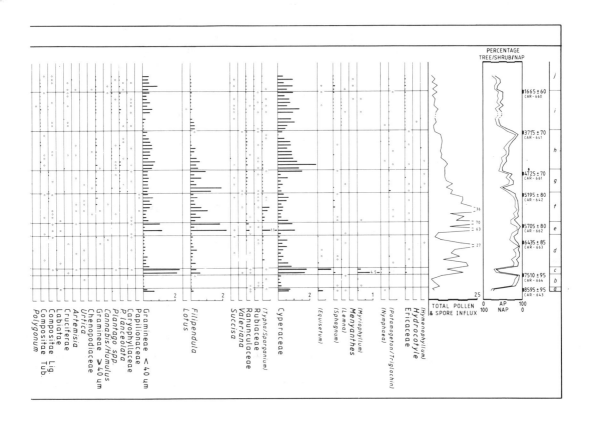

disturbance could be attributed by some to beaver activities (cf. Coles & Orme 1983), charcoal dust noted at the phase *b/c* boundary indicates the influence of fire here.

High *Alnus* pollen influxes are recorded from the start of phase *d* (*ca.* 7270 B.P.) to the close of phase *f* (*ca.* 5140 B.P.), indicating the re-establishment and local persistence of alder carr. An episode towards the close of phase *d* has somewhat lower influx of *Alnus* pollen coinciding with increased silt content, suggesting further disturbance or migration of the stream close to the sampling point, but these *Alnus* influx values are still as high as those in phase *b*. Despite a recession in phase *g*, alder apparently persisted in the valley-mire until the end of phase *h*, when a stratigraphic change from carr peat to *Sphagnum*-sedge peat is recorded, dated at 3715 ± 70 B.P. A predominantly open valley-bog flora was then established which has persisted to the present, with increasing representation of Ericaceae and *Sphagnum*, whilst some streamside *Salix* also probably grew in phases *i* and *j*.

Human influence upon vegetation: the Mesolithic

Evidence for human influence upon vegetation by Mesolithic peoples derives mainly from low AP/NAP ratios and from records of charcoal in the pollen-preparation residues, which together suggest local disturbance involving fire in phases *a* and *c* (Figs. 2, 3). Whilst natural fires cannot be ruled out, repeated natural fires in a damp fen or carr environment seem unlikely. Increased silt content, charcoal records and reduced AP/NAP ratios of phase *d* might also be explained by a non-human agency, but the overall stratigraphic indications of repeated disturbance in the valley mire lead to at least circumstantial evidence for the effects of continual activity of Mesolithic peoples in this area (see below).

Neolithic, Bronze Age and Iron Age environments

At the beginning of phase *g* (*ca.* 5140 B.P.), the reduced AP/NAP ratios are mainly due to the decline in *Alnus* values, but also to reduced pollen influx of other arboreal taxa, particularly *Ulmus* (cf. Figs. 2, 3), coinciding with the first appearance of *Plantago lanceolata* pollen. These features strongly indicate Neolithic forest clearance and farming in the area early in the phase. AP/NAP ratios later recover, but decline again late in phase *g*, ^{14}C-dated to 4725 ± 70 B.P.

Rising AP/NAP ratios in phase *h* mainly reflect the recovery of alder carr, but a dramatic impact upon the landscape is indicated by pollen and stratigraphic changes towards the close of the phase, dated to 3715 ± 70 B.P. Whilst the percentage pollen data (Fig. 2) highlight the demise of the alder carr (reflected in the change from wood peat to *Sphagnum*-sedge peat), the influx data (Fig. 3) also show a decline of most arboreal taxa.

The pollen spectra and low AP/NAP ratios of phase *i* (Fig. 2) indicate therefore that Late Bronze Age and Iron Age peoples in the study area would have inhabited a landscape which was already predominantly open as a result of the activities of earlier cultures. The significance of this is considered more fully in the context of the archaeological evidence discussed below.

Discussion

In Figure 4, a summary of the above interpretation is combined with dated and conjectured archaeological evidence to facilitate discussion.

Settlement and land-use by Mesolithic cultures

Models of Mesolithic land-use in the British uplands have stressed hunting, particularly of red deer; gathering, notably of hazel nuts; seasonality of upland land-use; and possible management of upland resources by periodic burning (Simmons 1975; Jacobi *et al.* 1976; Simmons *et al.* 1981). Dennell (1983) remarked that these models of Mesolithic land-use are untested, and the available data can sometimes be equally well explained by a variety of models, whilst R. B. White (personal communication) has described the present pattern of finds in Wales as "the archaeology of exposure", as coastal processes and forestry deep-ploughing are the main agents of discovery of Mesolithic remains.

For Wales, Jacobi (1980) linked known upland and lowland sites in a single economic model with seasonal occupation of each. He speculated on the utilization of higher ground inland from the nearest 'winter site' at Aberystwyth (Fig. 1a), but the corresponding upland 'summer camp sites' remain to be found. Meirionnydd is devoid of Mesolithic sites and artefacts (Livens 1972; Wymer 1977), but red-deer antlers found in submerged forests suggest plentiful game in extensive wooded coastal lowlands (Kelly 1983) where 'winter sites' may lie undiscovered, submerged by eustatic sea-level rises.

Charcoal records from MYG in the early phases (Fig. 2) provide circumstantial evidence of Mesolithic upland land-use in Meirionnydd. Periodic burning of upland forests could have encouraged seasonal browse for deer, whilst the immediate MYG locality could have been opened up for water fowling or fishing (cf. Smith 1984). The early *Alnus* pollen suggests opportunist colonization of burnt wetland by alder (McVean 1956; Smith 1984) to create the earliest recorded upland alder carr in postglacial Britain (Chambers & Price 1985). Later records of charcoal (phases *c, d*) suggest continual Mesolithic disturbance in this locality and imply that discoveries of Mesolithic sites and artefacts may be expected in Ardudwy.

Settlement and land-use in the Neolithic and Early Bronze Age

Fig. 4. Summary of archaeology and environment in the study area.

There are several Neolithic chambered tombs in the district and more numerous burial cairns, stone circles and other monuments of the Bronze Age (Bowen & Gresham 1967). The pollen data (Figs. 2, 3)

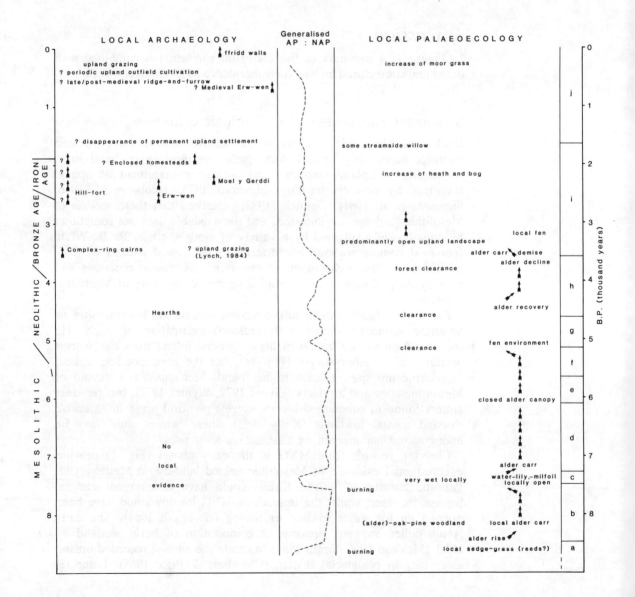

are in concordance and suggest marked impacts upon woodland in the Neolithic *ca.* 5195 and *ca.* 4725 B.P., but a more dramatic impact in the Bronze Age from *ca.* 3715 B.P., indicating the development by late-prehistoric times of a predominantly open, upland landscape.

The pollen evidence from MYG for Neolithic cereal cultivation is equivocal; although several large Gramineae grains >40 μm were recorded in phases *f* and *g* (*ca.* 5640-*ca.* 4625 B.P.), Cerealia pollen was not certainly identified in these horizons. During excavation of Moel y Gerddi, 2 Neolithic hearths were detected underneath the late-prehistoric settlement, demonstrating local Neolithic presence (Fig. 4). The hearth charcoal dates of 4590 ± 80 B.P. and 4540 ± 70 B.P. (Dresser 1985) correspond with phase *h*, and it may be noted that the influx of Gramineae grains >40 μm was not exceeded subsequently in the mire section (Fig. 3). However, the agricultural significance of these records is difficult to assess (cf. Caseldine 1981).

Charcoal samples from one of a pair of complex-ring cairns near Moel Goedog (Fig. 1b), excavated in 1978 by Frances Lynch, were ^{14}C-dated between 3635 and 3445 B.P. in the Early Bronze Age (Lynch 1984). This is immediately after the pollen evidence for major forest clearance at the start of phase *i* (Figs. 2, 3).

The subsequent low pollen influxes of the major tree taxa suggest that local tree cover had become scarce. The nearest woodland today is the *garth* (Fig. 5) which is too steep for arable use and has remained tree-covered in recent centuries. The *garth* is apparently too distant from the MYG core to contribute more than a background AP level today, and AP influxes were similar in phase *i*. What is perhaps the most significant aspect of the reduced AP levels of phase *i* is the possibility that phase *g/h* settlement structures (Neolithic or Bronze Age) could have been of wood rather than stone, and will have left little trace, making it very difficult to assess the density of settlement in the study area for those periods.

Settlement and land-use in the late-prehistoric period

Circular enclosures

Later in phase *i*, the settlements of Erw-wen and Moel y Gerddi were constructed. Radiocarbon dates from Erw-wen range from 2660 ± 60 to 2405 ± 60 B.P. for successive timber round-houses and later a stone building. Radiocarbon dates from the late-prehistoric settlement at Moel y Gerddi range from 2345 ± 65 to 2245 ± 110 B.P., suggesting more recent construction of this site, although contemporaneous occupation of the sites was likely for a time.

A small number of charred *Hordeum* and *Triticum* seeds were recorded from the excavations and a quern was found at Moel y Gerddi, all suggestive of local cereal growing, although grain could have been brought in from elsewhere. Cerealia pollen, recorded in low frequencies from o.g.s. samples (Table 2), might be interpreted as confirming local cultivation (cf. Vuorela 1973). However, structural evidence suggests both sites had a pastoral bias to their agriculture. Animal bone evidence in these excavations was lacking due to the acidity of the soils.

The succeeding stone phase of Erw-wen might be interpreted as a shift in constructional material due to resource shortage, but it could equally indicate an improvement in or a more appropriate technology of construction. Both the later phase of Erw-wen and the building of Moel y Gerddi employed some wood, and the latter site may have been abandoned when extensive refurbishment and rebuilding of its timber structures became necessary.

The structural evidence suggested permanent (i.e. year-round) settlements, with later rebuilding and reinforcing of walls with stone. Both appear to have been abandoned well before the establishment of Roman rule.

Enclosed homesteads

Enclosed homesteads have been regarded as developments of earlier more circular forms and have been assigned to the Roman Period (Bowen & Gresham 1967), although recent excavations in Arfon suggest their origins may be in the pre-Roman Iron Age and their occupation may outlast Roman rule (R. B. White personal communication), which effectively ended in this area *ca.* 383 A.D. The hey-day of such settlements may perhaps have corresponded with a period of supposed climatic warming (Lamb 1982).

The distribution of hill-forts in Ardudwy implies that they were focal points for each block of territory (cf. Hogg 1960) with every hut settlement "within a day's walk to and from" a hill-fort (Kelly 1982a). It is likely that the hill-fort of Moel Goedog was once the permanent territorial centre for Erw-wen and Moel y Gerddi and probably continued as a centre after their abandonment, perhaps providing the focus for the enclosed homesteads. However, it cannot be assumed the enclosed homesteads were later, or permanent, settlements without detailed excavation, though their extensive field systems strongly suggest a permanent base.

Settlement and land-use in historical times

Permanent upland settlement near Moel y Gerddi seems to have ceased. At Erw-wen a small, largely unexcavated, Medieval settlement might be a *hafod* (summer dwelling) in a system of seasonal stock movement, or, in view of former arable fields nearby, a year-round dwelling. Its later abandonment, plus evidence for late- or post-Medieval cultivation in the form of preserved ridge-and-furrow to the south of Moel y Gerddi, attest to the study area's historical marginality. Although the Medieval climatic optimum may have stimulated upland settlement and marginal arable agriculture (Jones 1965; Parry 1975), Jones (1964, 1972, 1973) has emphasized that periodic upland outfield cultivation would still have been practised in later centuries, when the settlement pattern in North Wales is traditionally considered to have incorporated a system of so-called 'transhumance' with seasonal movement from *hendref* (home base) to upland *hafod* (summer dwelling). Kelly (1982b) argues that this so-called 'transhumance' is misnamed, and that the pattern may have been more variable than the literature suggests, with not all farmsteads being involved, some *hafod* settlements at lower altitudes, and daily rather than seasonal stock movements to some pastures.

Although establishment of the *hendref*-based *gwely* system was claimed to have created a "veritable agrarian revolution" in North Wales between 1100-1300 A.D. (Pierce 1961), Davies (1979) seemed prepared to accept functional parallels drawn between 16th century documentary records of occupied summer huts in large upland pastures (*hafoddydd*) and the unexcavated remains of earlier undated upland hut settlements with their associated paddocks.

The antiquity of so-called 'transhumance' in upland Ardudwy is now considered in the light of local field, archaeological, dating and pollen evidence.

The hendref/hafod model of 'transhumance'

In the traditional 'transhumance' model (Fig. 5) the *ffridd* (rough pasture) and *mynydd* (mountain) would have provided grazing for cattle and sheep in the area of summer dwellings (*hafotai*), whilst the lower-altitude *gwaun* (meadow, moor, pasture), *rhos* (moorland) and *gwern* (place where alders grow) would have been exploited from *hendrefi* (permanent dwellings) situated on the drier slopes and margins which would have been cultivated. The *garth* hillside supplied wood; the drier areas of the low-lying *gwaelod* would have been settled and cultivated, but the *morfa* (salt marsh, dunes) was not enclosed until early in the 19th century (Thomas 1963).

Hut-circles as hafotai?

Although Moel y Gerddi lies beyond the historical *hendref* area (Fig. 5), the excavation evidence indicated a more substantial, year-round settlement there with mixed farming (though with a pastoral bias), rather than merely a summer dwelling concerned mainly with pastoralism (Kelly unpublished). Only unenclosed huts with their small diameter and relatively flimsy construction could reasonably be accepted as clear *hafotai*; none was found in the study area, but one small hut is located just off Figure 1b north-east of Moel Goedog.

There is insufficient evidence to confirm that permanent upland settlement was contemporary with the hill-fort and it ended when the hill-fort was abandoned. One could also postulate that, while the hill-fort was occupied, any temporary summer dwellings would have either been at higher altitudes on the *ffridd* and the lower *mynydd* slopes or, at any altitude, but at some distance from the hill-fort in less centrally located areas. Seasonal stock movement may then have been practised, but its pattern would have been disrupted in the vicinity of inland hill-forts. Hence in this interpretation, it it suggested that the traditional, coastal-zoned 'transhumance' model could not apply in the study area whilst the hill-fort was in occupation.

'Transhumance' earlier in prehistory

Jones (1979) has argued that the territorial organization of Early Medieval Gwynedd, involving permanent lowland settlement and the integration of lowland and upland resources, was probably established in prehistoric times. For the Mesolithic, only circumstantial evidence has been found from the MYG core to support Jacobi's (1980) model of seasonal upland land-use. For the Neolithic, Moore (1981) found scant evidence for upland cereal growing in mid-Wales. He suggested that transhumance was probably widely practised in Neolithic Wales, with pastoral exploitation of the uplands. The hearths of Neolithic date at Moel y Gerddi neither confirm nor deny this hypothesis, and the mire pollen evidence is equivocal on upland cereal cultivation. For the Bronze Age, Lynch (1984) presented evidence from the study area which may bear on the question of 'transhumance' in prehistory. Mineralogical analysis of soil adhering to cremated bone excavated from a Bronze Age ring cairn on the slopes of Moel Goedog indicated that this soil was not immediately local (Conway in Lynch 1984). The ring cairn is located beside an ancient trackway (Fig. 1b) of suspected Bronze Age date which runs south-west to Llanbedr on the coast. Here, there are soils of similar mineralogy to the material adhering to the cremated bone, suggesting reburial of bone brought from the lowlands. Lynch (1984) concludes that the upland Bronze Age monuments "... should therefore relate to a population who were living some distance away from them, although they may have been using the surrounding land for grazing. We know as yet too little of the settlement patterns and economy of coastal Ardudwy to speak with confidence in a Bronze Age context, of transhumance or of the traditional *hafod* and *hendref* systems of later centuries; but the reburial in Circle I of bones originally buried in lowland ground hints

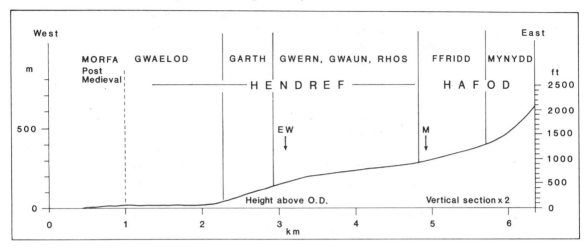

Fig. 5. The traditional *hendref/hafod* model (see text) showing land-use zonation for Ardudwy on a generalized west-east transect (Kelly forthcoming).
EW = Erw-wen; M = Moel y Gerddi (hut-circle sites); O.D. = sea level. There is some doubt as to the post-Medieval origin of the *morfa* due to recent finds of older artefacts (P. Crew personal communication), but its drainage and cultivation commenced in the 19th century.

at perhaps such a system".

It is contended that *if* there ever were such a system in the study area by the Bronze Age, then in later prehistory it became disrupted for a time by permanent upland settlement and possibly remained disrupted until at least the demise of the hill-fort.

Summary

(1) The abundance of stone-built hut settlements in Gwynedd, North Wales has prompted much speculation about their age, function and environmental setting, exacerbated by a lack of detailed archaeological excavations and their integration with palaeoecological work.

(2) An area of ancient landscape in Ardudwy was threatened with upland pasture improvement, so rescue excavations of two late-prehistoric hut-circle sites were made. Their environmental setting was investigated through pollen analysis.

(3) Data are presented as percentage and influx ^{14}C-dated pollen diagrams from a peat site (MYG) in a nearby valley mire. The development of the late-prehistoric cultural landscape is inferred from pollen analyses, archaeological fieldwork and site excavations.

(4) The valley mire was dominated by *Alnus* soon after 8500 B.P.; charcoal records suggest the establishment of alder carr may have been facilitated by Mesolithic clearances, although the area is devoid of known Mesolithic sites or artefacts. Neolithic interference episodes are recorded, but the major decline of forest and alder carr occurred in the Early Bronze Age. Predominantly open conditions prevailed subsequently.

(5) Late-prehistoric settlement included upland farmsteads, originally timber, but later in stone and timber. Pollen indications of timber

shortage in late prehistory hint at earlier, unrecorded, timber dwellings.

(6) A model of historical land-use zones, based on so-called transhumance between *hendref* (winter base) and *hafod* (summer dwelling) settlements, is evaluated for the study area and seems not to apply there in late prehistory.

Acknowledgements

P. Q. Dresser supplied the ^{14}C-dates. Excavations, pollen analyses and ^{14}C-dates were financed by grants from GAT Ltd. through funding from *Cadw*: Welsh Heritage. Thanks are due to R. B. White for comments on the Mesolithic in Wales and on Romano-British hut settlements; to Frances Lynch for information on ring cairns near Moel Goedog; and to G. R. J. Jones for sources concerning the *hendref/hafod* system.

Nomenclature

Plant nomenclature follows A. R. Clapham, T. G. Tutin & E. F. Warburg (1962) *Flora of the British Isles* (2nd edition), Cambridge University Press, Cambridge.

Pollen Analytical Studies of Mountain Summer-Farming in Western Norway

Mons Kvamme

Introduction

The transhumance practice of summer farming is mentioned as a well established and familiar phenomenon in the sagas and old Norwegian laws (Hougen 1947; Reinton 1961; Sølvberg 1976). Discussions concerning the history of summer farming have, so far, been mainly based on archaeological material, historical records or place-name evidence. Both the area of summer farming and the associated buildings are referred to as *støl* or *seter* (Indrelid this volume).

In recent years, archaeological investigations in western Norway have been concerned with prehistoric human activity in mountain summer-farm areas (e.g. Magnus 1986; Bjørgo 1986; Moe *et al.* this volume). These studies were often initiated because of proposed hydro-electric power development, which will eventually flood large mountain areas. As part of this work, pollen analysis has provided information about the history of summer farming, supplementary to and independent of the archaeological material. These historical insights have come from elucidating the development of the cultural landscape around summer farms. This chapter presents some results from these investigations, and considers both the methodology and the vegetational and landscape development. A more comprehensive account (including several additional localities) will be published elsewhere.

The cultural landscape of the summer-farm region

Summer farming almost exclusively took place at higher altitudes than the area of all-year occupation (Indrelid this volume). Its main purpose was to keep cattle away from the infields and outfields during summer time, so that as much as possible of the produce of the permanent farm could be collected for winter fodder (Reinton 1955). In addition, the system allowed the rich grazing potential of the mountains to be utilized. Even though summer farming has lost its economic importance during the last 50 years, its earlier importance is still easily understood

Fig. 1. Former summer farm at Hovden, locality 4. The site is marked by an arrow.

in most agricultural areas of western Norway (Indrelid this volume). The cultural landscape of summer farms is a characteristic feature of the sub-alpine and alpine zones (Figs. 1, 2). In general, this cultural landscape has 3 main elements: (1) There is a treeless area. The absence of trees primarily results from exploitation of the forest for timber and wood. In particular, collection of fire-wood for cheese production did much harm to the forest. Cattle grazing prevented regeneration. (2) The field-layer is dominated by grasses, sedges and herbs favoured by grazing. Where utilization has been strongest, a grass-dominated meadow is often found, with many weeds. However, the influence of grazing can also readily be seen in the surrounding forest, where the floristic composition and an often park-like structure indicate past and present utilization. (3) There were some houses and often also some buildings for cattle. Due to their declining economic importance, the old buildings of many summer farms are disappearing from the cultural landscape, even if the related vegetation types still survive. These areas are still used as pastures, but dairy operations are no longer practised. Some buildings have been converted into holiday cottages.

There is a considerable variation in the vegetational composition of summer-farm cultural landscapes in different geographical districts and at different altitudes. In addition, land-use practices other than grazing, such as mowing or leaf-fodder collection (Austad this volume; Hauge this volume), have contributed to the diversity. However, most of the vegetation types found have one feature in common: they have, until recently, been directly influenced by and have thus reflected

Fig. 2. The Sunndalsætra summer farm, locality 3. The site is marked by an arrow.

human utilization. Studies of their development through time provide a means of reconstructing the past land-use of an area. Vegetation types can, in this way, turn out to be as informative as archaeological material.

The use of local pollen diagrams

Traditionally, conclusions concerning vegetational and cultural landscape development have been based upon changes in regional pollen diagrams from lake sediments (e.g. Berglund 1969). For my investigations into the history of summer farming, this approach is not possible, partly because of the lack of suitable lakes, and partly because of the limited areal extent of the intensively utilized vegetation. In regional diagrams, indications of summer farming would be slight and ambiguous. The alternative is to study cultural landscape development from changes in local vegetation, as recorded in local pollen diagrams (e.g. Fægri 1954; Hafsten 1965 in the Norwegian mountains; Kaland 1979, 1986 in the Norwegian coastal-heath region; Andersen this volume; Odgaard this volume).

For the present study, peat deposits have been selected from within the summer-farm cultural landscape. Most sites are situated in, or are closely connected with, intensively used areas, and several of them are between the summer-farm houses. Such sites reflect a limited spatial scale (Birks 1986a). The relevance of the results from such sites for the reconstruction of regional vegetational history is limited

Depth below surface in cm.	Layer no.	Sediment type	Colour	Physical properties				Component deposit elements										Notes
				Nig	Strf	Elas	Sicc	Tb	Th	Tl	Dl	Dh	Dg	Ld	Ag	Ga	Gs	
Loc. 1: Frettestøl																		
0-9	6	Sph.-peat	Yellowish brown	2	1	4	2	1_3	1_1	+	+							laminated
9-13	5	Fibrous peat	Brown	3	0	2	2	3_1	2_2					2_1	+			
13-17	4	Fibrous peat, minerogenic	Greyish brown	2	0	1	2		2_2					3_1	1			
17-35	3	Peat, sand	Greyish brown	2	0	1	2		2_1					3_2	1	+	+	
35-38	2	Peat	Brown	3	0	1	2	+	2_2					3_2	+			
38-54	1	Peat	Dark Brown	3-4	0	1	2		2_2		+		1	3_1				bark
Loc. 2: Seltuftene																		
0-6	7	Fibrous peat	Dark brown	3-4	0	3	2	+	2_1	0_1				3_2	+			roots
6-12	6	Peat	Brown	3	0	2	2	+	3_1					2_3	+			
12-23	5	Felted peat	Brown	3	2	3	2	3_1	3_1					3_2	+			
23-36	4	Felted peat	Brown/black	3	2	2	2	3_1	+			1		3_2	+	+		laminated
36-41	3	Peat	Light brown	2	1	3	2	3_1	3_1			1		3_1	+			
41-45	2	Peat, minerogenic	Greyish brown	2	0	0	2	+	+					3_2	1	1		stumps
45-80	1	Peat	Dark brown	3-4	0	2	2		3_1		1			3_2	+	+		branches
Loc. 3: Sunndalsætra																		
0-4	4	Felted peat	Brown	3	0	2	2	2_1	0_2	0_1			+	+	++			
4-8	3	Peat, minerogenic	Yellowish brown	2	0	1	2	+	1_1	+				3_2	1	+		
8-17	2	Peat	Dark brown	3-4	1	2	2	2_1	2_1				+	3_2	+			
17-50	1	Felted peat	Light brown	2	0	1	2	1_2	2_1		+		+	2_1				
Loc. 4: Hovden																		
0-12	9	Sph.-peat.	Yellow	1	0	4	2	0_4		+								
12-18	8	Sph.-peat	Yellowish brown	2	0	3	2	1_3	+	1_1				+				
18-22	7	Fibrous peat	Reddish brown	3	0	3	2	3_1	1_2	+				2_1	+			
22-32	6	Peat, minerogenic	Greyish brown	2	0	2	2		2_2	+				2_1	++	1		
32-36	5	Peat, minerogenic	Greyish brown	2	0	2	2		2_2					2_1	1	+		
36-44	4	Peat, minerogenic	Greyish brown	2	0	2	2		2_2	+				2_1	++	1		
44-53	3	Peat	Dark brown	3-4	0	2	2		2_1				1	3_2	+	+		
53-72	2	Peat	Brown	3	0	1	2		2_1		+			3_3				
72-85	1	Peaty humus	Black	4	0	0	2		+					3_4	+			

because of their local nature and because of the danger of site disturbance by nearby human activity. However, as these sites are primarily studied to document such 'disturbances', the latter feature is regarded as an advantage rather than a problem.

Here, 4 examples from summer-farm areas demonstrate how human utilization is recorded by different patterns of vegetational development in local pollen diagrams. Methods follow Fægri & Iversen (1975) and sediment descriptions Troels-Smith (1955) (Table 1). Seven ¹⁴C-datings were obtained (Table 2).

Evidence for intensively used vegetation: locality 1

My example of this type is from Etne, southern Hordaland (Fig. 3). Frettestølen is situated in an open, treeless valley bottom, dominated by extensive mires. The presence of scattered birch forest on the mountain-side *ca.* 100 m away demonstrates that the area lies below the timberline in the sub-alpine vegetational zone.

Local summer-farming was abandoned *ca.* 100 years ago, and today the buildings are in ruins. However, until recently, grazing maintained a grass-dominated vegetation. The summer farm lies on a slope, and the pollen samples were collected from a *Scirpus caespitosus*-covered mire, a few metres below the transition between grassland and peat.

In the pollen diagram (Fig. 4), zone 1 represents an open, birch-dominated forest. Grazing is indicated by grass pollen percentages of *ca.* 20%, and by the occurrence of pollen of, for example, *Artemisia*, *Urtica* and *Plantago lanceolata*. In a sub-alpine birch forest, which the pollen assemblage presumably reflects, it is unlikely that all these NAP result from long-distance transport.

Local deforestation is recorded at the transition to zone 2, together with temporary maxima in the values of *Potentilla* and *Melampyrum*.

◀

Table 1. Lithostratigraphical description of the sediments at the different sites using the system of Troels-Smith (1955). Depths and layer no. correspond to figures in the pollen diagrams (Figs. 4-7). Sph. = *Sphagnum*.

Table 2. List of ¹⁴C-dates. All peat samples were sieved according to K.-D. Vorren (1979) and the >1 mm fraction removed.

Lab. no.	Locality name (no.)	Material	¹⁴C-age B.P.	Calibrated age, MASCA
T-5559	Frettestøl (1)	Peat	2330±100	555±155 B.C.
T-5560	Frettestøl (1)	Peat	930±100	1065±105 A.D.
T-5561	Frettestøl (1)	Peat	Recent	Younger than 1690 A.D.
T-4205	Seltuftene (2)	Branches	2880±90	1155±145 B.C.
T-5607	Sunndalsætra (3)	Peat	1460±80	525±85 A.D.
T-5245	Hovden (4)	Peat	1090±70	890±90 A.D.

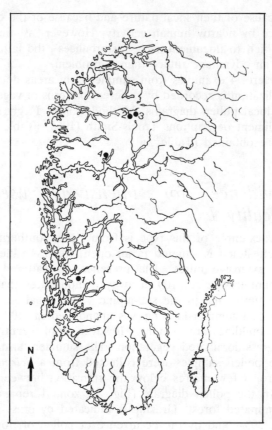

Fig. 3. Map of western Norway showing the location of the sites (1-4) discussed in the text.

These reflect an opening of the forest, but hardly the intensive use of the area. Most likely, this phase represents wood- or leaf-fodder collection as part of the existing use of the area.

The formation of the present cultural landscape is recorded at the transition to zone 3. Pronounced increases in the values of grass, *Rumex*, *Ranunculus* and Asteraceae pollen, together with the assemblage of other herb taxa present, indicate that the vegetation was now intensively grazed. This is supported by most of the pollen curves remaining unchanged up to modern times, suggesting that the past vegetation was similar to the present grazed area. At the same level, high percentages of microscopic charcoal particles indicate regular, human habitation not far from the site.

The intensified utilization can also be detected in the peat (layers 3, 4) from its increased minerogenic content (Table 1). This is probably due to erosion initiated by grazing (cf. locality (loc.) 4).

In zone 3, records of 2 cereal pollen (*Hordeum*-type, *Triticum*-type) merit special attention. The natural dispersal of cereal pollen (except for *Secale*) is so poor that these grains may indicate local cereal cultivation. Neither the altitude of the site nor the local edaphic conditions make this impossible. Even though cereal cultivation was never a major activity at summer farms, it was not uncommon in suitable areas during historical times (Reinton 1957). Its importance in prehistoric times is not known.

The frequencies of weed pollen are very low, however. With cereal cultivation near the site, one would expect a higher representation of pollen such as Brassicaceae, *Rumex longifolius*-type and *Stachys*-type (probably *Galeopsis*), but the values of these and other weeds are no higher than are found at any intensively grazed summer-farm. In addition, weeds typically associated with cereal cultivation in western Norway, such as *Spergula arvensis* and *Caltha palustris* (Kvamme 1982), are missing. Furthermore, there are no definitive traces of local cultivation, such as lynchets or ancient fields.

On the other hand, traditional summer-farming implied frequent contact with lowland areas. Dispersal of cereal pollen by anthropogenic transport, as demonstrated by Vuorela (1973), thus seems to be more probable than local cultivation.

At the transition to zone 4 decreasing values of grasses and other herbs, increasing values of AP and Cyperaceae pollen, and a sediment change to fresh *Sphagnum* peat, all indicate a less intensive utilization of the area around the site. This probably represents the end of traditional summer-farming activities, which ceased *ca.* 100 years ago. The ¹⁴C-date gives a recent age, with a maximum age of 260 years, thereby confirming this interpretation. The curve of charcoal particles does not decrease, possibly because of inwash from dry ground (with minor peat growth) just above the site. As the area is mostly treeless today, the AP increase must derive from long-distance transport. As there is no reason why such transport should increase around the turn of the century, the increased AP percentages are probably a result of reduced local pollen production, possibly because of a lower pollen production today in the Cyperaceae-dominated mire, than in the earlier grass-dominated grazed vegetation.

Figure 4 illustrates several phases of different intensities of utilization during the development of the summer-farm cultural landscape. The ¹⁴C-date at the bottom of the diagram (not the bottom of the peat-deposit), gives a minimum age for extensive grazing in the area of 2330 B.P. (MASCA calibrated age ≈ 555 B.C.). This is around the Bronze Age-Early Iron Age transition. The start of intensified utilization at the summer farm is ¹⁴C-dated to 930 B.P. (1065 A.D.). This also dates the beginning of the recent cultural landscape at this summer farm.

The pattern of vegetational development at this site is typical of sites situated in the centre of summer farms. The 3 main elements of summer farming mentioned above are easily recognized. The decreasing AP curves reflect the opening of the forest, the rises in the values of grasses and other herbs record grazing effects, and the charcoal particles document the presence of people staying regularly at the summer farm.

Evidence for grazed forest: locality 2

This example is from Erdalen, Stryn, Sogn og Fjordane (Fig. 3). The sequence is from a small peat deposit in sloping terrain at Seltuftene (Kvamme & Randers 1982). The site is situated within grazed sub-alpine birch forest, next to a grass-dominated, partly open pasture-area. In this area, about 300 m away from the site, remains of house foundations have been found that date from the Viking or Early Medieval Period (Kvamme & Randers 1982).

The pollen diagram (Fig. 5) does not reflect any major local forest changes. As today, sub-alpine birch forest has dominated the area as far back as the diagram extends. However, distinctive changes in the field-layer vegetation can be seen. In zone 1 there are high percentages of Asteraceae, *Rumex* and *Filipendula* pollen, and a diversity of other NAP, together with large amounts of fern spores (70-90%). The abundance of ferns and low grass values suggest that the high values

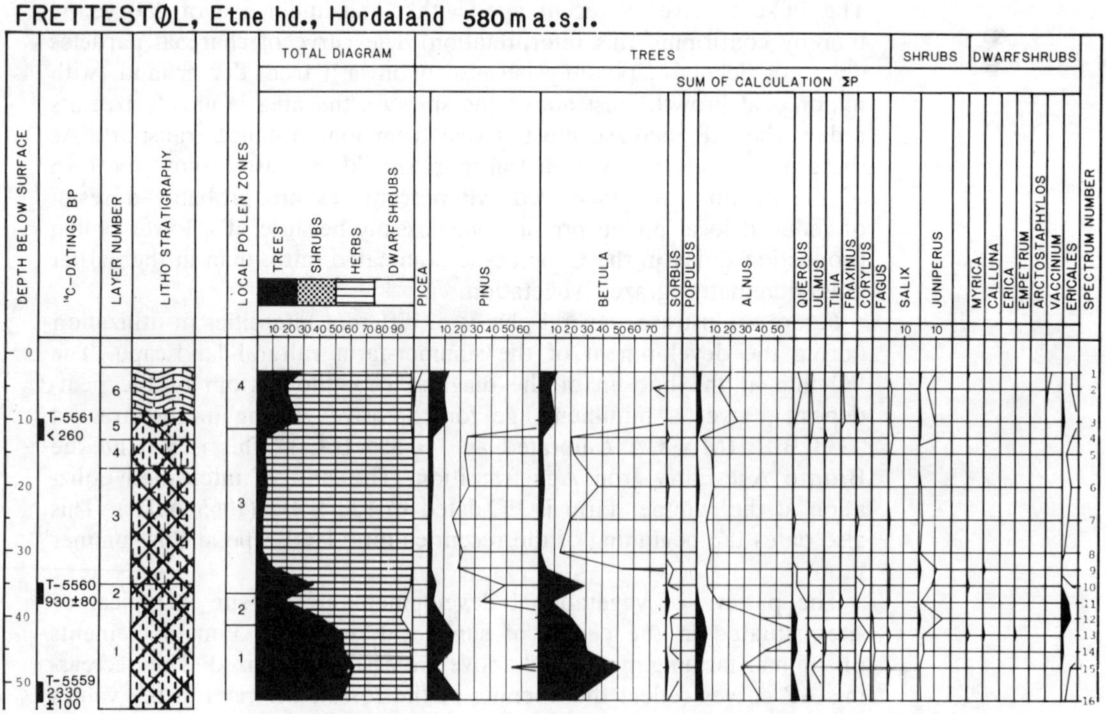

Fig. 4. Pollen diagram from Frettestøl, locality 1.

FRETTESTØL, Etne hd., Hordaland 580 m a.s.l.

of herb pollen are unlikely to represent weeds or be connected with grazing.

The high values of Asteraceae sect. Cichorioideae pollen are of special interest. The only species within this group that could be expected in such quantities in this vegetation type is *Cicerbita alpina*. This species is typically found in sub-alpine 'tall-herb' vegetation. The other NAP types found in zone 1, (e.g. *Rumex, Ranunculus, Filipendula*) probably also represent species in this vegetation type. Tall-herb vegetation usually occurs on fertile soils with adequate supplies of moisture, often in association with tall-fern vegetation (Nordhagen 1943).

At the transition to zone 2 a significant decline in Asteraceae sect. Cichorioideae occurs, together with decreases in several other NAP types and fern spores. *Cicerbita alpina* is preferred by grazing animals (Lagerberg *et al.* 1958; Høeg 1976), and together with other tall-herbs and tall-ferns, it is intolerant of grazing and trampling (Nordhagen 1943; Aune 1973; Kielland-Lund 1975). Therefore the changes at the transition to zone 2 probably represent the onset of extensive grazing in the forest. This interpretation is supported by the parallel rise in grass percentages from this level. The intolerance of many tall-herbs to trampling and grazing explains why this type of vegetation in grazed sub-alpine forest today is mainly restricted to river banks and other

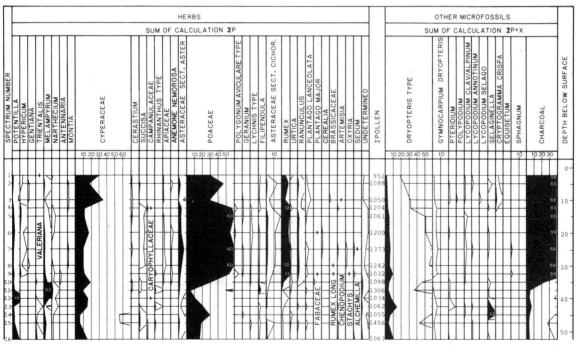

Analysis: Svenn Sivertsen & Mons Kvamme-84

inaccessible places (Odland 1981).

The onset of grazing is dated to 2880 B.P. (1155 B.C.). This accords with a burial find further down the valley, at the main farms that traditionally used this summer farm (Fett 1961). The slight increase in charcoal frequencies higher up in the diagram is probably associated with the nearby house foundations of Viking or Early Medieval Age.

Compared to the results from Etne (Fig. 4), evidence for grazing is much weaker in this diagram. This is not only due to the greater distance of the site from the centre of utilization, but also because the forest around the site, although grazed, was never cleared. However, similar changes have been found in several diagrams prior to the onset of summer farming, suggesting that extensive grazing was a prelude to summer-farm establishment.

Fig. 5. Pollen diagram from Seltuftene, locality 2.

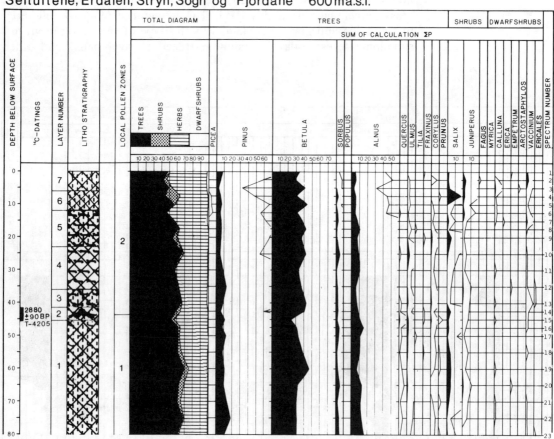

Seltuftene, Erdalen, Stryn, Sogn og Fjordane   600 m a.s.l.

Evidence of anthropogenically induced changes in forest composition: locality 3

This type of vegetational change is illustrated in a diagram from Sunndalsætra, Stryn, Sogn og Fjordane (Fig. 3). The samples were collected from a small peat deposit situated between the summer-farm houses (see Fig. 2). The vegetation at the site consists of *Deschampsia caespitosa*, *Nardus stricta*, *Potentilla erecta* and *Vaccinium myrtillus*, and is next to a grass-dominated 'setervoll'. The vegetation of the valley is dominated by pine forest, which extends upward to its presumed climatic limit at *ca.* 600 m elevation (Skogen 1984). The summer farm lies at 460 m, relatively lower than the preceding two sites.

The lower part of the diagram (Fig. 6) is dominated by *Pinus* pollen,

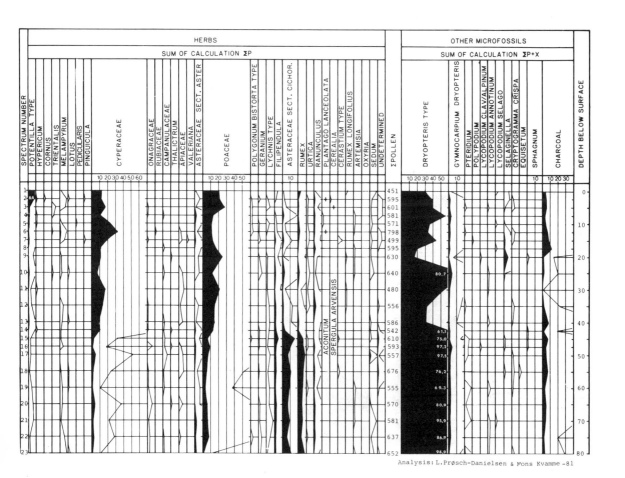

Analysis: L.Prøsch-Danielsen & Mons Kvamme -81

suggesting that the summer-farm area was, at this time, covered by pine forest. A few finds of pollen possibly indicative of agricultural activity, such as *Artemisia* and *Rumex longifolius*-type, are not unambiguous evidence for local utilization at this time.

The development of the cultural landscape at this summer farm occurs in two stages. The first is the transition to pollen zone 2, where *Pinus* values decrease rapidly, accompanied by an increase in *Betula*. This change in forest dominants can hardly have been caused by climatic change, as pine forest grows today at higher altitudes without any significant pollen values at the surface of the sequence. Furthermore, increases in pollen values of *Juniperus*, *Ranunculus*, *Rumex* and *Urtica*, together with continuous occurrences of *Artemisia*, *Plantago lanceolata* and *Rumex longifolius*-type, document the anthropogenic nature of this vegetational change. This flora indicates local grazing. A marked increase in the charcoal curve at the beginning of zone 2 reflects human occupation near the site.

Fig. 6. Pollen diagram from Sunndalsætra, locality 3.

The continuous curve of *Plantago lanceolata* warrents discussion. Stryn is situated far east of the main distributional area of this species

SUNNDALSÆTRA, STRYN HD. SOGN OG FJORDANE   460 m a.s.l.

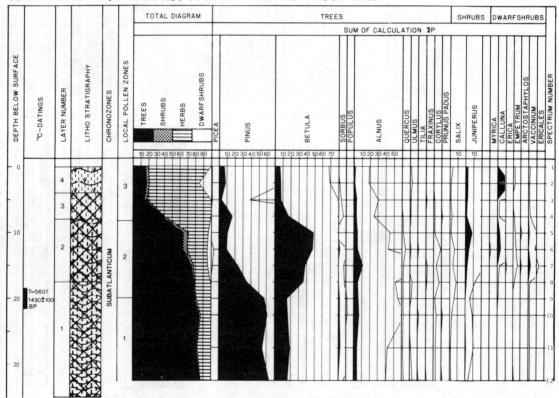

today (Fægri 1960). It is uncommon, even in the agricultural districts of the main valleys. It is therefore unlikely that its pollen occurrences represent long-distance transport following agricultural expansions there. More probably, the pollen reflects local, sporadic growth of the plant in response to grazing. Due to its high pollen production, it is well represented locally.

The second stage of cultural landscape development occurs at the transition to zone 3. The fall in *Betula* values demonstrates that, at this time, the birch forest was cleared as well, and the increases in grass, *Rumex*, *Potentilla* and *Ranunculus* percentages suggest intensive grazing close to the site. These changes probably reflect the establishment of the present-day cultural landscape at the summer farm. This is supported by the large increase in the frequencies of charcoal particles at the same level, suggesting regular human habitation nearby.

The change in forest dominance from pine to birch at the transition to zone 2 was probably due to selective exploitation of pine in the area for timber, tar-production or other purposes. Utilization of local vegetation at the site was not as intensive as later, and thus birch

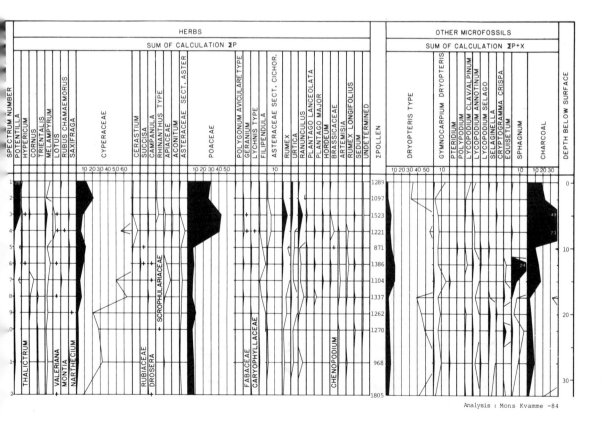

Analysis : Mons Kvamme -84

forest was able to succeed. However, the percentages of charcoal particles and the weed-flora composition indicate that an area nearby was intensively used. Apparently the site at this time occupied a peripheral situation within the cultural landscape, compared to later. In many areas in western Norway, there is today a zone of birch forest strongly influenced by utilization between the summer farm and the dominant regional forest type, and zone 2 may represent this. The transition to zone 3 may thus reflect a moving of the centre of utilization to its present location.

The transition to zone 2 is ^{14}C-dated to 1430 B.P. (525 A.D.). Even if this does not date the onset of the summer farm in its present situation, it dates the beginning of the type of utilization of which it has been a part.

Fig. 7. Pollen diagram from Hovden, locality 4.

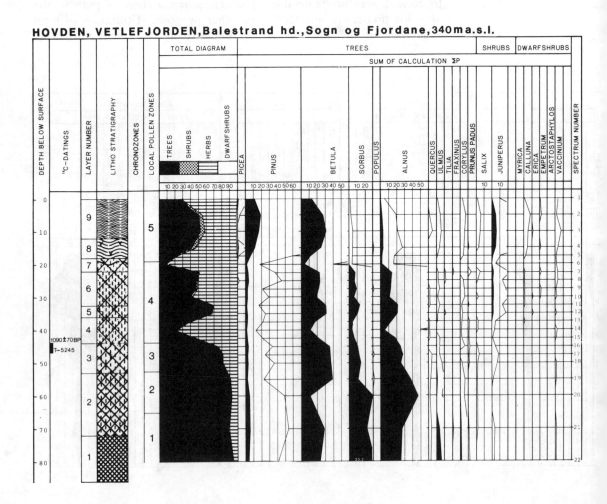

Evidence for erosion due to intensive grazing: locality 4

Evidence for this type of utilization comes from a pollen profile in the former summer farm at Hovden, Vetlefjorden, Sogn og Fjordane (Fig. 3). The samples are from a small mire at the transition to a flattened shoulder in an otherwise steeply sloping mountain-side (Fig. 1). The summer farm (abandoned earlier this century) was situated on this shoulder, where the vegetation is still grass-dominated with common occurrences of species such as *Ranunculus acris*, *Rumex acetosa*, *Potentilla erecta* and, in particular, *Urtica dioica* and *Juniperus communis*. The vegetation on the mire consists of *Eriophorum angustifolium*, *Scirpus caespitosus*, *Juncus filiformis* and species of *Carex* and grasses. The mountain-side surrounding the summer farm is dominated by fern-rich birch forest (Fremstad & Moe 1982).

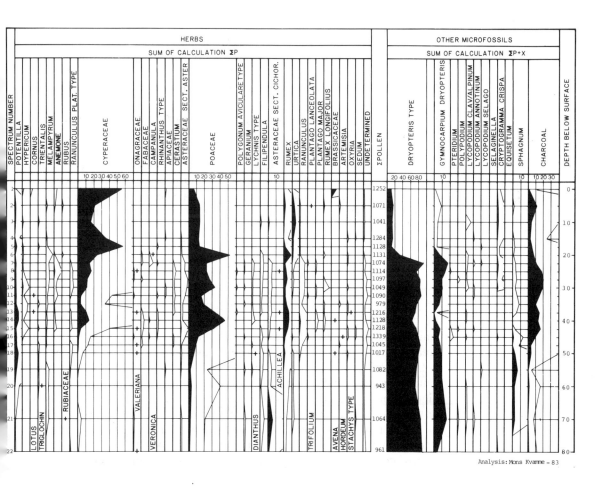

Analysis: Mons Kvamme – 83

The pollen diagram (Fig. 7) is divided into 5 pollen zones. The lowest one reflects a fern-rich, deciduous forest, not unlike the type of vegetation found in the area today, but with *Sorbus* and *Alnus* (probably *A. incana*) as prominent associates with birch. The unusually high percentages of the entomophilous *Sorbus* probably reflect local stands on the slopes just above the site. *Ulmus* reaches values of *ca.* 5%. Today this tree occurs regularly in the valley, but is hardly recorded in the top of the diagram. Elm was probably commoner at this time.

At the transition to zone 2, there are distinctive decreases in *Ulmus* and Asteraceae sect. Cichorioideae (probably *Cicerbita alpina*, cf. loc. 2) values. The forest composition seems to have become more variable. These changes may be due to the beginning of human utilization including leaf-fodder collection (Nordhagen 1954; Troels-Smith 1960). Apart from the *Ulmus* curve, indicators for this are few and weakly developed. Taking the local representation of the diagram into consideration, they give no certain evidence for local cultural landscape development at this time.

In zone 3, decreasing AP and increasing *Juniperus* and *Potentilla* values suggest that the forest became more open. Increasing grass and *Rumex* values and a higher number of herb taxa indicate grazing of the forest. A small increase in charcoal provides further evidence for an anthropogenic cause for these vegetational changes, as do 2 cereal pollen grains. But, as at locality 1, these grains do not prove that there was any local cereal cultivation.

At the transition to zone 4 AP falls abruptly by more than 50%. There are maxima in zone 4 of grasses, *Rumex*, *Ranunculus*, Asteraceae sect. Asteroideae and charcoal particles, and nearly continuous records of *Rumex longifolius*, *Plantago major* and *Artemisia*. The zone clearly reflects the start of intensive local utilization, probably summer-farm establishment.

The trends in the AP and fern-spore curves in zone 4 apparently contradict this interpretation. The AP curve soon increases again, and in spite of the intolerance of ferns to grazing and trampling, the fern-spore curve continues more or less unchanged during the zone (cf. loc. 2). However, the lithostratigraphy shows that layers 4, 5 and 6 (corresponding to zone 4) contain considerable amounts of minerogenic material (Table 1). This could be due to flooding by small streams, deposition during snow-melt, avalanches, or erosion. None of the first 3 processes occur here today, and this part of the diagram contains strong indications of intensive grazing. The presence of minerogenic material therefore probably results from erosion caused by the effect of heavy grazing and trampling on the vegetational cover on the slopes above the site. Redeposition of minerogenic material is a common feature in this type of profile (cf. loc. 1).

Erosion has, in this case, not only affected the minerogenic content of the sediment, but has also introduced reworked organic material containing pollen from the previous vegetation of the area. The increase in zone 4 of the dominant taxa of zones 1 and 2 (*Betula*, *Alnus*, *Sorbus*, fern spores) most probably results from this redeposition.

At the transition to zone 5 the curves of grazing indicators rapidly decrease, such as grasses, *Rumex* and *Ranunculus*, and the present-day colonization of derelict pastures by *Urtica* and *Juniperus* is clear. Another indication of reduced utilization is the decrease in charcoal particle values. The minerogenic content of the peat decreases at the same level (layer 7). The topmost sediment (layers 8, 9) consists mainly of weakly decomposed *Sphagnum* peat, probably formed in this century. The curves of presumed redeposited AP fall immediately, and the increase in *Betula* and *Pinus* values in zone 5 represent the reforestation of the area that is occurring today. Simultaneously, different fern species rapidly expand into previously grazed areas. However, the values of fern spores fall considerably at the transition to zone 5, and remain low. This implies that most of the fern spores in zone 4 may be secondary.

The upper 3 cm of layer 3 is ^{14}C-dated to 1090 B.P. (890 A.D.). This sample was taken just below the lithostratigraphic boundary, to avoid contamination from redeposited material. This gives a maximum age for the intensified utilization related to the summer-farm establishment. As the dated sample represents a third of zone 3, it also gives a minimum age for the local cultural landscape development, which started at the beginning of this zone. This can be estimated to *ca.* 700 A.D., assuming the ^{14}C-date is correct.

Conclusions

The results of these investigations demonstrate the value of local pollen diagrams as a tool for elucidating the development of the cultural landscape at summer farms. Due to the restricted pollen-source areas of the sites investigated, the pollen assemblage must represent the vegetation type exploited locally. The kind of information obtained is thus directly related to the scale of the site of interest. When diagrams are made from the intensively exploited parts of the summer farms, evidence for intensive grazing and regular human occupation is usually obvious (loc. 1 zone 3; loc. 3 zone 3). The palynological evidence for anthropogenic impact rapidly decreases when the sites are located some distance away (cf. Hicks this volume). Here, in most cases, utilization has been of a more extensive character, resulting in other vegetation types with different kinds of human impact (loc. 2 zone 2; loc. 3 zone 2). It is thus possible to document the length of time the present ecological effect of summer farming has had on the local landscape.

This does not mean that all human impact on vegetational development recorded at this kind of site necessarily implies summer farming. It is, however, the most likely explanation when interpretations are based upon similarities between present and past vegetational composition, and the results conform to patterns of vegetational development recognized at most summer farms, such as those described here.

At existing summer farms, archaeological material often turns out to be sparse or absent, even after days of surveying (e.g. loc. 3, loc. 4). In such cases, pollen analysis not only supplies information independent and supplementary to the archaeological material, but also provides the only easily accessible record of the history of individual summer farms. Furthermore, when archaeological material is present, indications of local grazing provided from vegetational history tend to extend further back in time (e.g. loc. 2). However it must be emphasized that pollen analysis alone does not give a complete cultural picture of summer-farming history. That can only be achieved by combining pollen analytical and archaeological data from several summer farms and the surrounding mountain areas within the same geographical district. This is currently being attempted in inner Sogn (Kvamme & Randers 1982; Magnus 1986; Bjørgo 1986; unpublished material by the same authors). Here, there is evidence for summer-farm expansion between *ca.* 1000 B.C. and 1000-1100 A.D.

The examples given above demonstrate that it is possible to obtain a picture of cultural landscape development at summer farms. The aims of the present studies have primarily been to date the last intensification of utilization at different summer farms. Hopefully, the development of the different types of cultural landscapes at summer farms can be further elucidated by future research.

Summary

(1) Because of proposed hydro-electric power development in mountain areas of western Norway, extensive archaeological surveys have been carried out prior to dam construction. As part of these surveys, pollen analytical studies have been made at or near summer farms.

(2) The work has been based on local pollen diagrams from sites usually situated in areas of intensive utilization of the local vegetation. The initiation of local summer-farming has been established by dating the origin of vegetation types characteristic of the present cultural landscape.

(3) Intensive utilization is recognized by (i) deforestation, indicated by decreasing arboreal pollen (AP) percentages; (ii) intensive grazing, indicated by increasing pollen values of grasses, *Rumex*, *Ranunculus* and other herbs; (iii) human occupation, indicated by the presence

of charcoal particles; and (iv) erosion resulting from disturbance, indicated by increased minerogenic content of the peat.

(4) Extensive utilization is recognized by anthropogenic changes in forest composition, indicated by a non-climatic change from pine to birch forest; and grazing effects on field-layer vegetation, indicated by decreasing values of Asteraceae sect. Cichorioideae pollen (here probably *Cicerbita alpina*).

(5) At several summer farms, palynological data provide the only available evidence for early utilization. Where archaeological material is present, the palynological results tend to suggest an older age for utilization.

Acknowledgements

These investigations have been financed by Haugesund Elektrisitetsverk, Norges Vassdrags- og Energiverk and Sogn & Fjordane Energiverk. The author also wishes to thank P. E. Kaland, K. Fægri, John Birks, Hilary Birks and collaborators at the Historical Museum, Bergen, for help, stimulating discussions, and constructive criticism during the work.

Nomenclature

Plant nomenclature follows J. Lid (1985). *Norsk-svensk-finsk Flora*, Det norske Samlaget, Oslo.

Development of the Cultural Landscape in the Lofoten Area, North Norway

Eilif J. Nilssen

Introduction

The geographic situation of the Lofoten islands (Fig. 1), extending out into the warm Gulf Stream, leads to mild winters. Lofoten could thus be expected to be one of the most suitable areas in Norway north of the Arctic Circle for the introduction of domestic animals and farming. Therefore this area may have suffered a considerable human impact on the local vegetation.

The islands of Vestvågøy and Gimsøy are situated in the centre of the Lofoten islands, at 68°03'-68°21'N, 13°27'-14°16'E. The climate is oceanic with a mean winter temperature (January-March) of -0.5°C, and a mean summer temperature (June-August) of +11°C. Annual precipitation is *ca*. 1100 mm.

Today Lofoten is the most densely populated region in northern Norway with *ca*. 21 inhabitants km^{-2}. About one third of the population of 26 000 is engaged in fishing and farming. Lofoten is known for its steep mountains rising abruptly from the sea. Only *ca*. 4% of the land area (8% at Vestvågøy) is cultivated. Today the cultivated areas produce grass for silage, pasture for cattle, sheep and goats, and potatoes. A couple of human generations ago, barley was also common on farms, when the barter economy dominated.

The landscape is anthropogenically influenced to varying degrees by farming and extensive grazing by sheep and goats. In areas where farming has continued for hundreds of years, only isolated trees of *Betula pubescens* survive as remnants of a presumed formerly continuous *Betula* woodland. In areas more distant from farming activity, the ericaceous dwarf-shrubs *Vaccinium myrtillus, V. vitis-idaea* and *Empetrum hermaphroditum* dominate in woodlands. In areas even further from present farms woodlands are dominated by ferns such as *Athyrium filix-femina, Gymnocarpium dryopteris, Thelypteris phegopteris* and even *Matteuccia struthiopteris* and tall-herbs such as *Geranium sylvaticum* and *Valeriana officinalis*.

Site	Own ref.	Lab. ref.	Age ^{14}C-years B.P.	Calibrated age (MASCA)	^{13}C o/oo
Tangstad	T 32	T-6152	2000 ± 60	50 ± 110 B.C.	-27.6
	T 57	T-6153	2910 ± 70	1165 ± 135 B.C.	-28.4
	T 222	T-6158	8900 ± 100		-28.0
Årstrand	Å2	T-2912	3990 ± 90	2635 ± 195 B.C.	
	Å3	T-2913	4280 ± 130	3020 ± 160 B.C.	
Rystad	RY 47	T-6154	1770 ± 60	190 ± 80 A.D.	-26.2
	RY 85	T-6155	2340 ± 70	535 ± 135 B.C.	-25.4
Justad Øvre	JØ 57	T-6157	580 ± 70	1335 ± 75 A.D.	-26.4
	JØ 150	T-6357	1240 ± 70	730 ± 90 A.D.	-28.1
	JØ 177	T-6156	1790 ± 50	180 ± 70 A.D.	-27.9

Table 1. Radiocarbon dates.

Three of the sites studied, Tangstad, Rystad and Justad, (Fig. 1) are mentioned as farms from 1567 A.D. in the oldest historical records from the area (Rygh 1905). The fourth site, Årstrand, lies *ca.* 2 km from the closest old farm Vik mentioned in 1567. The recent Årstrand farm was established during the last century. According to archaeological finds (Bertelsen 1985), Rystad was established as a dwelling site between 3000 B.C. and 0 A.D., and Justad and Vik (near Årstrand) were established as farms in the (Norwegian) Late Iron Age (600-1050 A.D.). There are no archaeological records relating to Årstrand itself. There is no evidence for settlement at Tangstad until the middle of the Medieval Period (*ca.* 1200 A.D.), but according to Rygh (1898, p. 77) and Olsen (1926, pp. 77-105) the name indicates an earlier establishment.

Recently published pollen diagrams from Bøstad (K.-D. Vorren 1979) and Moland (Nilssen 1983) reconstruct the ecological history of these farms. Two other diagrams from Vestvågøy, Stormyr (Moe 1983a) and Dønvold (Nilssen 1983) also provide general information about vegetational history.

Materials and methods

Pollen profiles, except Årstrand, were taken as near the farm centres as possible. This is important for detecting small vegetational changes caused by early human activity (K.-D. Vorren 1979). The hypothesis is that vegetation changes caused by human activity started in the centre of what is a farm today, and that the natural vegetation gradually changed through time and in space outwards from this centre if the activity was maintained. The profiles, Tangstad, Rystad and Justad Øvre, were collected less than 150 m from the supposed farm centre.

The Årstrand profile was collected *ca.* 2 km from the nearest old farm Vik, and 1 km from the Årstrand farm.

All profiles were collected from bogs in continuous 10 cm-diameter plastic pipes. In the laboratory the pipes were cut open and samples for pollen analysis and ^{14}C-dating taken. Material for dating was 2 cm thick or less, except for the Årstrand profile where the dated material was *ca.* 7 cm of peat. Peat was washed through sieves, and material between 1.0 and 0.063 mm size was dated at the Radiological Dating Laboratory, Trondheim (Table 1). The dates are uncorrected values based on a half-life of 5570 years.

Pollen preparation methods follow Fægri & Iversen (1975). The pollen diagrams are based on a minimum of 500 pollen per sample.

Fig. 1. The investigation area with the pollen sites marked by dots.

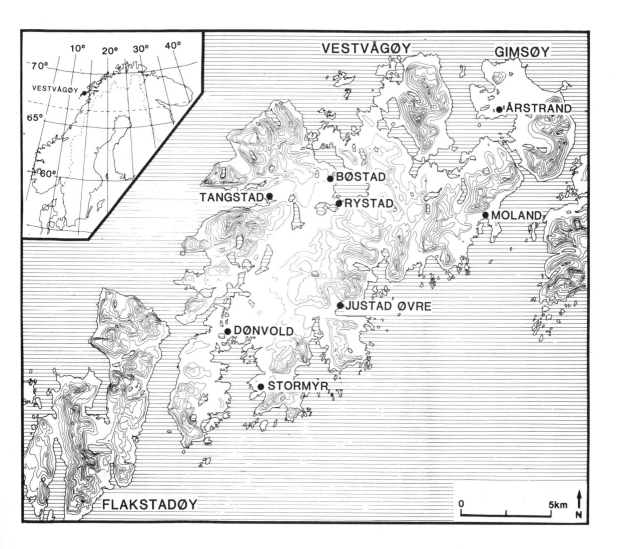

Pollen from local bog plants are excluded from the pollen sum and presented as percentages of total pollen plus the individual taxon. *Calluna vulgaris* is not included in Ericales. Grouping of taxa into ecological groups follows Vorren & Alm (1985).

The suggested time-scale is based on the sedimentation rates interpolated between ¹⁴C-dates. In the Årstrand diagram the top of the profile is assumed to be modern. In the Rystad diagram a characteristic peak in the *Pinus* curve dated in another diagram from the area (Nilssen 1983), was used in constructing the suggested time scale. A uniform sedimentation rate and an absence of any hiatuses between fixed time points are thus assumed in constructing these chronologies.

Pollen identifications follow Fægri & Iversen (1975) and, for cereals, Beug (1961).

The sites

Fig. 2. Simplified pollen diagram from Tangstad, Vestvågøy.

Pollen diagrams from the 4 sites are presented in Figures 2-5. The oldest deposits are at Tangstad, (*ca.* 9000 B.P.), but deposits predating any human impact on vegetation are also found at Rystad and

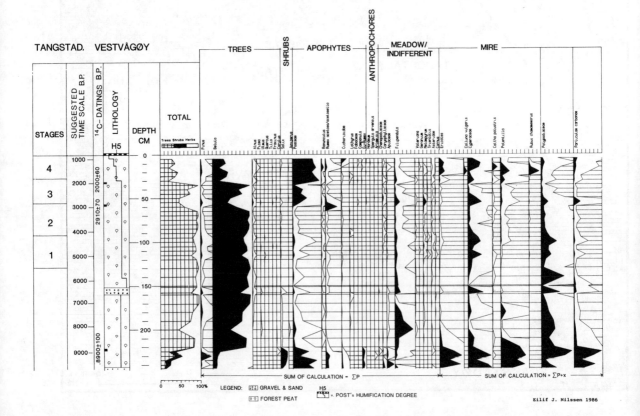

Årstrand. The pollen assemblages from these deposits reflect the natural vegetation of Lofoten at that time. This was dominated by *Betula* woodland, rich in ferns and herbs such as *Filipendula* and *Cornus suecica*.

The history of human impact on the vegetation as reflected in the pollen diagrams can be conveniently divided into 4 stages. These are indicated on the left side of the pollen diagrams.

Tangstad

The lowermost part of the diagram (Fig. 2) suggests an open vegetation, dominated by Ericales, *Salix*, Poaceae, *Rumex* and Eu-*Rumex*, typical for Preboreal times in this area. At 222 cm (8900 ± 100 B.P.) *Betula* starts to expand. It became totally dominant in the area, with *Filipendula* and Polypodiaceae in the understorey. A short interruption in peat accumulation (158-151 cm) is due to a layer of sand and gravel deposited by a marine transgression at that time (Møller 1986).

Stage 1. From 120 cm (*ca.* 5200 B.P.) there is a small increase in Poaceae, *Ranunculus*, *Rumex*, Cichorioidae and charcoal. These changes are interpreted as the first signs of anthropogenic influence on the vegetation. Increased values for long-distance transported *Pinus* pollen suggest a reduction of the local *Betula* woodland. Introduction of domestic animals and permanent all-year settlements may have also occurred.

Before stage 2, from 90-60 cm (*ca.* 4000-3000 B.P.), anthropogenic indicators show reduced values and *Betula* regenerates. This suggests the settlements recorded in stage 1 were abandoned.

Stage 2. From 60 cm (*ca.* 3000 B.P.) there is an abrupt increase in Poaceae. In the levels above, the first occurrences of *Hordeum* pollen are found. These changes are interpreted as the establishment of the farm with cereal cultivation and development of grass heaths. At the end of this stage (*ca.* 2100 B.P.) there are high values of *Betula* and a corresponding reduction in Poaceae. Although this is recorded in only one spectrum, it may represent abandonment of the farm.

Stage 4. From 45 cm (*ca.* 2000 B.P.) increased values of Poaceae and other apophytes, occurrences of *Hordeum* and *Spergula arvensis*, and *Betula* values below 40% indicate greater woodland clearance and a more intense utilization of the area around the site.

The suggested time-scale based on the sedimentation rate between the uppermost two dates indicates no sediment after 1000 B.P. This may be due to peat-cutting which is common in this area.

Årstrand

Peat accumulation began at Årstrand (Fig. 3) at 4280 ± 130 B.P. (T-2913). *Betula* woodland appears to have been well established. The values of *Betula* are above 70% at the bottom of the profile, and they remain more or less constant (*ca.* 60%) throughout the rest of the diagram, up to the present day. This suggests that the local extent of *Betula* woodland has not changed much during the last 4000 years. The relatively high values of mire plants not included in the pollen sum indicate large areas of mires around the site. It would appear that the landscape around Årstrand has changed little throughout the

Fig. 3. Simplified pollen diagram from Årstrand, Gimsøy.

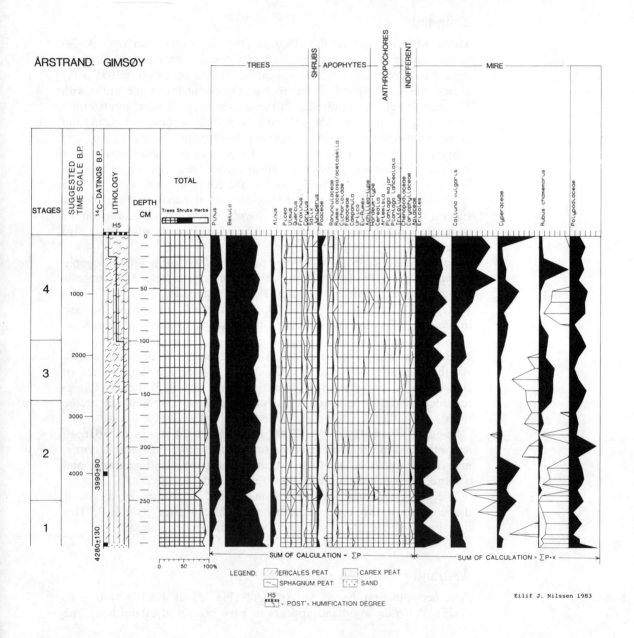

ÅRSTRAND. GIMSØY

Eilif J. Nilssen 1983

late Holocene. Large areas around the site were covered with mires as they are today, and forests were primarily growing on the hillsides to the south-east and north.

The only indication of human activity present in the diagrams is in stage 2, at 250 cm, interpolated to 4150 B.P. and some spectra above. A depression in the smooth *Betula* curve is concomitant with an increase in Poaceae and *Rumex*. Pollen of *Hordeum* and one *Plantago major* grain were recorded. In addition some pollen with sculpturing and annulus features of the *Hordeum*-group, but with a diameter within the *Triticum*-group (*sensu* Beug 1961) were recorded. These are marked as Cerealia in the diagram. This phase is interpreted as a short agricultural period, lasting no more than 100 years.

Fig. 4. Simplified pollen diagram from Rystad, Vestvågøy.

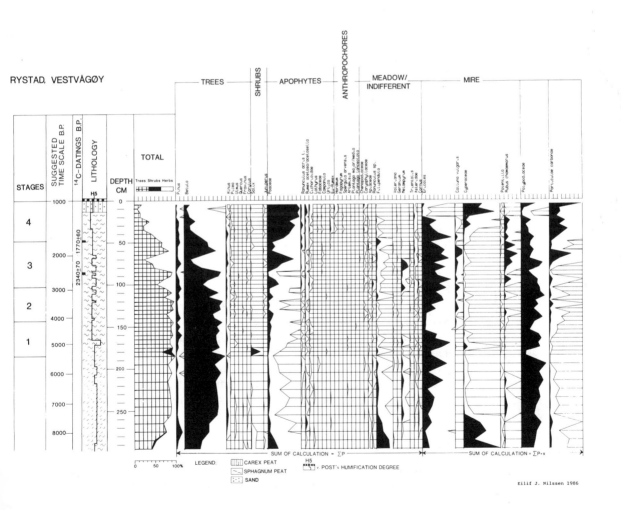

Rystad

The lowest part of the Rystad diagram (Fig. 4) shows increasing *Betula* and decreasing values of *Salix, Juniperus*, Poaceae and *Filipendula*. These reflect a succession during the establishment of the natural *Betula* woodland in this area. From 255 cm this succession ended, and the natural vegetation cover stabilized.

Stage 1. From 185 cm (*ca.* 5500 B.P.) there is a marked increase in the curves of Poaceae, *Pinus* and, later, charcoal. These changes probably reflect clearance of the *Betula* woodland to form pastures. Long-distance transported *Pinus* pollen increases as the *Betula* woodland decreases. Few other anthropogenic indicators occur at these levels except *Urtica, Artemisia* and, later in this stage, *Plantago lanceolata. Urtica dioica* is often associated with dung heaps. *Artemisia vulgaris* is a weak competitor and may be found along field margins today. *Ranunculus, Rumex* and Cichorioidae are accidental and show no marked increase.

Stage 2. From *ca.* 145 cm (*ca.* 4000 B.P.) increasing values and more frequent occurrences of Caryophyllaceae, Apiaceae and charcoal are recorded. Possible species included in these families are the field-weed *Stellaria media* and *Anthriscus sylvestris*. The latter is common in natural vegetation and increases with agricultural activity. Charcoal is a good indicator of human activity. In periods with human activity within a certain distance of the sampling site, the charcoal curve parallels curves for apophytes, especially Poaceae. *Calluna vulgaris* becomes continuous and *Rubus chamaemorus* becomes more frequent and increases. *Calluna* is favoured by intensive grazing, and *Rubus* by bog disturbance.

Stage 3. From 105-50 cm (*ca.* 2800-1800 B.P.) there are great fluctuations in the Poaceae, *Betula* and charcoal curves. The high values of Poaceae and the corresponding decline in *Betula* are interpreted as a more intense encroachment on the natural vegetation. These fluctuations may indicate that the farmers were moving from place to place, or that the farm-sites were deserted for some time and were re-established later.

Stage 4. From 50 cm (*ca.* 1800 B.P.), values for Poaceae and charcoal show a pronounced increase as *Betula* decreases to values *ca.* 10%. *Hordeum* and the field-weed *Spergula arvensis* are recorded here for the first time. This is interpreted as reflecting more intensive farming activity, where cereal harvesting was more common. Heavier utilization of the surrounding area reduced the *Betula* woodlands to about today's level. Expansion of grasslands resulted.

Justad Øvre

The lowermost part of the diagram (Fig. 5) shows well-established *Betula* woodland with slight anthropogenic influence on the vegetation from the start of the sequence *ca.* 2350 B.P. From 180 cm (1790 ± 50 B.P.), curves for most apophytes and charcoal increase. At the same time there are records of *Artemisia* and *Plantago major/media*. This is interpreted as farm establishment. Decreasing values are also observed for *Filipendula* and Polypodiaceae. These changes suggest that the establishment of the farm caused a reduction in the *Betula* woodland and associated plants.

The increased *Betula* values at 150 cm (1240 ± 70 B.P.) and from 57 cm (580 ± 70 B.P.) to 35 cm indicate ephemeral regenerations of the woodland caused by farm abandonment. At 57 cm there is a ^{14}C-date of 1335 ± 75 A.D., which supports historical information that this farm was deserted during the time of Black Death. Unfortunately, historical records do not extend far enough back to verify any abandonment phase at 730 ± 90 A.D. (1240 ± 70 B.P.).

Fig. 5. Simplified pollen diagram from Justad Øvre, Vestvågøy.

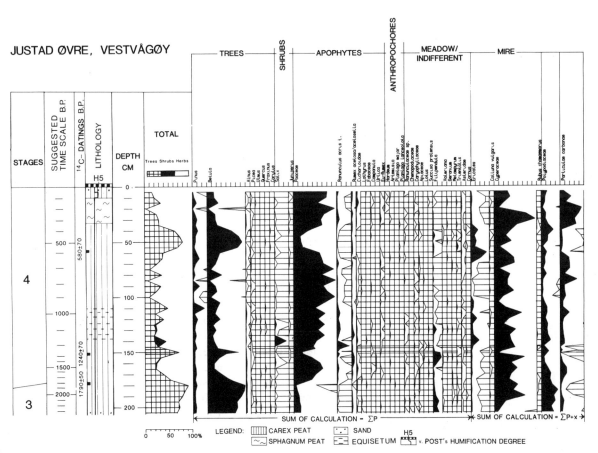

JUSTAD ØVRE, VESTVÅGØY

Eilif J. Nilssen 1986

Regional synthesis

The pollen diagrams show that after *Betula* had expanded in the area by 8900 B.P., a rich *Betula* woodland was established and lasted almost undisturbed for *ca.* 3000 years. There were then 4 stages in the develpment of the cultural landscape.

Stage 1. Increased values of Poaceae and charcoal from *ca.* 5500 B.P. in diagrams from Rystad and Tangstad are interpreted as the first forest-clearance by humans, to form pastures within the forest for domestic animals, or to obtain fire-wood and timber for house constructions in all-year settlements. This accords with pollen-analytical results of K.-D. Vorren (1979).

Stage 2. In the Årstrand diagram, *Hordeum* occurs as early as *ca.* 4150 B.P. A more uncertain record of cereals was made at Moland (Fig. 1) at this time (Nilssen 1983) and also at 3740 ± 40 B.P. (T-2223) at Bøstad (Fig. 1) (K.-D. Vorren 1979). As these records are sporadic and of short duration, the farmers may not have been successful with barley growing or they may have shifted from place to place.

Archaeological evidence for a crop-harvesting culture in northern Norway has been found in the cave Stiurhellaren in Rana (Hultgren *et al.* 1985). Charred seeds of *Hordeum* and bones of goat or sheep in waste layers dated between 4380-4170 B.P. were discovered. At Moland, Vestvågøy, Johansen (1982) found ard furrows, but the age of these are uncertain.

Stage 3. From *ca.* 2800 B.P. there are marked changes with increased values for apophytes, especially Poaceae, and a corresponding decline of *Betula* at Tangstad and Rystad. This is interpreted as the establishment of farms. However, the farms were probably not permanent, because, especially at Rystad, there are such great fluctuations in the pollen curves that several abandonment phases must have occurred during this period. According to the chronology, both the farming and the abandonment phases lasted more than one human generation. There are also some minor and shorter depressions, especially in the Poaceae curve, at Tangstad that may indicate abandonment phases. The sedimentation rate at Tangstad is much slower than at Rystad, and each pollen sample represents a longer time interval. Thus short-time events cannot be detected with certainty. A similar clearance and farming episode is also recorded in the pollen diagram from Moland (Nilssen 1983).

At Elgsnes in southern Troms, Vorren (1986) found a leached podsolic horizon within an agricultural field-layer. This is interpreted as a phase when the field was fallow for several hundred years. This event occurred in the centuries after 2050 B.P., and before *ca.* 1700 B.P. according to another diagram from Elgsnes (K.-D. Vorren personal communication).

Thus it seems that during the centuries around 2000 B.P. there was an abandonment phase of some regional extent as it is recorded in the diagrams from Rystad and Tangstad on Vestvågøy and even as a field podsol in southern Troms.

Stage 4. From *ca.* 1800 B.P. farming expansion is seen at all sites except Årstrand, and at Justad farming becomes established in an area marginal to farming today. Deforestation accelerates and pastures and ericaceous heaths develop. The extant cultural landscape is primarily a development of the events which occurred at that time.

The Black Death caused dereliction of *ca.* 80% of the farms on Vestvågøy (Nielssen 1977), but most were subsequently recolonized. The potential for regeneration of *Betula* woodland is clearly seen at Justad that was abandoned at this time. The change of farming activity and techniques with cultivated pastures for cattle and a large reduction in the number of sheep and goats which has taken place during recent times, will probably lead to a regeneration of the *Betula* woodland and a change in the cultural landscape that had developed 1700-1800 years ago and has been maintained in a more or less stable condition since that time.

Summary

(1) The aim of this study is to reconstruct the development of the cultural landscape in the central part of Lofoten, North Norway.

(2) It is suggested that the first vegetational changes began at sites most suitable for farming and that changes spread in time and space, as increasingly marginal areas for farming were required.

(3) Standard pollen analysis was carried out on peat collected near farms with different agricultural potential.

(4) Development of the cultural landscape occurred in four stages: stage 1, *ca.* 5500 B.P., there was a minor reduction of *Betula* woodland due to human activity at sites suitable for farming; stage 2, *ca.* 4150 B.P., there was a sporadic attempt at cereal growing at a site with no previous farming history; stage 3, *ca.* 2800 B.P., farming activity intensified with pasture development; stage 4, *ca.* 1800 B.P. intensive agricultural activity resulting in the creation of the present cultural landscape.

(5) Abandonment phases are also recorded. An abandonment phase at *ca.* 2000 B.P. may be of regional significance. The present cultural landscape has developed since *ca.* 1800 B.P. It will soon change with the introduction of modern farming methods.

Acknowledgements

I thank T. Nilssen for help with the fieldwork, K.-D. Vorren for discussions and critical reading of the manuscript and A. M. Odasz for correcting the English text. The investigations were supported by the Norwegian Research Council for Science and the Humanities (NAVF).

Nomenclature

Plant nomenclature follows J. Lid (1985) *Norsk-svensk-finsk Flora*, Det norske Samlaget, Oslo.

Agriculture in a Marginal Area - South Greenland from the Norse Landnam (985 A.D.) to the Present (1985 A.D.)

Bent Fredskild

Introduction

When Red Eric returned to Iceland around 985 A.D. after his 3 years as an outlaw he named the land where he had stayed Greenland, because, as is told in the saga, "he argued that men would be all the more drawn to go there if the land had an attractive name" (Jones 1986). He persuaded a number of people to follow him back to Greenland, and 25 ships with man and beast sailed westwards. Some ships perished, more were forced back, but 14 arrived safely. At that time, 3-4 generations after the Icelandic landnam, it was only the very poorest parts of Iceland that had not been settled, and moreover this was only 10 years after a famine year in all of north-western Europe, so maybe it was not difficult to persuade people. What then did they meet?

Norse landnam 985 A.D.

Judging from palaeoclimatological evidence from the Greenland Ice Cap cores the climate at that time was similar to, or slightly warmer than that of the mid-20th century. It is therefore concluded that the vegetation resembled what can be found today in areas far from grazing sheep and woodcutting people.

The outer coast areas do not look too hospitable. Dwarf-shrub heaths, dominated by *Empetrum hermaphroditum*, rich in mosses and lichens cover most of the gneissic peninsulas and islands. Today the climate is cool, with a mean summer (June-August) temperature of 5-6⁰C, and often during summer a cold fog, caused by the 20-30 km broad drift-ice belt coming from the Polar Sea with the East Greenland current around Kap Farvel, prevents the sun from reaching the ground. However, during the first part of the Norse era this drift-ice did not

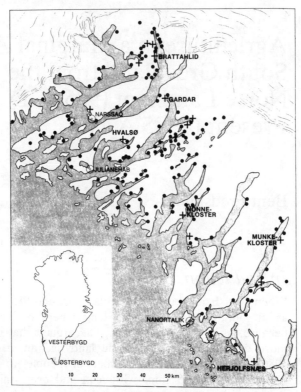

Fig. 1. Map of the central part of the Eastern Settlement (Østerbygd). Only a few farms were situated outside the map. Dot = Norse farm; cross = church; Nonnekloster = nunnery; Munkekloster = monastery; ring = present-day town (after Meldgaard 1975).

reach south Greenland. When sailing in through the fjords the vegetation is seen to gradually change and around the head, subarctic types take over. Most conspicuous is the open, mostly 3-4 m tall forest of *Betula pubescens* and *Salix glauca*, which in favoured sites like the Qingua valley at Tasermiut includes birches up to 10 m tall. However, most of the lowlands are covered by heaths with *Betula glandulosa* and *Salix glauca* dominating, interspersed with low willow copses and open grassland communities. Fens and marshes are numerous around lakes and ponds and along streams in the lowland. On south-facing slopes, supplied by water throughout the summer, luxuriant herb-slopes can be seen. The summer in the interior is much warmer (*ca.* 10⁰C) and the precipitation is lower than at the outer coast.

No wonder that the first farmers settled here (Fig. 1). The ruins are concentrated around Brattahlid - Red Eric's farm - and in Vatnahverfi south of Gardar. Some time later the landnam expanded to the Godthåbsfjord area 300-400 km to the north-west. There in the Western Settlement, the summer is as warm as in the Eastern Settlement, in fact the highest Greenland July average temperature (10.9⁰C) is measured in the interior, but the growing season is shorter, only 5

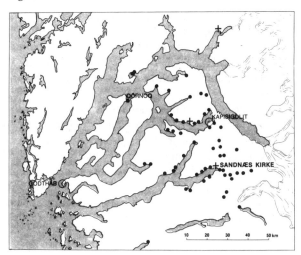

Fig. 2. Map of the Western Settlement (Vesterbygd) (after Meldgaard 1975). Signatures: see Fig. 1.

months against 7 with temperatures above 0°C and there is no forest. Here, Norse ruins are only found in the interior (Fig. 2).

The landnam is very marked in pollen diagrams from the Eastern Settlement, the most prominent feature being the introduction of *Rumex acetosella* and, less common, the subarctic *R. acetosa* ssp. *lapponica*. Both are included in the same pollen curve in the diagram from a shallow lake a few km from Brattahlid (Fig. 3). The *Salix* curve decreases as does that of *Betula*, mainly caused by a severe reduction in the amount of tree *Betula*. Approximately 0.5 km from the Norse church in Brattahlid is a former pond, the water of which was used for irrigation in the Norse period and again half a century ago. In the diagram from here (Fig. 4) decreases in *Betula*, again mainly tree *Betula*, *Salix* and the fern *Gymnocarpium dryopteris*, which grows in mossy scrub, are evident, and the emergence of *Rumex* and *Achillea millefolium* leap to the eye. But just as important is the re-occurrence or re-flourishing of a number of herbs from the pioneer vegetation, which invaded the land after deglaciation 7-8 millennia earlier, such as *Sagina, Silene acaulis, Plantago maritima, Rhodiola rosea* and/or *Sedum, Thalictrum alpinum, Thymus praecox* and one or more species of *Saxifraga*.

The explanation seems obvious: clearing of the copses by fire or axes, the trampling of people and animals, grazing and peeling-off of the vegetation mat for the construction of the turf-and-stone-buildings gave these open-ground pioneer plants another chance which they grasped immediately. The uncovering of the minerogenic soil is also reflected in the content of sand washed or blown into the basin (Fig. 6). Especially in the Western Settlement burning of the vegetation in connection with the landnam is recorded as a charcoal layer under the ruins (Iversen 1934) and in a peat profile below natural vegetation (Fredskild unpublished) as well as by an explosive increase in the amount of microscopic charcoal in lake sediments (Fredskild 1985).

Their fields were rich in introduced weeds and apophytes - native plants favoured by cultivation - as seen, for example, in a profile from a moist depression between the farms at Brattahlid (Fig. 5). Most conspicuous are the numerous seeds of introduced annual weeds: *Poa annua, Stellaria media, Capsella bursa-pastoris* and *Polygonum aviculare*. The native *Montia fontana*, growing on damp ground in fens and along brooks, is extremely favoured by human activity, growing today in every Greenland town and settlement except the northernmost ones. Pollen and seeds of flax (*Linum usitatissimum*) have been found in the middens of two Western Settlement farms together with pollen of *Spergula arvensis* (Sørensen 1982; Fredskild unpublished). Flax is not grown in Greenland today, and only three finds of introduced *Spergula*, two from gardens and one from a chicken yard, are known in modern time, which may indicate the growing of both species in Norse time.

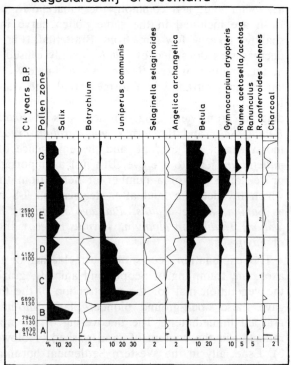

Fig. 3. Selected pollen curves from Comarum Sø near Brattahlid, Eastern Settlement (after Fredskild 1973). The landnam occurs at the pollen zone border F-G.

But what else do we know about their agriculture? They could not grow cereals. Maybe they tried - one single pollen of *Hordeum* has been found at Brattahlid (Fredskild 1978) - but the sagas tell that many people there in the 13th century had never seen bread. A few small querns have been found but they may have been used in grinding imported cereals or maybe *Elymus*. The low-arctic species *Elymus mollis* is widespread along the coast northwards to 72^0, whereas the north European, boreal *E. arenarius* and a possible hybrid with intermediate chromosome number is only found at a couple of Norse ruins in the Eastern Settlement (Fredskild 1973). In Iceland *Elymus* was much utilized for making flour.

The economy was based on sheep and cattle. In addition they had some pigs and goats, and also some horses, mainly as a means of transportation. They had fenced homefields, conscientiously manured, for haymaking. They drained waterlogged areas and, conversely, dammed streams and created a system of irrigation channels as seen, e.g., at Gardar (Krog 1982). Some small upland buildings are interpreted as being saeter cottages. However, as evidenced by the numerous bones of seal, fish and caribou in the middens right from the beginning, they would hardly have survived without this supplement to the daily fare.

GALIUM KÆR B   (61°10'N,45°31'W)
Qagssiarssuk, S.Greenland

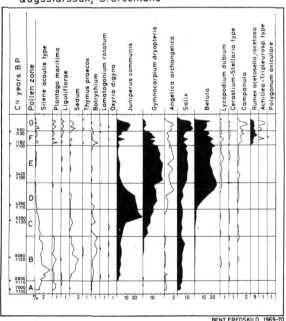

BENT FREDSKILD 1969-70

Fig. 4. Selected pollen curves from Galium Kær (after Fredskild 1973). The landnam occurs at the pollen zone border E-F.

QAGSSIARSSUK Section QDB
MACROFOSSIL Diagram

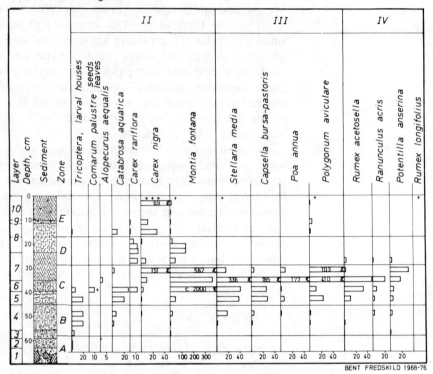

Fig. 5. Some of the macrofossils in a series of samples from a moist depression at Brattahlid. II includes plants (plus caddis fly larval houses), indicating moist conditions, III includes annual weeds, and IV perennial weeds (after Fredskild 1978).

We know that the Western Settlement was abandoned *ca.* 1350 and the Eastern Settlement presumably late in 15th century, but what happened during these centuries, and why did the Norsemen disappear from the scene?

The explanations are manifold, ranging from some that can be rejected as purely speculative or of minor, local effect only, to some that may be seriously considered. Among the first may be mentioned: inbreeding, the Black Death, plundering and murdering by pirates, attacks by larvae of the butterfly *Agrotis occulta*, eating everything green in the pastures and hayfields, battles with, or the contrary, gradual mixing with the Eskimoes, who expanded their hunting grounds to south Greenland during the later part of the Norse era, or the gradual recession back to Iceland or Scandinavia.

The more important factors causing the extinction of the Norsemen are the following that are closely connected in a very complex manner: climatic change, overgrazing, and changes in the structure of the community. As to the latter, I have only a few comments. Obviously the first immigrants took the best land at the head of the fjord, leaving only marginal sites closer to the outer coast or in unfavourable areas.

In the beginning of the 12th century Greenland got a bishop, tithes had to be paid, and in 1261 Greenland became a part of the Norwegian Kingdom, which meant further tax-paying. At the same time a concentration of fortune, including cattle, was seen. Thus, at Gardar, the bishop's farm, two byres with room for no less than *ca.* 100 cows were built. This mirrors a change in the social structure of the community. If we suppose the same development as in Early Medieval Iceland then the number of freeborn farmers and tenants decreased, both groups having some influence on political decisions, while 'lower cast' people increased in number. During the later part of the era, trade with Europe decreased and finally stopped.

Overgrazing is, and beyond doubt was, one of the most severe threats to agriculture in Greenland. Apart from the peaty soil in fens and swamps most soils are coarse textured with only a low clay and humus content and therefore with a low water-carrying capacity. Periodically the upper soil is very dry because of the frequent foehn-winds, and the vegetation cover is easily broken. Once this happens, an often irreversible wind- or water-erosion starts. As seen in the Galium Kær diagram (Fig. 6) the amount of sand increased markedly immediately after the landnam. Because of a contemporary change in sedimentation rate, the increase of annually deposited sand was even more drastic than indicated by the curve. Further proof is seen under a hayfield at Sdr. Igaliko in the Eastern settlement. Under 0.5 m of

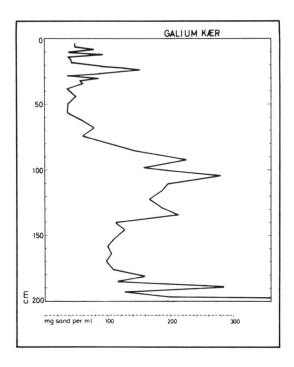

Fig. 6. Content of sand in the sediments of Galium Kær at Brattahlid. The decrease at 100 cm is connected with the spreading of *Betula glandulosa* dwarf-shrub heath to the surrounding slopes, whereas the increase at *ca.* 35 cm is contemporaneous with the Norse landnam (after Fredskild 1973).

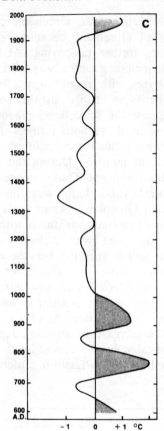

Fig. 7. Fluctuations in the mean temperature in the past 1400 years in Greenland, determined by isotope measurements on the ice core from Crête on the centre of the Greenland Inland Ice. Fluctuations shorter than 120 years are smoothed out (after Dansgaard 1975).

sandy loess an 'old' zonal soil profile is seen with a dark humic stripe with charcoal on top, resting on a podzol (Jacobsen & Jakobsen 1986). Pollen analysis of this stripe showed that it was formed under an open willow copse, rich in ferns and grasses. A ^{14}C-dating gave an age at the beginning of the Norse era, indicating that wind deposition of any importance - apart from late-glacial time - did not occur until the introduction of agriculture. Similar soil profiles are common in the same area (Jakobsen 1986), where a Norse farm was also covered by drifting sand, even during the habitation period.

Another effect of grazing is the change in species composition in the pastures. In some grassy vegetation types, controlled grazing may result in increasing productivity by favouring grasses, e.g. *Poa pratensis* propagating by subterranean runners. In other communities, like the highly productive herb-slopes, species composition rapidly changes if they are grazed early or even in the middle of the growing season, when translocation is still from underground reservoirs (roots, rhizomes, etc.) to the above-ground parts of the plants. Likewise, dwarf-shrubs, most important for grazing sheep during winter, are vulnerable, and therefore those species are favoured that are not utilized by animals or at least only to a smaller extent. Unfortunately, the more important of these, e.g. *Nardus stricta, Kobresia myosuroides*, cannot be traced by pollen analysis, as their pollen cannot be separated from other grasses and sedges, respectively.

As happened in Iceland, Greenland was undoubtedly filled up with farms to the ultimate limit after a few generations. If the Norsemen had been so clever or lucky that they had found an ecological equilibrium as regards carrying capacity - this would have been first and so far the only example in the history of mankind - what would happen to this marginal area if a climatic deterioration set in? In other words, if the period in which the cattle were depending on winter fodder was extended by *x* weeks, then the vegetation on the hayfields, yielding the fodder just sufficient for a 'normal' winter, would have to produce more feed units per ha in a *x* week shorter and besides colder summer season, which is of course impossible. In fact, the half millennium of the Norse era saw some severe deteriorations as shown by isotope analysis of ice cores (Fig. 7). It seems as if the Viking Age expansion westwards took place during a warm period. During the first centuries only a slight and gradual decrease in temperature is registered, but then, around 1300 the Little Ice Age started, the temperature reaching a first minimum contemporary with the extinction of the Western Settlement. After a short-lived amelioration the subsequent drop in temperature was contemporary with the close of the Norse period in Greenland.

Present-day agriculture

The present-day agriculture started around the turn of the century as a hobby, based on a stock of Faroese sheep. As it turned out to be a success, the Sheep Breeding Experimental Station at Uperniviarssuk was established in 1915, and Icelandic sheep were introduced. In 1924 the first full-time breeder started at Brattahlid, today's Qagssiarssuk, close to Red Eric's farm, and more breeders followed. Today 60-70 families have sheep breeding as their chief occupation. However, for most of them, fishing and hunting occupy a not unessential part of their time. Cattle breeding has been tried without success, whereas the introduction of some yak-oxen was a great success - they were too tasty! Initially, sheep breeding was rather extensive, and the sheep had mainly to serve themselves all year round. However, some catastrophic winters, e.g. 1966-1967, when the number of sheep was reduced from 48 000 to 20 000 in spite of imported fodder spread by boat and even by helicopter, made an alteration of the management essential. In 1975 cooperative work with the Agricultural Research Institute in Iceland was initiated. As a result of joint field-work with the Greenland Experimental Station during 1977-1981, a report was published (Thorsteinsson 1983) which formed the basis of a rather ambitious "Detail Plan" in 1982 operating with the establishment of a number of new farms and an expansion of the infrastructure, including the building of roads, small harbours, slaughterhouses, etc.

This 10-year plan was to have been financed by the EEC and Denmark, but after Greenland's resignation from membership of the EEC, the Greenland Home Rule Authorities in 1985 passed a revised plan according to which only a few farms should be built and the better ones of the already existing farms should be consolidated, whereas the smaller, uneconomic farms should be closed down.

As the crucial point in Greenland sheep-breeding is winter fodder, the hayfield area is being extended, partly by cultivating, but with as little mechanical soil preparation as possible, partly by fertilizing natural plant communities. The grasses cultivated for hay are mainly foreign plants: barley, oats, rye and timothy (*Phleum pratense*), partly local species like *Poa pratensis, Festuca rubra* and *Agrostis* spp., some of which are now being improved by breeding. Unfortunately, the harvest of timothy is often reduced by up to 50% as a result of the activity of 2 native species of mites, being in clover in such monocultures (P. Nielsen 1984). And weeds - the same species as in the Norse period - are still a serious problem, and reduce the yield.

Exploitation of upland pastures is now being attempted by a kind of mountain pasturing that allows the lowland pastures close to the farms to recover fully during the summer. In the autumn the sheep are collected, the animals to be slaughtered are shipped away and the remainder kept close to the farms in big folds with sheepcotes, being regularly foddered throughout the winter. If sufficient locally produced winter fodder is at hand, sheep breeding may pay one day - at present the average yearly support to each family is 100 000 Dkr.

In 1983 the Greenland Home Rule Authorities and the Danish Ministry for Greenland appointed a working group of three scientists - a botanist, a geographer and a zoologist - with an administrative chairman, to follow the impact of sheep breeding on the environment. One way to do this is by carefully describing the vegetation and soil in fenced and unfenced reference areas, of which several have already been laid out, and more are planned.

As in the Norse period, soil erosion is a major threat. Degradation of soil and vegetation is now clearly seen around the first farms, and in the western part of Vatnahverfi, an erosion border, many km long, locally in the form of a 3-4 m high wall, divides the landscape into two contrasting parts (Fig. 8). On one side of the border there is still luxuriant vegetation, whereas on the other side bright, polished rocks, not even with epilithic lichens, alternate with sand dunes and stony abrasion flats. The question is how to stop the erosion which has already started and prevent it happening in the future. The Research Station and the farmers are fully aware of the problem, at the same

Fig. 8. Erosion front south of Sdr. Igaliko. Photo: N. Kingo Jacobsen.

time being under political pressure to expand sheep breeding, thereby creating more work in a community with a very high unemployment rate. No wonder that the recommendations to the Greenland Government from the working group do not always harmonize with those of the sheep breeders' organization.

Conclusion

In a lecture to the American Anthropological Association Meetings in Washington D.C., McGovern raised the question: "If Norse land-use practices were indeed destructive of such vital economic resources, why did Norse farmers not perceive and correct the problem", or, in other words "why did environmental feedbacks not trigger appropriate management response?" (McGovern *et al.* 1985). In discussing this, they mention 6 factors causing what they call "information flow pathologies", which are, of course, as relevant today as they were in connection with the Norse period:

1. False Analogy - managers' cognitive model of ecosystem characteristics (potential productivity, resilience, stress signals) may be based on the characteristics of another ecosystem whose surface similarities mask critical-threshold differences from the actual local ecosystem.
2. Insufficient Detail - managers' cognitive model is overgeneralized, fails to sufficiently allow for the actual range of spatial variability that underlies surface similarity of an ecosystem whose patchiness is better measured in resilience than initial abundance.

3. Short Observational Series - managers lack a long enough memory of events to track or predict variability in key environmental factors over a long period and are subject to chronic inability to separate short-term and long-term processes.
4. Managerial Detachment - managers are socially and spatially distant from agricultural producers who both carry out managerial decisions at the lowest level, and who are normally in closest contact with local-scale environmental feedbacks.
5. Reactions Out of Phase - partly as a result of the last 2 factors, managers' attempts to avert unfavourable impacts are too little and too late, or apply the wrong remedy.
6. S.E.P. - (Someone Else's Problem), managers at many levels may perceive a potential environmental problem, but do not feel obligated to take action as their own particular short-term interests are not immediately threatened.

The first three problems are most relevant in a pioneer situation, whether in 11th or 20th century - there is (was) still a chance that decision-makers at all levels may learn by more experience and longer observation periods. On the other hand, the final one unfortunately has a universal character. It is to be hoped, however, that the modern Greenland community has the right medicine to cure these pathologies - if not, sheep breeding is heading for ruin, and on top of the social and economic tragedies, part of the south Greenland landscape will be devastated.

Summary

(1) Agriculture, based on sheep and cattle, was introduced to south Greenland by the Norse landnam *ca.* 985 A.D.

(2) Climatic deterioration, overgrazing and possibly changes in social structure first caused the fall of the Western Settlement in the interior Godthåbsfjord region *ca.* 1300 A.D., and less than 2 centuries later, of the Eastern settlement further south in the Julianehåb-Narssaq region.

(3) Sheep were reintroduced at the beginning of this century, and today about 70 families have sheep breeding as their chief occupation. Cattle breeding was abandoned after a brief attempt.

(4) The threats to sheep breeding and the environment are the same as during the Norse era; overgrazing with subsequent soil erosion and catastrophic winters with too thick a snow-cover preventing grazing. They are met by a combination of mountain pasturing in summer and stall-feeding in winter. Hay is harvested either from fertilized natural grassland, or fields sown with timothy, barley, oats, rye or other grasses.

Nomenclature

Plant nomenclature follows T. W. Böcher, B. Fredskild, K. Holmen & K. Jakobsen (1978) *Grønlands Flora*, P. Haase & Søn, Copenhagen.

Changes in Agricultural Practices in the Holocene Indicated in a Pollen Diagram from a Small Hollow in Denmark

Svend Th. Andersen

Introduction

Pollen diagrams from lakes or bogs integrate pollen produced from a large area; discrete communities cannot be distinguished, and the effects of various human activities cannot be separated or localized. Due to the special aerodynamic conditions in forests (Raynor *et al.* 1975), arboreal pollen (AP) deposited on the forest floor is derived mainly from trees within a short distance (30 m), as shown by pollen analyses within present-day woodlands (Andersen 1970; Bradshaw 1981a). Pollen diagrams from very small ponds and bogs or soil sites that were shaded by the tree canopy, therefore make it possible to distinguish single tree communities in a time sequence (Andersen 1973, 1978, 1984; Baker *et al.* 1978; Bradshaw 1981b; Birks 1982). If the trees around such a site were felled, this impact would manifest itself clearly in the pollen diagram. However, we have little experience yet as to how various human activities are reflected in pollen spectra from such sites.

The site investigated

Næsbyholm Storskov on Zealand (Fig. 1) is unusually rich in prehistoric monuments: a few megalithic graves and many Bronze Age barrows occur, and the whole forest area is covered by field systems (V. Nielsen 1984). Hence, we know that people were active in the area during Neolithic and Bronze Age-Iron Age times. The area contains numerous undisturbed small hollows. One was selected and called *Glyceria*-hollow (not to be confused with the *Glyceria*-hollow from the Løvenholm forest in Andersen 1984). This hollow is 18 x 12 m in size (Fig. 2), and is situated in an area of many small hills with Bronze Age barrows and prehistoric field terraces delimited by lynchets (Fig. 3). A flat area west of the hollow adjoins it by a steep slope 1

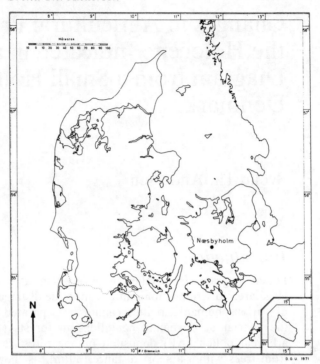

Fig. 1. Map of Denmark with the position of Næsbyholm Storskov.

m high, which proved to be a lynchet. A colluvial layer extends from this slope across the hollow, the amount of stones decreasing with distance (Fig. 4). The flat area was used as a field in prehistoric time, and tilled soil flowed into the hollow.

Pollen analyses

Samples were collected from an excavation through the colluvial layer and from a boring. The sediment above and beneath the colluvial layer consists of strongly decomposed wood and leaves that were deposited in the small pond. Below is late-glacial clay (Fig. 4).

Pollen preservation was moderately good. Percentages of AP are calculated as percentages of the AP-pollen sum (Andersen 1970, 1978), as shown in the pollen diagram (Fig. 5). A summary diagram of non-tree pollen types is presented in Figure 6. Their percentages are based on the transformed AP-sum and terrestrial non-tree pollen (with wind-pollinated non-trees divided by 2). These percentages are generally similar to the untransformed percentages, but artificial changes due to changes in total AP output are minimized. Pollen percentages of local plants (*Salix*, wet ground taxa, aquatics) are not assumed to indicate areal percentages but rather to indicate whether these plants were common or scarce in or near the hollow. Curves for pollen concentration, clay and sand content and charcoal dust are also presented in Figures 5 and 6.

Radiocarbon dating

Two dates were obtained from pieces of wood after removal of soluble humus with alkali (K-2579, 33-35 cm and K-2580, 60-62 cm). The organic sediment was not dated because of a large content of soluble humus, except for one sample just below the colluvial layer (K-3952, 50-55 cm), where infiltration by younger humus is unlikely. The dates are shown on Figures 5 and 6 (in calibrated calendar years, Clark 1975). Other ages were inferred by comparison with dated levels from other sites.

Development of vegetation and landscape

The AP curve (Fig. 5) shows that trees were scarce around the *Glyceria*-hollow in the Younger Dryas and again in late Subboreal-early Subatlantic time. The lowermost treeless period corresponds to the basal clay. The uppermost minimum for trees is from the colluvial layer and reflects cultivation of the field next to the site. At other times, the area was covered by forest, still with a rather open tree cover in Preboreal and Boreal time, in the early Subboreal and in a period just after the field-cultivation stage. *Salix*, wet-ground plants and aquatics were common at times when the tree cover above the hollow was scarce.

Fig. 2. The *Glyceria*-hollow in Næsbyholm Storskov, seen from the south. The distance between the vertical poles is 10 m. (September 1986).

There are maxima for clay and sand and charcoal in the Late-Glacial and Preboreal, the early Subboreal and late Subboreal-early Subatlantic time (Fig. 6).

The vegetation development

The vegetation development from Younger Dryas to the *Ulmus*-decline at the end of the Atlantic reflects the invasion of late-glacial park-tundra by trees (*Betula*, *Pinus*) and the establishment then of *Corylus* and later *Tilia-Corylus*-dominated forest. *Quercus* expanded in middle Atlantic time but remained a subordinate member of the forest. This development reflects a natural succession of tree communities on high and well-drained ground with increasing shade reflected by decreasing percentages for *Betula*, *Pinus* and non-tree pollen. Percentages for moist and wet-ground trees such as *Ulmus* and *Alnus* are much lower than in regional pollen diagrams; their pollen was probably transported from other sites. Hence, the *Ulmus*-decline is likely to reflect the *Ulmus*-decline common in all regional pollen diagrams from Denmark at *ca.* 4000 B.C. (Andersen *et al.* 1983).

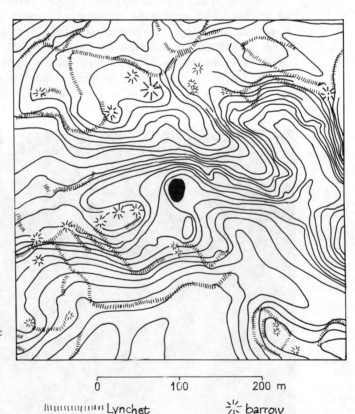

Fig. 3. Map with surface contours (1 m) and prehistoric barrows and lynchets around the *Glyceria*-hollow (black). Redrawn after original map by V. and G. Nielsen (from Andersen 1985, see also V. Nielsen 1984).

0   100   200 m

ıllıllıllıllıllıllı Lynchet   ☀ barrow

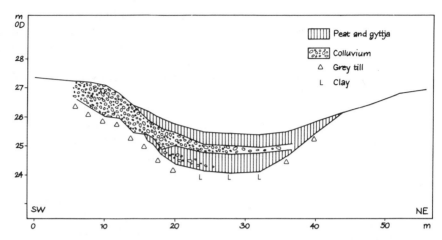

Fig. 4. Cross section of the *Glyceria*-hollow (from Andersen 1985).

The decrease of clay and sand washed into the hollow from the local soils reflects the increased denseness of the vegetation. The high frequency of charcoal dust in the Preboreal was probably connected with the predominance of *Pinus*, as natural fires caused by lightning are frequent in pine forests (Uggla 1958). Charcoal became scarce as deciduous trees (*Corylus*, *Tilia*) became dominant. Hence, it can be assumed that the clay, sand and charcoal-dust maxima recorded later were due to human disturbance.

Tilia forest. Shredding stage. Subboreal

Ulmus declines permanently at 128 cm depth. At the same time *Tilia* increased and all other plants except for wet-ground herbs virtually disappeared. High clay and sand content and high frequencies of charcoal dust indicate local disturbance. The only possible explanation seems to be that humans cleared all the vegetation except the *Tilia* trees and burned the litter. *Tilia* continued to flower, as shown by only a small decrease in its pollen concentration (Fig. 6). *Plantago lanceolata* pollen occurs continuously from this level suggesting that the *Tilia*-maximum coincides with the oldest Neolithic, when agriculture and cattle breeding were introduced in Denmark, but nothing indicates arable or pastoral activity near the site. It would appear that the *Tilia* around the site was favoured for harvesting of leaf-fodder for cattle. Evidence for Early Neolithic harvesting of leaf-fodder has also been found in Switzerland (Guyan 1981; Troels-Smith 1981).

Harvesting of leaf-fodder from trees does not necessarily prevent the trees from flowering. Rackham (1976) described two methods of harvesting twigs from trees: pollarding and shredding (Fig. 7). With pollarding, the tree trunk is cut at a height of 2-3 m and all the new twigs are harvested regularly. This prevents flowering if the pollarding is sufficiently frequent. With shredding, by contrast, the whole trunk and tree top are preserved, and only adventitious twigs along the trunk are cut off. By this method, accordingly, the tree top permanently delivers organic nutrients to the root system, and flowering continues and may be stimulated by increased light. For shredding, it is necessary to climb the tree, but this difficulty can be helped by preserving the bases of major side branches (see Fig. 7).

Pollarding has been common in Scandinavia up to the near present (Austad this volume), whereas shredding has been used widely in central and southern Europe and eastern Asia. In Denmark, nearly all trees depicted in Medieval church fresco-paintings have been shredded. In many cases, *Tilia* is the obvious species (Fig. 8), and in one case (Fig. 9) the process itself is depicted. Shredding must have been widespread in Denmark in former times. It seems to be a bold step to interpret the palynological changes in the *Glyceria*-hollow as Early Neolithic shredding of *Tilia*. However, at present, there is no other interpretation that fits the data so well. It would be desirable

Fig. 5. Pollen diagram from the *Glyceria*-hollow, with a curve for pollen concentration (transformed numbers of land plant pollen (P) mg⁻¹ organic matter) (from Andersen 1985).

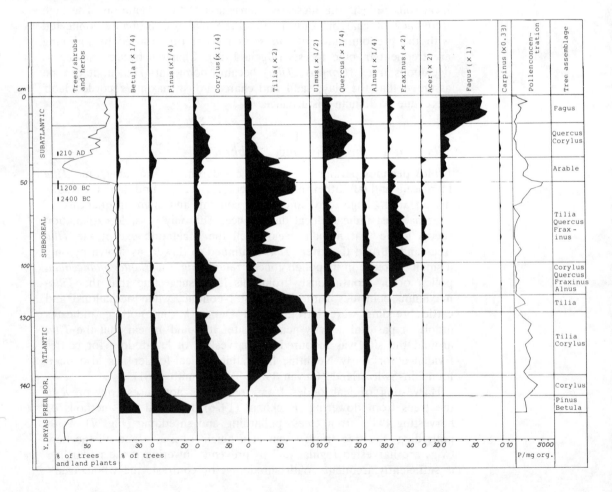

to look for similar events at other sites which reflect local processes in a similar way. Until more evidence is found, it may be too early to say how the use of shredding may have affected regional pollen diagrams. Judging from the thickness of the sediment, the *Tilia* maximum at *Glyceria*-hollow probably lasted a few hundred years.

Corylus-Quercus-Alnus-Fraxinus forest. Pastoral stage. Subboreal

The *Tilia* forest was succeeded abruptly by a forest of *Corylus*, *Quercus*, *Alnus* and *Fraxinus*, whereas *Tilia* was strongly suppressed. This change was accompanied by maxima for open-ground herbs (mainly Poaceae, *Plantago lanceolata*, *Rumex acetosella*, *Artemisia*, Chenopodiaceae, *Hypericum*) and humus plants, particularly *Pteridium*. *Crataegus* occurred sparsely. The amount of charcoal dust decreased steadily, whereas clay and sand continued to be deposited for a while and then abruptly decreased. Open forest with light-demanding trees

Fig. 6. *Glyceria*-hollow. Curves for clay and sand content, charcoal particles (larger than 0.01 mm) per numbers of land plant pollen, and ecological groups of non-tree pollen.
A = *Avena*;
H = *Hordeum*-type.
The white silhouette for open-ground herbs indicates wild grasses. The black silhouette for uncertain herbs indicates Compositae sect. Cichorioideae (from Andersen 1985).

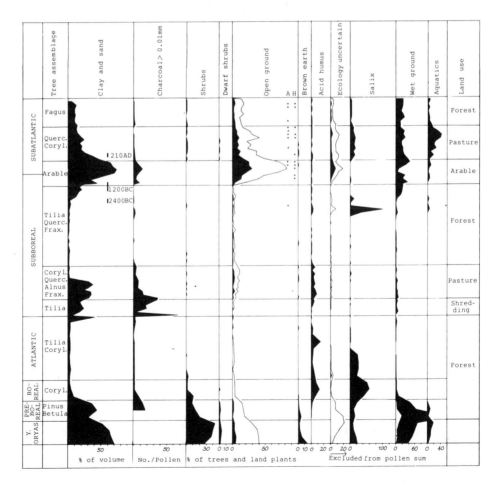

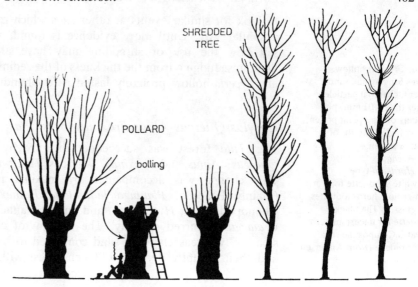

SHREDDED
TREE

POLLARD

bolling

Fig. 7. Pollarded and shredded
trees (from Rackham 1976,
reproduced by permission
from O. Rackham).

and grassy glades with *Pteridium* are indicated. *Alnus* and *Fraxinus* probably grew on moist flushed soil at the foot of the steep slope north-east of the hollow.

Human influence is unmistakable. Shredding of *Tilia* was discontinued, and *Tilia* was suppressed to remove its shade. The open forest was maintained, probably with the help of fire, for browsing and grazing by cattle or sheep. The weed *Pteridium* was difficult to suppress. This pastoral stage probably lasted 500 years, judging from the sediment thickness.

Two very distinctive methods of agriculture thus succeeded each other at the *Glyceria*-hollow: a stage with shredding of *Tilia* was followed abruptly by a pastoral stage. It cannot be decided, however, whether this change was due to a change of land-use by the same population or to immigration of new people.

Tilia-Quercus-Fraxinus forest. Subboreal

The pastoral stage was followed by *ca.* 2000 years with very high AP frequencies and low amounts of clay and sand and charcoal dust, indicating that humans had abandoned the area. Between the *Ulmus*-decline and 2400 B.C. the pollen concentration was low, indicating rather low decomposition of organic matter in the hollow. Between 2400 B.C. and 1200 B.C. the pollen concentration is much higher, indicating a much higher decomposition of organic matter, perhaps due to some drying out of the hollow. As a result, the proportion of residual clay and sand increases. This does not indicate increased inwash to the site.

The regenerating forest suppressed *Corylus* and dry-land herbs. The distribution and abundance of the different trees in this Subboreal forest were controlled by natural ecological factors unaffected by people. *Tilia* would probably have favoured drier soils, whereas *Quercus*, *Alnus* and *Fraxinus* may have favoured the increased moisture in flushed soils on the slopes near the hollow.

Fig. 8. Shredded *Tilia*. Scene from fresco-painting of *ca.* 1475 A.D. in Elmelunde Church, Denmark. The National Museum of Denmark.

Fig. 9. Shredding of *Tilia*. Scene from fresco-painting of *ca*. 1475 A.D. in Elmelund Church, Denmark. The National Museum of Denmark.

Arable cultivation. Subboreal and Subatlantic. 1000 B.C.-200 A.D.

Tree pollen decreases dramatically over 10 cm to a minimum of 16%. It was replaced by pollen from open-ground plants. Clay and sand and charcoal dust also increase at this level, which corresponds to the lower border of the colluvial layer. The field next to the site was now cleared and tilled. The field must have been made all at once, but the gradual changes of the curves over 10 cm suggest that the lower-most part of the colluvial layer was mixed with the topmost part of the peat. Hence, the organic matter dated in the ^{14}C-sample at 50-55 cm (1200 B.C.) is slightly older than the clearance of the field, which accordingly must have occurred at *ca*. 1000 B.C.

Trees, particularly *Tilia*, remained in uncultivated places, probably on the northern side of the hollow. *Fagus* is also represented. Its pollen increase may be a result of a regional increase, as documented at Holmegaard Bog, Zealand (Aaby in Andersen *et al*. 1983, this volume).

The peak of the herb pollen curve depicts conditions during the cultivation of the field. Poaceae, *Plantago lanceolata*, *Rumex acetosella* and *Artemisia* are the most frequent dry-land herbs and numerous other perennial weeds are represented, whereas annual weeds are scarce (mostly Chenopodiaceae, *Polygonum aviculare*). The pollen of insect-pollinated weeds is probably poorly dispersed, as is cereal pollen (*Hordeum*-type, *Avena*); the pollen analyses therefore may give a somewhat distorted picture of conditions in the field.

Wet-ground plants (*Carex*, *Glyceria*) and aquatics (*Ranunculus*, *Hottonia*, *Lemna*) were frequent in and around the hollow during the arable stage.

Cultivation of the field must have stopped at a time which corresponds to the level where AP increases and the clay and sand content of the sediment decreases. This level is dated to just before 225 A.D. The field was thus used for *ca.* 1000 years.

Quercus-Corylus forest. Pastoral stage. Subatlantic, after 200 A.D.

Corylus and *Quercus* became common and *Tilia* became scarce. *Fagus* increased slightly, but remained uncommon. The field was probably colonized by shrubs, indicated by finds of *Crataegus, Rhamnus catharticus, Viburnum* and *Sorbus*, and by *Corylus* and *Quercus*. *Tilia* and *Fagus* were apparently suppressed, and open patches with shrubs, and open-ground herbs (mainly Poaceae, *Plantago lanceolata*, *Artemisia*, Chenopodiaceae) persisted. This suggests that the forest was open and was used for browsing and grazing by cattle and sheep. Pig husbandry may also have been important. Although the land-use changed around the hollow, it is possible that other fields in the area were still cultivated.

Wet-ground plants and aquatics were common in and around the hollow indicating a sparse tree cover.

Fagus forest. Subatlantic

The pastoral phase at the *Glyceria*-hollow was interrupted by an increase in AP frequencies to *ca.* 90%. *Fagus* replaced *Corylus* and *Quercus*. This indicates that grazing was reduced, allowing *Fagus* to become established. The date is uncertain. At Holmegaard Bog, the *Fagus* expansion is dated to *ca.* 400 A.D. (Aaby in Andersen *et al.* 1983, this volume). A few records of pollen of dry-land herbs and shrubs such as *Crataegus, Rhamnus* and *Sorbus* suggest that glades apparently occurred, and were used for cattle grazing. *Fagus* was not suppressed, and grazed *Fagus* forest continued, as was well known in Denmark, up till *ca.* 1800 A.D. when grazing in the forest was abandoned.

Tilia became very scarce, and it may have been removed in order to promote pig husbandry. *Salix* and aquatics became scarce, indicating increased shade over the hollow.

The detection of different farming practices by pollen analysis

Four phases with human impact are recorded in the *Glyceria*-hollow: shredding, and later pasture in an early part of the Neolithic, arable

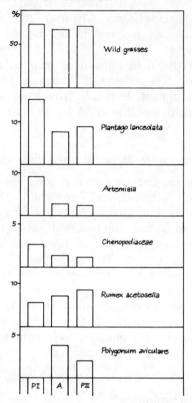

Fig. 10. Average frequencies of the most common herbs, as percentages of the total numbers of pollen grains from open-ground herbs.
PI = Neolithic pastoral phase;
A = arable phase from Roman Iron Age;
PII = pastoral phase from Roman Iron Age (from Andersen 1985).

cultivation in the Late Bronze Age and Pre-Roman Iron Age, and grazing by cattle in the Roman Iron Age.

Only *Tilia* was present in the shredding phase. All other vegetation was removed by felling and burning, and there are no traces of other activities near the site at that time.

Open forest occurred in the two pastoral periods: *Corylus* and *Quercus* were dominant, whereas the shade trees, *Tilia* and later also *Fagus*, were suppressed. There were extensive glades with shrubs and herbs.

During the cultivation of the field next to the *Glyceria*-hollow there were few trees in the vicinity, and pollen of perennial herbs and wild grasses predominate over that of annual weeds and cereals.

Figure 10 shows average frequencies for the most frequent herbaceous plants, as percentages of total open-ground herbs and wild grasses in the 2 grazing stages mentioned above (P I and P II) and the arable stage (A). Wild grasses dominate and have the same frequencies in the 3 periods. *Plantago lanceolata*, *Artemisia* and Chenopodiaceae are slightly more frequent in the Neolithic grazing stage, and *Rumex acetosella* in the Bronze Age-Iron Age periods. Only one species, *Polygonum aviculare* is clearly most frequent in the arable stage. The pollen records of the herbaceous plants are therefore very alike in prehistoric pastoral and arable stages even at this type of site, presumably because pollen from the same weeds are best dispersed in both cases. A similar picture is found from lakes or bogs more distant from cultivated sites (Behre 1981; Aaby in Andersen *et al.* 1983).

Summary

(1) *Glyceria*-hollow is a 18 x 12 m depression in an area rich in Bronze Age barrows and field lynchets in Næsbyholm Storskov. The pollen diagram from its sediments reflects very local vegetation.

(2) A natural forest succession is recorded from the Preboreal (*Pinus-Betula*), Boreal (*Corylus*) and Atlantic (*Tilia-Corylus- Quercus*).

(3) Just above the *Ulmus*-decline at 4000 B.C. the dominance of *Tilia* pollen together with the occurrence of charcoal dust is probably a response to shredding of *Tilia* near the site by Early Neolithic people. This was followed by a pastoral phase lasting *ca.* 500 years.

(4) The area was abandoned for *ca.* 2000 years, and natural forest regenerated.

(5) From 1000 B.C. to 200 A.D. a field was cultivated next to the site. Soil washed into the hollow. Pollen analyses show weed, grass and a few cereal pollen.

(6) The field was abandoned, and was colonized by secondary forest which was maintained for animal grazing.

(7) The area was almost abandoned, and was colonized by *Fagus* *ca.* 400 A.D.

(8) Even from such a local pollen catchment it is difficult to distinguish the herb flora connected with pastoral and arable activities by pollen analysis.

Nomenclature

Plant nomenclature follows *Flora Europaea* except for Poaceae (= Gramineae).

Land-use History during the Last 2700 Years in the area of Bjäresjö, Southern Sweden

Marie-José Gaillard and Björn E. Berglund

Introduction

The study of Bjäresjö is part of the Ystad research project entitled 'The cultural landscape during the last 6000 years' the main purpose of which is to understand the interactions between humans, vegetation and climate from Neolithic time until today (Berglund & Stjernqvist 1981; Berglund this volume). The project area is in southern Scania and includes a coastal zone, an outer hummocky zone and an inner hummocky zone (Berglund 1985b, this volume; Fig. 1).

Bjäresjö lake was chosen as a reference site to describe the land-use history of the outer hummocky zone, which today is intensively exploited. As part of the same project, thorough archaeological investigations were performed close to the lake and in Bjäresjö village. Since the lake and its catchment are very small, it was expected that the sediments would record mainly local events, and would give a picture of the palaeoenvironment closely related to the settlement history. Moreover, some levels in the sediment sequence could be ^{14}C-dated to provide a time-scale. Therefore, Bjäresjö possessed many advantages for the study of cultural landscape history.

We present the preliminary results of pollen analysis as part of a larger, multidisciplinary investigation of the lake's history. This includes the elucidation of vegetational and land-use history, and the evaluation of soil erosion, trophic changes and water-level fluctuations, based on a chronology of ^{14}C- and ^{210}Pb-datings. Hopefully, it will eventually be possible to evaluate the interactions between humans, vegetation and climate through time (Gaillard 1985a).

This chapter discusses the site, a preliminary time-scale and the first pollen-analytical results and makes a tentative interpretation of land-use, particularly in the Iron Age and Middle Ages. Water-level changes are also discussed.

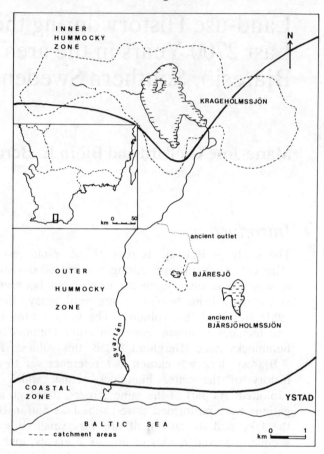

Fig. 1. Location of the study area showing Bjäresjö and Krageholmssjön and their catchment areas and the ancient lake of Bjäresjöholmssjön studied by Nilsson (1961). The different geographical zones of the Ystad area are also indicated.

The site

Bjäresjö lake is 5 km north-west of Ystad, in the outer hummocky zone of the project area, at 48 m above sea level (Fig. 1). It is a small, more or less circular lake, with a maximum diameter of 200 m, a surface area of *ca.* 2 ha, and a maximum depth of 1.7 m. It was probably formed by dead-ice melting and it lies between hills of clayey till. Its catchment area is very small, *ca.* 10 ha. A series of maps (1699, 1809, 1812, 1888, 1915, 1970) of the area show that the shape and size of the lake have hardly changed since 1699. An outlet drained to the river Svartån probably until about 1900.

Bjäresjö village, on a low hill north of the lake, consists today of *ca.* 20 farms and houses, and one church. Bjärsjöholm Castle is 1.5 km south-east of Bjäresjö village. The sediments of the ancient lake nearby (Bjärsjöholmssjön), studied by Nilsson (1961), provide a standard pollen diagram for the cultural landscape in southern Scania (Berglund 1969).

Archaeological investigations show a more or less continuous settlement on Bjäresjö hill since the Iron Age. There were Roman Iron Age and Viking settlements at the eastern edge of the modern village. The remains of a large farmhouse were discovered north-west of the church and dated to 1000-1100 A.D. This farmhouse was enlarged to a spacious and rich stone-house during the Middle Ages (1200-1350 A.D.) and was probably the main building in the village until 1350 when it was destroyed by fire (J. Callmer personal communication). It is probable that the main farmhouse was moved then to the site of Bjärsjöholmssjön (Skansjö 1987b). The Bjäresjö landowners built Bjärsjöholm Castle in 1576.

Methods

The field work, sediment description, laboratory techniques and pollen analysis follow methods described earlier for the Ystad project (e.g. Gaillard 1984). 1000-2000 AP + NAP were counted at intervals of 5 to 10 cm.

Stratigraphy and sediment description

A transect of 12 cores was used to study the stratigraphy of the basin (Fig. 2). The initial peat deposit was interrupted by a water-level rise and subsequent deposition of clayey gyttja. This is more silty and sandy near the shores, and includes a layer of coarse-detritus drift gyttja across the whole basin.

The curves of loss-on-ignition at 550^0C, carbonate content and water content are presented (Fig. 3) for the main core, C3. The compact, coarse-detritus gyttja (layer 7) is well differentiated from the clayey fine-detritus gyttja by its very high loss-on-ignition values. There is no evidence for any hiatus at its transition to the clayey gyttja (layers 5, 6, 8). In the top m (layers 1, 2), carbonate, water and organic content increase, and carbonate is particularly high in the upper 50 cm.

Preliminary chronology

Radiocarbon-dating is unreliable in south Scania sediments because of hard-water effects produced by their high carbonate content (Olsson 1979). Therefore dating must rely on correlation with the ^{14}C-dated pollen zones of Ageröds Mosse in north Scania (Nilsson 1964). This can be difficult because Nilsson's zone boundaries are not always readily recognizable in south Scania. The boundaries between Nilsson's zones SB2/SA1 and SA1/SA2, dated respectively at 2140 ± 85 B.P. and 1250 ± 85 B.P. in Ageröds Mosse, and defined by an increase in *Fagus* and a decrease in *Alnus*, are very difficult to place in the Bjäresjö diagram, owing to the low AP percentages. The rational limit of *Fagus* is especially uncertain. Therefore, it is particularly important to date the coarse-detritus gyttja. The ^{14}C-datings were performed by Håkansson (1986) and the ^{210}Pb-dating by El-Daoushy (1986).

The results are given in Figures 2 and 4. The upper part of the peat (carr peat and moss peat, layers 13, 14) gives two identical dates, 2680 ± 50 B.P. and 2690 ± 60 B.P., whereas slightly clayey coarse-detritus gyttja at the base of the lake-sediment sequence (layer 12) is slightly older, viz. 2760 ± 50 B.P. The time-depth diagram (Fig. 4) shows a fairly constant accumulation rate of 0.66-0.76 mm yr[-1] for the coarse-detritus gyttja (layers 5, 6, 7). The accumulation rate of the underlying clayey gyttja (layers 8-11) is 0.66 mm yr[-1], when the date of the moss peat is taken into account. Nevertheless, the possibility of a hiatus at the peat/gyttja transition cannot be excluded, as long as no absolute dating is available for the clayey gyttja. The same argument applies to the upper boundary of the coarse-detritus gyttja. The only way to clarify this problem would be to ^{14}C-date seeds and fruits from the clayey gyttja by accelerator mass spectrometry.

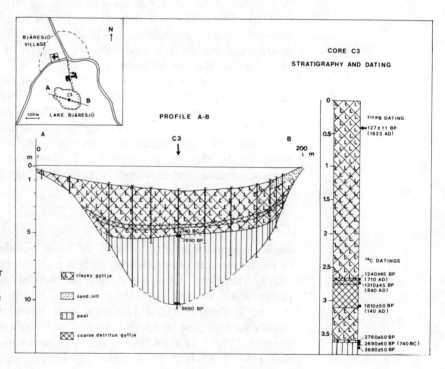

Fig. 2. Stratigraphic profile for Bjäresjö, showing simplified stratigraphy and dating. In the main core (C3), the sediment thickness is 9 m. The upper 3.7 m of lake sediments were studied by pollen analysis.

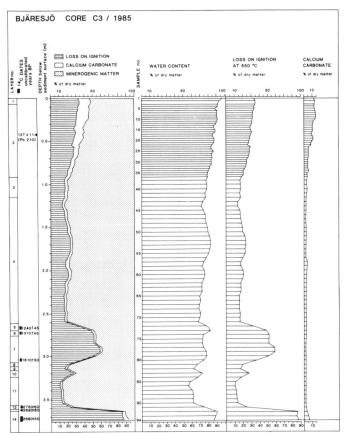

Fig. 3. Sediment analysis of the main core (C3) in Bjäresjö. Loss on ignition, calcium carbonate and water content are given as percentage of dry matter. [14]C- and ^{210}Pb-datings are also indicated. The detailed description of sediments (layers 1-14) will be published elsewhere.

The ^{14}C-dates obtained for the layer of coarse-detritus gyttja made it possible to fix the boundaries SB2/SA1 and SA1/SA2 with greater certainty. They also show that our tentative correlation with Nilsson's pollen zones was reasonable (Fig. 5).

Our preliminary chronology is based upon the assumption that the sediment sequence is continuous and that the accumulation rate is more or less constant between 1200 and 2700 B.P. (0.66 mm yr[-1]) and between 127 (^{210}Pb-dating) and 1200 B.P. (2.3 mm yr[-1]). It is shown to the left of each pollen diagram (Figs. 5-9). The archaeological periods were delimited using this preliminary time-scale, and are therefore independent of palynological evidence. The ages used for the boundaries of the archaeological periods are those generally accepted for south Scandinavia (Hedeager & Kristiansen 1985). The pollen-analytical results are considered using this preliminary time-scale, which means that our discussion of land-use history and chronology is liable to change in the future.

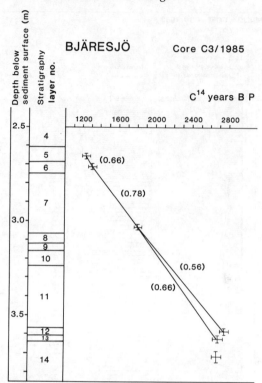

Fig. 4. ¹⁴C-dates plotted against depth. The accumulation rates are given for each segment of the curve, with two alternatives for the oldest one.

Pollen-analytical results

Presentation of results

Bjäresjö sediments are characterized by very high percentages of pollen from local aquatic and lake-shore vegetation, and from the immediate surroundings of the lake, such as fields, meadows and pastures. This is shown by the very low AP/NAP ratios for the whole profile (Fig. 9.1) and by an exceptionally rich herbaceous pollen flora. The complete list of identified pollen and the taxon groupings will be published later. The aim here is to discuss the land-use history on the basis of the most common or significant taxa. These taxa are divided into 6 groups in order to simplify discussion: (1) trees, (2) cultivated plants, (3) weeds, (4) ruderal plants, (5) pasture land indicators, and (6) aquatics and lake-shore vegetation.

Since trees are rare, and aquatics and herbaceous plants are abundant, we used 5 different sums to facilitate discussion and minimize numerical artefacts due to percentage calculations:

- Σ AP used for trees (Fig. 5)
- Σ NAP - (lake-shore plants + aquatics + Cannabaceae), used for cultivated plants, weeds, ruderal plants and pasture indicators (Figs. 6, 7).
- Σ aquatics + lake-shore plants, used for lake vegetation (Fig. 8).
- Σ AP + NAP - (lake-shore plants + aquatics + Cannabaceae), used for the summary diagram (Fig. 9.1).
- Σ shrubs + NAP, used for the summary diagram (Fig. 9.2).

Fig. 5. Bjäresjö, core C3. Pollen diagram for the common trees. The time-scale is based on the [14]C- and ^{210}Pb-dates given in Figures 3 and 4. The archaeological periods are presented for comparison. The pollen-zones and chronozones of Nilsson (1964) are also indicated (for further explanation, see text).

Regional vegetation and forest history

Figure 9.1 shows that AP percentages never exceed 45% after the Late Bronze Age, and remain below 30% from the beginning of Viking time until today. This feature is unusual for the region, when compared with diagrams from Bjärsjöholmssjön (Nilsson 1961), Krageholmssjön (Gaillard 1984) or Fårarps Mosse (Hjelmroos 1985b), where trees are much commoner. As the small size of Bjäresjö lake and its

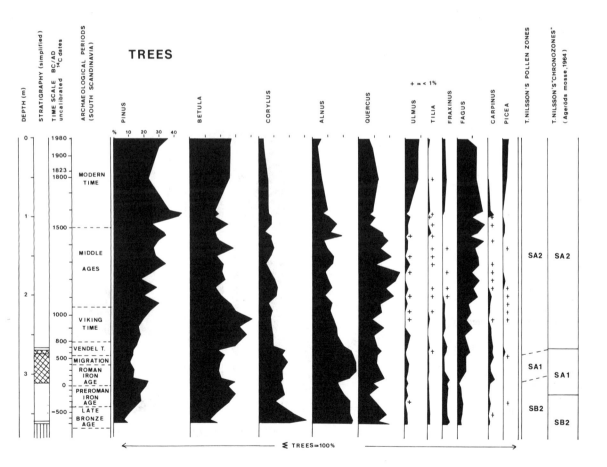

catchment imply a local pollen source, this shows that its immediate surroundings have been almost treeless since the later part of the Late Bronze Age. The AP diagram (Fig. 5) shows a progressive increase in *Fagus* and *Pinus*, as well as a gradual decrease in *Corylus*, all typical of Nilsson's zones SA1 and SA2. The high values of *Carpinus* during SA1 and the low frequences of *Ulmus*, *Tilia* and *Fraxinus* are also characteristic. The late increase in *Ulmus*, *Fraxinus* and *Picea* is also shown in both diagrams from Krageholmssjön (J. Regnéll personal communication; Gaillard 1984). It reflects 18th and 19th century plantations (see below). The *Pinus* rise after 800 A.D. is probably due to long-distance pollen dispersal in an open landscape.

The fluctuations in the *Betula* and *Alnus* curves correspond mainly to local events. The peak of *Alnus* is associated with the coarse-detritus gyttja and probably reflects the expansion of *Alnus* swamp. The rapid increase in *Betula* around 700-800 A.D. (Vendel/Viking Time) may correspond to a period of forest clearance (see below).

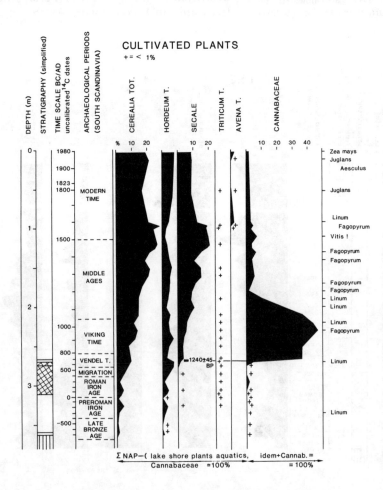

Fig. 6. Bjäresjö, core C3. Pollen diagram for cultivated plants, common weeds and ruderal plants. The diagram is divided into 7 zones corresponding to different phases of cultivation history. The time-scale and archaeological periods are presented for comparison (cf. Fig. 5).

It is clear that forested areas were very small around Bjäresjö since *ca.* 500 B.P. *Fagus* was not dominant as it was elsewhere in the region (e.g. Krageholmssjön, Bjärsjöholmssjön), but *Quercus*, at least from Viking times until *ca.* 1500 A.D., was the commonest tree.

Human-impact indicators and land-use history

In an initial attempt to characterize the cultivation history and to try to estimate areas occupied by different types of land-use, 'human impact indicators' have been grouped into 4 categories: (1) cultivated plants, (2) weeds, (3) ruderals, and (4) pasture-land indicators, following Behre (1981), our ecological knowledge of the different taxa and information from, for example, lists of weeds reported by Linné in 1749 from farmlands in Scania (compiled by G. Olsson personal communication).

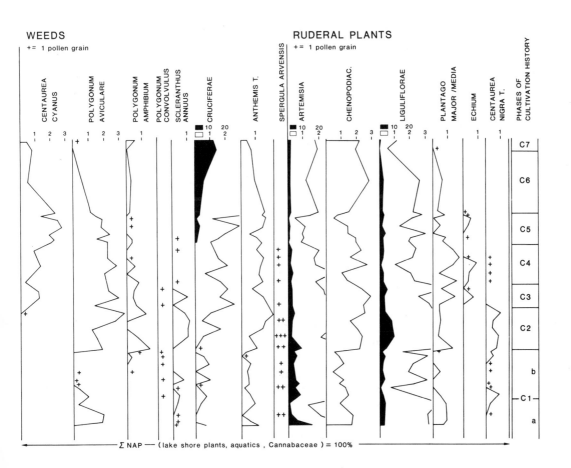

Cultivated plants

In Figure 6, the separation of the different cereal pollen types is based on Beug (1961), our own observations, and comparison with recent material. Attempts at separating *Cannabis sativa* from *Humulus lupulus* using Punt and Malotaux (1984) showed that more than 50% of the Cannabaceae pollen could, with certainty, be ascribed to *Cannabis sativa*.

Cereals consist mainly of *Hordeum*-type until Vendel Time. Since *ca*. 700 A.D., *Secale* is regularly represented and becomes more and more frequent. This is common in pollen diagrams from Scania (Nilsson 1961, 1964; Digerfeldt 1974; Gaillard 1984). It is the dominant cereal pollen type from the beginning of the Middle Ages. *Triticum*-type already occurs during the Pre-Roman Iron Age, but as low percentages throughout. *Avena*-type was observed in the upper sediments only.

Cereal history in Sweden is quite well known, thanks to investigations of cereal impressions in pottery from prehistoric time (Hjelmqvist 1955, 1963, 1969, 1979). The oldest find of *Secale* impressions is from the Pre-Roman Iron Age in Västergötland, north of Scania. Grain impressions of *Secale* are rare during the Early Iron Age, but increase markedly during Viking Time. Since then the cultivation of *Secale* expanded considerably.

Our pollen-analytical results show the same trends. However, it is widely recognized that *Secale* has a higher pollen representation than other cereals. Therefore, the high percentages of *Secale* pollen during the Middle Ages, and particularly around 1450-1550 A.D., do not necessarily imply that *Secale* was the dominant cereal. *Avena* and *Triticum* are usually very poorly represented in pollen diagrams, as at Bjäresjö. Nevertheless, the cultivation of *Avena sativa*, *Triticum dicoccum* and *T. monococcum* is known from the Bronze Age and Iron Age. But these cereals were of minor importance compared with *Hordeum* (Hjelmqvist 1963, 1979).

The high frequencies of Cannabaceae between 700 and 1150 A.D. are of particular interest. The occurrence is known from other pollen diagrams for Scania, e.g. Krageholmssjön (Gaillard 1984), and particularly in another small lake, Bysjön (G. Digerfeldt personal communication). There the increase in pollen values also occurs just above the boundary SA1/SA2. *Cannabis* cultivation in Sweden has been discussed in detail by Fries (1958) for the Varnhem area (Västergötland, southern Sweden) where very high pollen percentages of Cannabaceae are recorded during Late Iron Age and Early Middle Ages. A recent survey of the *Cannabis/Humulus* pollen curves in some Swedish pollen diagrams (Påhlsson 1981) shows that the culminations of the curve range from the Migration Period to Medieval times (*ca*. 500-1500 A.D.). The very few macroscopic finds of *Cannabis* are from Early Iron Age and Viking Time (Hjelmqvist 1955), suggesting that hemp was introduced into Sweden by the first centuries A.D.

At Bjäresjö, the increase in Cannabaceae percentages unfortunately occurs at the transition between the coarse-detritus gyttja and clayey fine-detritus gyttja, immediately above the upper level dated by ^{14}C (Fig. 6). The possibility of a hiatus at that level has been mentioned above. However, using all the information available, the culmination of the Cannabaceae curve at Bjäresjö can be dated to 700-1150 A.D. Its decrease in pollen percentages since *ca.* 1150 A.D. may be due to either a much reduced cultivation of hemp or retting in another basin.

Linum usitatissimum and *Fagopyrum esculentum* are both palynologically under-represented which makes reconstruction of their history difficult. *Linum* pollen was found at 6 levels at Bjäresjö, scattered between the Pre-Roman Iron Age and Modern times (*ca.* 1600 A.D., Fig. 6). Macrofossils and grain impressions have been found from the Late Bronze Age, and Early and Late Iron Age in Sweden and Denmark (Hjelmqvist 1979; Jessen 1933, 1951). In Scania, the oldest find is from Vendel Time. The pollen of *Linum* found in the Pre-Roman Iron Age sediments of Bjäresjö may be the first proof for the early cultivation of flax in Scania.

Fagopyrum pollen is recorded in 6 levels, from Viking times onwards. It is generally accepted that buckwheat was introduced during the Middle Ages along the routes of the Mongols. Pollen records from earlier times (e.g. Neolithic) may result from confusion with the very similar pollen of wild *F. tataricum*. The Bjäresjö results show that *Fagopyrum* was possibly introduced and cultivated in Scania during the Early Middle Ages, perhaps as early as 1000 A.D. However, no real proof for the early cultivation of buckwheat is forthcoming until macroscopic remains are found.

Weeds

The diagram for weeds is also presented in Figure 6. They consist mainly of weeds of winter cereals, particularly *Secale*. This could explain why most of them increase together with the first increase in *Secale*. *Centaurea cyanus* increased only when *Secale* cultivation expanded at Bjäresjö, from the beginning of Middle Ages *ca.* 1050 A.D. The first peak of *Polygonum aviculare* (Late Bronze Age and Pre-Roman Iron Age) seems to be unrelated to cereal cultivation. It may have been growing in natural or ruderal communities. The upper curve, however, seems to be intimately related to *Secale* cultivation. *Polygonum amphibium* is included in the weed diagram, as it is sometimes common in cereal fields on wet clayey soils. Nevertheless, it is probable that it also grew in the lake, and that the peaks in its curve are, instead, associated with the water-level rise deduced from the succession of aquatics (between 700-1200 A.D., Fig. 8). *Polygonum convolvulus* and *Scleranthus annuus* are mainly characteristic of winter cereals.

Cruciferae pollen probably originates mainly from cereal fields, but also from ruderal communities, particularly since *ca.* 1600 A.D., when pollen percentages increase markedly. The high values in the two upper levels probably reflect the cultivation of *Brassica napus*.

The curve of *Anthemis* type (Moore & Webb 1978) is probably also a composite one. According to Behre (1981), Tubuliflorae are particularly common on fallow land and in wet meadows, pastures and ruderal communities. They also grow in cereal fields and dry pastures. Since the curve of *Anthemis* type has similarities with that of *Polygonum aviculare* from Vendel Time to *ca.* 1600 A.D., it is possible that many of the grains originate from *Anthemis arvensis*, typical of winter rye fields.

Ruderal plants

Traditional indicators of ruderal communities are *Artemisia*, Chenopodiaceae and *Plantago major/media*. To those we add Liguliflorae,

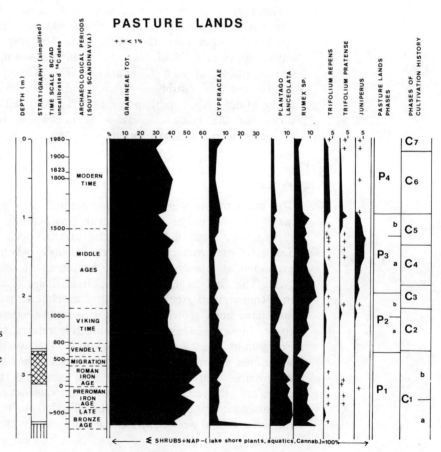

Fig. 7. Bjäresjö, core C3. Pollen diagram for common pasture-land indicators. The diagram is divided into 4 zones corresponding to the different phases of pastoral history. The phases of cultivation history are also indicated (cf. Fig. 6). The time-scale and archaeological periods are presented for comparison (cf. Fig. 5).

Echium and *Centaurea nigra* type. All are difficult to classify in a particular group.

Liguliflorae can also be common on fallow land, meadows, pastures and grazed forest (Behre 1981). In the Bjäresjö diagram, Liguliflorae reach maximum percentages during the Cannabaceae culmination together with *Centaurea nigra* type and *Scleranthus annuus*. This suggests that these taxa were particularly well represented in the *Cannabis* fields.

Echium vulgare is a very characteristic ruderal, growing along roadsides and field margins. It is striking that it is restricted in the pollen diagram to between 1050-1500 A.D., and seems to be related to the high *Secale* and *Centaurea cyanus* values. This species may thus also have been associated with the *Secale* fields.

Chenopodiaceae and *Plantago major/media* were probably most common in ruderal communities. The latter expanded with the development of cereal cultivation around 700 A.D. *Artemisia* also occurs regularly throughout the profile. It reaches higher values from Late Bronze Age to Vendel Time, where it may consist mainly of *A. campestris* growing in meadows and pasture land. Note that all ruderal species are also indicators of fallow land, together with weeds and grasses.

Pasture-land indicators

There are few, if any, good indicators of pastures. We have grouped in Figure 7 those taxa that are often related to pastoral farming, such as Gramineae, Cyperaceae, *Plantago lanceolata*, *Rumex* and *Juniperus*, along with *Trifolium repens* and *T. pratense*.

Plantago lanceolata is certainly the best pastoral indicator, but it is under-represented in pollen spectra (Berglund *et al.* 1986). At Bjäresjö, the curves of *Plantago lanceolata* and *Juniperus* are considered together. *Plantago lanceolata* has high percentages until *ca.* 600 A.D., when it decreases at the first increase in *Secale* and Cannabaceae. A second significant decrease occurs around 1100 A.D., after which it diminished gradually until 1500 A.D., whereas *Juniperus* expands (until 1400 A.D.). These trends can be interpreted in the following way: pasture covered large areas during the period 700 B.C.-600 A.D. With the development of cereal and *Cannabis* cultivation, they were partly replaced by fields. Around 1100 A.D., pasture began to be invaded by *Juniperus*, which, together with decreasing percentages of *Plantago lanceolata* may signify that grazing diminished progressively. The drop in *Juniperus* values between 1400 and 1500 A.D. probably reflects the replacement of pastures by cereal fields (cf. the *Secale* curve). Since 1500 A.D., *Juniperus* has almost completely disappeared from the region.

Rumex acetosa, characteristic of wet meadows and pasture land, is probably the main component of the *Rumex* curve during the period 700 B.C.-500 A.D. The increase in its values between *ca*. 900 and 1100 A.D. closely parallels the *Secale* curve, which would suggest that *Rumex acetosella*, a frequent weed in winter-cereal fields, is involved. The subsequent decrease between 1100 and 1500 A.D. may be due either to a decrease in pastoral species, due to overgrowing by *Juniperus*, or to some factor unfavourable to *Rumex acetosella* in the fields.

Trifolium repens and *T. pratense* are particularly abundant since 1500 A.D. *T. repens* is common in pastures, grazed forests and ruderal communities. At Bjäresjö, it ought perhaps to be included in the ruderals, at least from 1500 A.D. until today. *T. pratense* is typical of wet meadows and pastures, but it may also be sown on fallow land, which could well be the case in Bjäresjö during the 16th century and even later.

Gramineae and Cyperaceae are sometimes used as indicators for pastures. However, in the Bjäresjö profile, about 30 to 50% Gramineae belong to *Phragmites*-type. Moreover, the highest percentages of Gramineae and Cyperaceae occur in the coarse-detritus gyttja, corresponding to a period of low water-level (Fig. 8).

Land-use history at Bjäresjö from the Late Bronze Age until today

We have divided the pollen diagram for cultivated plants, weeds and ruderals into 7 zones (C1-C7) that are interpreted in terms of cultivation history (Figs. 6, 9.3). Similarly, the diagram of pasture indicators is divided into 4 zones (P1-P4, Fig. 7). The following reconstruction of land-use history at Bjäresjö is based on these different zones.

C1. Late Bronze Age-Vendel Time: 700 B.C.-700 A.D.

This long period of 1400 years is characterized by the dominance of pastures, which is well illustrated by the *Plantago lanceolata* curve (zone P1) and by the summary diagram (Fig. 9.2). Trees have their highest values in the profile. Fluctuations in the AP/NAP curve are difficult to interpret (Fig. 9.1). The two peaks in AP percentages during the Pre-Roman Iron Age and Migration Period do not correspond to any significant decrease in human-impact indicators.

Subzone P1a corresponds to the later part of the Late Bronze Age and the transition to the Pre-Roman Iron Age. Cereals (mainly *Hordeum*) have very low pollen values. They may have been cultivated in the area, but very sparsely. Subzone P1b has higher percentages of cereals since *ca*. 200 B.C. Therefore, a first development of cereal cultivation occurred in the Pre-Roman Iron Age. *Linum* may also have been cultivated then. At the beginning of Vendel Time *Hordeum*-type increased, *Secale* occurred regularly and *Plantago lanceolata* decreased. Pastoral farming is progressively replaced by agriculture.

Fig. 8. Bjäresjö, core C3. Pollen diagram for common aquatics and lake-shore plants. The interpretation of the diagram in terms of water-level changes is given together with the results obtained for Krageholmssjön (Gaillard 1984). The time-scale and archaeological periods are presented for comparison (cf. Fig. 5).

C2. Vendel Time/Viking Time: 700-1050 A.D.

At *ca.* 700 A.D. a great change occurs in the pollen diagram due to the increase in *Secale*, Cannabaceae, weeds and ruderal plants (Figs. 6, 9.2). During the whole Viking Time, *Hordeum, Secale* and *Cannabis sativa* are common crops. *Linum* pollen may indicate flax cultivation already at this time. Fields increase in area, whereas forests (Fig. 9.1) and pastures (zone P2) diminish.

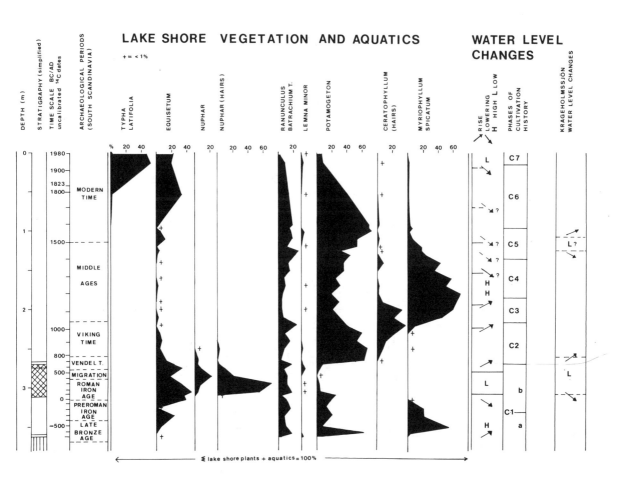

C3. Early Middle Ages: 1050-1200 A.D.

Secale cultivation becomes more and more common. Weeds of winter cereals expand. The decline of Cannabaceae suggests a decrease in *Cannabis sativa* cultivation. *Linum* is most probably cultivated during this time. *Juniperus* begins to invade pastures where grazing is less intensive (zone P2b).

C4. Middle Ages: 1200-1400 A.D.

Secale and *Hordeum* are the dominant crops, and *Fagopyrum* is definitely cultivated. The progressive increase in *Juniperus* (zone P3a) reflects the dereliction of pastures as grazing diminished in intensity.

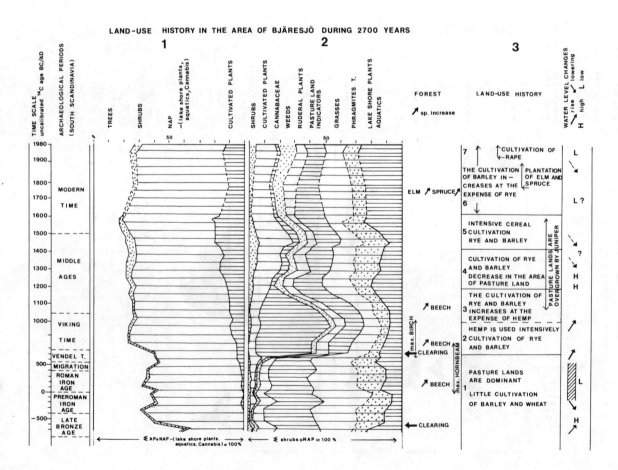

LAND-USE HISTORY IN THE AREA OF BJÄRESJÖ DURING 2700 YEARS

C5. Middle Ages/Modern Times: 1400-1600 A.D.

Secale cultivation reaches its maximum. *Fagopyrum* is still cultivated. *Trifolium pratense* is particularly well represented in the pollen diagram (zone P3b), perhaps sown on fallow lands. Part of the pastures is converted to cultivated fields, which would cause a decrease in *Juniperus*. Possibly, the last 'outfield' area (common grazing land) is reclaimed by this time, but its name Gållfäladsvång ('fälad' means outfield area) is retained on the map from 1699 (cf. Olsson this volume). Historical documents show that about 20% of the area of meadows was converted to cereal fields between 1570 and 1699 (Persson 1987).

C6. Modern Times: 1600-*ca.* 1930

Secale cultivation decreases somewhat in favour of *Hordeum* and *Avena*. *Ulmus* and *Fraxinus* were probably planted during the 18th and 19th centuries along roads and around farms and castles. Historical documents show that this commonly occurred in southern Scania (Emanuelsson *et al.* 1985). Around 1800 A.D., weeds such as *Centaurea cyanus*, *Polygonum aviculare* and *Echium* are much less common. Pastures are now almost nonexistant (zone P4). It seems that all areas previously colonized by *Juniperus* are reclaimed. From the end of the 19th century, *Picea* plantations are common in Scania (Emanuelsson *et al.* 1985).

C7. Modern Times: *ca.* 1930 until today

Secale and *Hordeum* cultivation is still dominant. The high percentages of Cruciferae may reflect the introduction of rape, *Brassica napus*. Weeds and ruderals such as Liguliflorae and *Plantago major/media* are now poorly represented. The complete pollen taxa show that the pollen flora is generally much poorer, which indicates the disappearance of several plant communities owing to intensive agriculture. The diversity of the landscape is reduced (see Ihse this volume; Birks *et al.* this volume).

◀

Fig. 9. Summary of the palynological results for Bjäresjö.
(1) AP/NAP diagram showing the percentages of trees, shrubs, cultivated plants and herbs. Note that lake-shore plants, aquatics, and Cannabaceae are excluded.
(2) Diagram for shrubs and herbaceous plants according to a preliminary grouping of all pollen taxa. Trees are excluded in order to give a better picture of the local vegetation.
(3) Summary of results in terms of forest and land-use history (simplified), as well as water-level changes. The time-scale and archaeological periods are presented for comparison (cf. Fig. 5).

Lake vegetation and water-level changes

It was mentioned above that pollen from lake vegetation was over-represented in the Bjäresjö pollen diagram. This fact has the advantage, however, of providing a detailed picture of lake vegetation through time and the possibility of following movements of littoral vegetation zones.

Lake-shore species and aquatics are presented in Figure 8. The diagram clearly shows successions that can be interpreted in terms of water-level changes:

(1) The peat-clayey gyttja transition is characterized by the sequence Cyperaceae - *Lemna minor* - *Potamogeton* - *Myriophyllum spicatum* and corresponds to a water-level rise during Late Bronze Age.

(2) The *Myriophyllum spicatum* - *Equisetum* - *Nuphar* sequence is due to a progressive water-level lowering that started at the beginning of the Pre-Roman Iron Age and culminated in the Roman Iron Age and Migration Period. It resulted in the deposition of coarse-detritus gyttja. The culmination of this event is dated to 140-640 A.D. Low water-levels have been recorded from other lakes of southern Sweden between 250 B.C. and 750 A.D. (Digerfeldt 1972, 1974, 1975, personal communication; Gaillard 1984). Therefore, this lowering may represent a regional event ascribable to climatic change (Gaillard 1984, 1985a, 1985b).

(3) The succession *Lemna* - *Potamogeton* - *Ceratophyllum* - *Myriophyllum spicatum* is interpreted in terms of a progressive water-level rise, taking place at *ca*. 700 A.D. (Vendel Time), progressing during Viking Time and Early Middle Ages, and culminating *ca*. 1200 A.D.

(4) From *ca*. 1300 A.D. until today, a new sequence is observed of *Myriophyllum spicatum* - *Potamogeton* - *Equisetum* - *Typha latifolia*. This may be due either to progressive water-level lowering, or to natural lake infilling by sediments. The second alternative is the most likely. Water depth decreased by a minimum of 1 m during 1300-1800 A.D., owing to sediment accumulation only. This succession of aquatic vegetation is partly also a result of increased minerogenic sedimentation and lake eutrophication.

The water-level rise dated at 2700 B.P. is difficult to understand, as is the enormous accumulation of peat between 9700-2700 B.P., which implies a gradual water-level rise of at least 5 m. The creation of Bjäresjö lake may result from forest clearance, causing changes in hydrological balance (cf. Moore 1986), increased humidity, or other palaeohydrological causes. A combination of human-induced and climatic changes is probably involved.

The second water-level rise seems to be directly related to cereal and *Cannabis sativa* cultivation. It is thus difficult to decide whether the rise is due to deforestation or further climatic change. Viking Time and the Early Middle Ages are characterized by intensive clearing in the whole area, so that the regional ground-water level may have been influenced significantly. However, a change towards a more humid climate cannot be excluded.

Conclusions

The Bjäresjö pollen diagram gives a local picture of vegetational and land-use history mainly within the Bjäresjö village area. The grouping of taxa makes it possible to distinguish temporal variations that may reflect changes in farming economy. Further grouping of pollen indicators, using all identified types and applying numerical analysis (e.g. Birks *et al.* this volume), could provide a more detailed and accurate interpretation.

Our tentative reconstruction of land-use history at Bjäresjö results in 7 different periods with changes in cultivation and utilization of pastures. If our preliminary chronology is correct, these periods have a direct connection with settlement history. The archaeological results demonstrate continuous settlement at Bjäresjö since Late Bronze Age. There is no evidence for any interruption during the Migration Period. A similar land-use development in a village environment is reported from Borup, Själland, in Denmark (Mikkelsen 1986).

The large main-farm building, dated to the end of the Viking Time around 1000 A.D., falls within the first period of intensive cultivation (zone C2, 700-1050 A.D.). Around 1100 A.D., the farm was altered into an even larger building. This period, also characterized by intensive cultivation, is transitional in land-use. *Cannabis* seems to be replaced progressively by *Secale* (zone C3, 1050-1200 A.D.).

Archaeological excavation has provided much material (e.g. coins, ceramics) dating from the Early Middle Ages (1200-1350 A.D.), which coincides with the expansion of *Secale* cultivation and the progressive surrender of pasture lands (zone C4, 1200-1400 A.D.).

Archaeologically, very little is known about the period 1350-1576. It is possible that the main farm had already been moved to Bjäresjöholm by 1380 (Skansjö 1987b). However, this period is distinguished palynologically as well (zone C5, 1400-1600 A.D.) and shows further agricultural development at the expense of pastoralism.

The beginning of zone C6 (1600-*ca.* 1930) may relate to the organization of an estate economy dominated by cereal and other crop production. Detailed study of historical documents from 1570 and 1699 shows that an important change in field and meadow distribution occurred between 1570-1699 (Persson 1987; Emanuelsson this volume). About 20% of the total meadow area was transformed into cereal fields. According to Persson (1987), this development of cereal cultivation corresponds to the expansion period generally recognized in Sweden and elsewhere in Europe during the last part of the 16th century and the beginning of the 17th century.

The palaeoecological investigations at Bjäresjö, including data on soil erosion and water trophy will, it is hoped, give a detailed picture of the environment in a small area of southernmost Sweden. It should provide a firm basis for discussing relationships between human activity and cultural landscape.

Summary

(1) Lake Bjäresjö, in the Ystad area of Scania is in an intensively exploited area. Its small size and limited catchment make it ideal for studying local land-use history.

(2) Six ¹⁴C-dates and one ²¹⁰Pb-determination provide a tentative chronology, and show that the lake sediments cover the period since the Late Bronze Age (2700 B.P.) to today.

(3) The results of pollen analysis show a very intensive land-use since the Late Bronze Age, in contrast to other sites in the area. The land-use phases through time agree well with the settlement history, described by thorough archaeological investigations close to the lake.

(4) From the Late Bronze Age to the end of the Roman Iron Age, pasture is predominant, and cereals were cultivated to a small extent (mostly *Hordeum*). Cereals then increase gradually until Modern Time (*ca.* 1500 A.D.). *Secale* becomes common after 700 A.D., reaching a maximum around 1400-1500 A.D. There is evidence for intensive *Cannabis* cultivation during Viking and Early Medieval times (*ca.* 700-1150 A.D.), which agrees with other results from southern and central Sweden. Pastures were overgrown by *Juniperus* through the Middle Ages before being reclaimed around 1500 A.D.

(5) Modern agriculture and landscape are characterized by the decrease in diversity of the flora and vegetation communities, by the introduction of new cultivated plants and by the planting of trees such as *Picea* and *Ulmus*.

(6) Water-level fluctuations have been reconstructed and related to land-use and climatic changes.

Acknowledgements

We are very grateful to our colleagues Anders Persson, Sten Skansjö, Gunilla Olsson and Johan Callmer for useful information concerning their investigations at Bjäresjö. We thank Gunnar Digerfeldt for reading the manuscript and giving interesting comments. We are much indebted to Thomas Persson and other colleagues for their invaluable help during field work. We also express our gratitude to Karin Ryde, Hilary Birks and John Birks who corrected the English text. Figures were drawn by Christin Andréasson. The study is part of a project supported by the Bank of Sweden Tercentenary Foundation.

Nomenclature

Plant nomenclature follows J. Lid (1963) *Norsk og svensk flora* (3rd edition), Det norske Samlaget, Oslo.

The Halne Area, Hardangervidda. Use of a High Mountain Area during 5000 Years - An Interdisciplinary Case Study

Dagfinn Moe, Svein Indrelid and Arthur Fasteland

Introduction

The Halne area is in the middle of the Hardangervidda plateau around the northern shore of Halnefjord at 1129 m elevation. All the archaeological and palynological sites are between 1129 and 1150 m elevation. The fieldwork was done as a part of 'The Hardangervidda project for interdisciplinary cultural research (HTK)' (Johansen 1973).

Climatic conditions are partly influenced by the oceanicity of western Norway and partly by the continentality of eastern Norway. In the western mountain valleys, precipitation is more than 1000 mm yr^{-1}, whereas to the east precipitation is less than 1000 mm. Mean temperatures also show differences from west to east. In the west (below the tree-line) at 746-780 m, the mean January temperature is -7.6^{0}C (Maurseth) and -9.8^{0}C at 1300 m (Slirå), whereas in the east, at 988 m, it is -9.6^{0}C (Hagastøl III). The mean summer temperature from west to east is 11.7^{0}C, 7.3^{0}C and 10.7^{0}C (Johannessen 1970; Brun 1962; Sterten 1974). Snow normally falls near the end of September and the vegetational period begins in June or July depending on snow-cover, exposure, etc.

The bedrock of the Hardangervidda is mainly of Precambrian and Cambro-Silurian origin. In the Halne area Cambro-Ordovician rocks dominate. The minerogenic soil contains a great amount of mica schist (T. Vorren 1979).

The Halne area is in the low-alpine vegetation zone above the present tree-line. The nearest trees are 17 km to the west at *ca.* 950 m, and 13 km to the east at 1050-1100 m. The local vegetation is mapped by Hesjedal (1974). The study area is dominated by soligenous mires, partly with *Salix* and tall-herb vegetation, partly dominated by *Sphagnum*. At Halnelægeret (locality (loc.) 434 in Fig. 1), the vegetation is mainly grass-dominated.

The vegetational history of Hardangervidda has been studied by Moe (1973, 1978, 1979). Forest immigrated into the mountain plateau

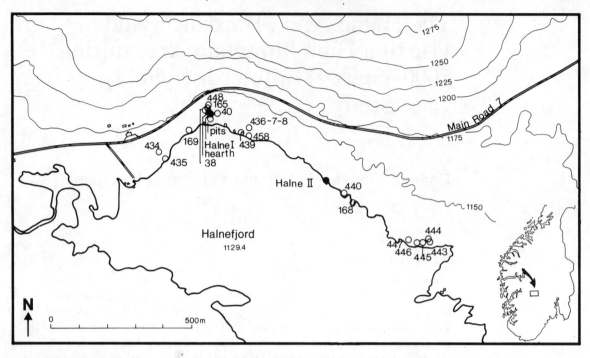

Fig. 1. Survey map of the
northern part of Halnefjord,
showing the location of the
archaeological and
palynological sites.

shortly after deglaciation, between 8500 B.P and 8000 B.P. It reached
1250 m with *Pinus* in the central and eastern parts. The forest retreated
at 8000 B.P. Tree-line probably passed below 1000 m in the west and
1100 m in the east. A temporary small rise of the tree-line started
ca. 5500 B.P. in the western area reaching 1050 m, but it retreated
again *ca.* 4700 B.P. to the present limit. The Halne area has thus
been unforested for the last 5000 years, and all tree-pollen recorded
there must originate by long-distance transport.

During historical time, the Halne area has been one of the boundaries in the South Norwegian mountains between east and west. The
name Halne itself means 'halfway' and the present county border
between Hordaland and Buskerud divides the area. One of the main
roads, R-7, crosses the mountains here, as did one of the earlier
main east-west paths 'Nordlige Normannslepa', before any roads and
railways were built.

Before the HTK project, the prehistory of the Hardangervidda had
been studied by, for example, Bendixen (1882, 1893), Negaard (1911)
and Bøe (1942, 1951). During the HTK project of 1970-74, systematic
surveys and excavations were carried out in the western, northern,
eastern and central parts of the plateau, and several hundred settlement
sites were found. Of these *ca.* 220 were older than 0 A.D. A further
30 had been found before HTK. Sixty-two of the 250 sites have been
partly or totally excavated (Indrelid 1986).

Younger sites, especially house-grounds, are numerous, but no complete overview exists. Relatively few have been excavated. Medieval house-grounds have yielded valuable information about the exploitation of the Hardangervidda between 1000 and 1600 A.D. (Negaard 1911; Fasteland 1971a, 1971b; Gustafson & Indrelid 1972; Blehr 1973).

Traditionally most sites older than 0 A.D. are found near lake shores. Therefore such areas were given priority during HTK surveys. However, recent finds from other mountain areas show that settlement sites from these periods may also occur far from lake shores (T. Bjørgo personal communication). Thus the distribution of pre-historic settlement sites on the Hardangervidda shown by the HTK project may not be representative of their real distribution. Sites from areas outside the shore zone (more than 100 m from the shores) are under-represented in our material.

Very few of the sites are stratified, and the cultural layer, which is usually 5-15 cm thick, is in most cases found just a few cm beneath the turf layer. In several cases the cultural layer has been strongly disturbed by freeze-thaw activity.

The excavated sites where charcoal was found in a reliable context have been ^{14}C-dated. In a few cases fire-cracked rocks have been dated by thermoluminescence (TL). Both excavated sites and sites where artefacts have been recovered from small test-pits have been dated typologically, but in many cases diagnostic artefacts are few, and it is only possible to give a chronological estimate within a wide range, up to 1000 years or more.

In the Halne area 21 archaeological sites have been found within a shore length of *ca*. 1 km (Fig. 1). In the same area two palynological sites, Halne I and II, were chosen. Dates from these sites are given in Table 1.

The ^{14}C-datings were done by Laboratoriet for Radiologisk Datering (NAVF), Trondheim. The TL dates were done by Atomenergikommisionens Forsøgsanlæg Risø, Denmark.

The aim of this chapter is to discuss the use of the area during the last 5000 years based on our interpretation of the available evidence.

Sites and datings

Archaeology

The 21 localities include 11 localities with house-grounds, of which 3 are associated with charcoal pits, 1 has charcoal-pits but no house-grounds, 1 hearth is unconnected with other structures, and 8 are open settlement sites.

The 11 house-ground localities show more or less visible traces of at least 14 houses and 3 charcoal-pits. One of the excavated house-grounds (loc. 447) is of Middle Mesolithic age; the rest are probably less than 2000 years old. Only 2 of these have been excavated (locs. 38, 165). Both are, according to artefact finds, of post-Medieval age, probably from the 17th-19th centuries (Fasteland 1971a).

Of the 8 unexcavated house-ground localities, 2 charcoal-pits outside the houses (locs. 446, 444) were dated to *ca.* 1800-1500 B.P. This is in good agreement with dates of similar charcoal-pits from other mountain districts of Central South Norway (Gustafson 1980; Magnus 1986; Prescott 1986). We may therefore assume that the unexcavated charcoal-pit locality (Table 1) is of the same age. The remaining 6 house-ground-localities (434, 438, 437, 445, 443, 436) are hard to date. None seems to be younger than 100-150 years, and they are probably not older than *ca.* 2000 years, but more exact dating is difficult without excavations. Locality 436 is probably of post-Medieval age.

Four open settlement sites have been excavated. One of them (loc. 169) has been TL-dated to *ca.* 3200-3800 B.P. using fire-cracked rocks (Indrelid 1986). Locality 40 has been ^{14}C-dated to *ca.* 1100-1300 B.P.

Table 1. Archaeological sites with thermoluminescence (TL) or radiocarbon datings, and estimated age by typology.

Locality	Type of structure	C/TL age (years B.P.)		Age estimated by typology (years B.P.)	Diagnostic artefacts
434	house-ground	-		2000 - 200	-
435	open site	-		5000 - 3800	flint flakes and blades
169	open site	TL: 3924 ± 300		5000 - 3800	tanged arrow-heads, blades
		TL: 3574 ± 200			
448	open site	-		3800 - 2500	leaf-shaped arrow-heads
-	charcoal pits	-		2000 - 1500	-
40	open site	C: 1210 ± 70	T-1954	3300 - 2500	leaf-shaped arrow-heads (triangular)
		C: 1140 ± 60	T-2195	300 - 150	fragments of clay pipe
				70 - 10	metal cartridge
165	house-ground			200 - 100	fragments of clay pipes, horse-nails
38	house-ground			350 - 100	fragments of clay pipes, horse-nails
439	open site			8000 - 2500	flint flakes
-	hearth	C: 3510 ± 90	T-999		-
458	open site	-		5000 - 3500	flint blades
436	house-ground	-		500 - 150	-
437	house-ground	-		2000 - 500	-
438	house-ground	-		2000 - 150	-
440	open site	-		8000 - 2500	flakes
168	open site	C: 4440 ± 130	T-1707	5000 - 4000	slate points, tanged arrow-heads, blades
447	house ground	C: 7520 ± 150	T-1798	8000 - 6000	microblades, transverse arrow-heads
446	charcoal pit	C: 1610 ± 110	T-1418	2000 - 1500	-
444	charcoal pit	C: 1710 ± 110	T-1641	2000 - 1500	-
445	house-ground	-		2000 - 500	-
443	house-ground	-		2000 - 500	-

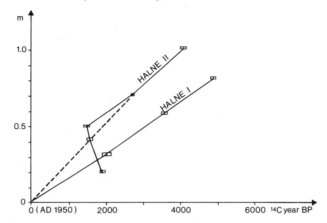

Fig. 2. Peat-accumulation rates from Halne I and Halne II using available dates (for interpretation see text).

(Indrelid 1986). The other sites are dated typologically. In some cases a general date to the period of Stone Age technology is the only possible estimate. These time ranges are extremely wide. Finds from other parts of the Hardangervidda indicate that this technology lasted until at least 2500 B.P. (Indrelid 1986), and in mountain areas to the north Stone Age technology lasted until 2000-2300 B.P. (Prescott 1986, p. 161). In Figure 6 a date of 2500 B.P. is used. It should be noted that the apparent break in cultural continuity in the second half of the first millennium B.C. may be a consequence of typological dating problems, rather than reality.

Some of the open settlement sites have provided indications of repeated settlements from different periods (e.g. loc. 40).

Vegetation history

The 2 palynological sites are *ca.* 500 m apart along the lake shore. Samples for pollen analysis and ¹⁴C-dating were taken from open peat walls. Laboratory procedures followed Fægri & Iversen (1975).

Radiocarbon dating

Eight ¹⁴C-dates have been obtained from the peat profiles (Table 2).

The accumulation rate in Halne I is based on 3 ¹⁴C-datings (Fig. 2). The dates give acceptable accumulation values and a more or less constant rate during the last 5000 years. The ¹⁴C-datings from Halne II are not acceptable. At least the 2 uppermost dates (T-5641, T-5644) are unacceptable and probably also T-5643, even when 2 standard deviations are considered.

Dating of peat material can, in some cases, give slightly younger

Locality	Material	Depth (m)	Age (C-years B.P.)	Lab. no.
Halne I	Peat	0.30 - 0.32	2050 ± 110	T-1430
Halne I	Peat	0.55 - 0.57	3510 ± 90	T-999
Halne I	*Betula nana*	0.80	4860 ± 90	T-1080
Halne II	Peat	0.19 - 0.21	1880 ± 80	T-5641
Halne II	Peat	0.39 - 0.41	1610 ± 80	T-5644
Halne II	Peat	0.48 - 0.51	1560 ± 80	T-5643
Halne II	Peat	0.69 - 0.71	2660 ± 40	T-5640
Halne II	Peat	0.99 - 1.00	4120 ± 80	T-5642

Table 2. Radiocarbon dates from the Halne peat profiles.

dates than expected, because younger roots infiltrate the lower sediments. However, our results also include older dates at younger levels, suggesting that some kind of redeposition must have occurred. Redeposited peat material has been discussed in other areas (e.g. Kaland 1986; Kvamme this volume). The main cause of redeposition of such material was probably human activity close to the site. The old path (Normannslepa) passes within 20-50 m of Halne II and erosion may have occurred. However, there are no obvious signs of this in the pollen or sediment stratigraphy.

The Halne area is well defined and is a suitable place for grazing and human habitation in the summer. It is therefore likely that one habitation influenced the whole area, most probably including the Halne I and Halne II sites. If this is not the case, the dispersal of lowland weed pollen and locally produced grass pollen is much more restricted than generally expected.

For further discussion, a chronological zonation is used for Halne I. For Halne II a bio-zonation (alternative I) is made based on dated local vegetation changes at Halne I. In the final discussion, an alternative II chronozonation is presented, based on the present-surface age and the 2 lowermost ^{14}C-dates only (Figs. 2, 6). Because of the unreliable datings at Halne II, only the lowest part of this diagram, up to 0.60 m, is used and discussed here.

Halne I

The diagram (Fig. 3) reflects vegetation changes during the last 5000 years, namely the end of the Atlantic, the Subboreal and the Subatlantic chronozones (Mangerud *et al*. 1974). The pollen spectra are dominated by herb pollen, e.g. Cyperaceae, Poaceae. Some shrub pollen occurs. All tree-pollen, except locally produced pollen of *Betula nana*, are long-distance transported. The pollen zones are summarized in Table 3.

Zone no.	Depth (m)	Estimated age (years B.P.)	Local pollen assemblage zone
HI-7	surface - 0.08	present - 450	*Juniperus*, Poaceae, *Rumex*
HI-6	0.08 - 0.175	450 - 1100	*Salix*, Poaceae, Apiaceae
HI-5	0.175 - 0.33	1100 - 2150	Poaceae, *Rumex*
HI-4	0.33 - 0.43	2150 - 2900	*Salix*, Apiaceae, *Betula (B. nana)*, *Dryopteris*
HI-3	0.43 - 0.67	2900 - 4050	*Juniperus*, Poaceae, *Dryopteris*
HI-2	0.67 - 0.78	4050 - 4800	Poaceae
HI-1	0.78 - 0.90	4800	*Betula (B. nana)*, Asteraceae, *Dryopteris*, *Rumex*

Table 3. Pollen zone characteristics for the Halne I profile (Fig. 3).

Halne II

Peat accumulation at Halne II also started *ca.* 5000 B.P. Most of the Subboreal and Subatlantic chronozones are represented. Pollen values are dominated by Cyperaceae together with *Salix* and *Betula (B. nana)* pollen. Tree pollen, except for possible *Betula nana*, is long-distance transported. The pollen zones are summarized in Table 4.

Discussion of the pollen diagrams

Moe (1973) discussed the possibility of local growth of plants interpreted as 'culture indicators' in mountain areas, against the alternative hypotheses of long-distance, regional wind transport and endo- and epizooic dispersal (Moe 1981, 1983b). All pollen of taxa such as *Plantago lanceolata*, *Urtica* and *Artemisia* recorded from this mountain area probably orginated from populations at lower altitudes and are mainly transported by humans and domestic animals. The interesting taxa found at Halne in the last 5000 years are *Plantago lanceolata*, *P. major*, *Urtica* and *Artemisia*. In addition, isolated grains of Chenopodiaceae and *Centaurea cyanus* occur. All these pollen taxa are subsequently called 'lowland-weed-pollen' (LWP). Moe (1973) showed that scattered or isolated occurrences of these taxa do not necessarily imply or demonstrate any human or cattle activity in the local area, but rather that they may reflect a small number of direct visits from nearby valleys where these taxa grew. Number of 'grazing days' (number of cattle x number of days cattle present) may be high at the same time as low or scattered LWP is found. Occurrence of LWP in spectra close to each other is an indication of more or less continuous traffic from the lowlands.

Therefore both the presence of LWP and changes in local vegetation at this altitude are of great interest in tracing 2 separate aspects of human activity, animal grazing and droving. In Halne I the first occurrence of LWP is in zone HI-3. Later, isolated grains occur in the oldest parts of zone HI-5, and then there is a maximum of *Artemisia* in late HI-5 and HI-6. In the most recent zone (HI-7) pollen of these taxa are again found.

Major changes in the local vegetation start in HI-2 with a decrease of shrubs, ferns and tall-herbs. A temporary rise in Poaceae suggests a change to grassland. Such a change most probably reflects increased local grazing by domestic animals.

Fern vegetation becomes re-established during HI-3 and HI-4. This may indicate reduced grazing. In HI-5 a rapid rise of grass and *Rumex* pollen occurs, together with reduced values of ferns and *Salix*. This probably reflects heavy grazing of the local vegetation.

Fig. 3. Pollen diagram from Halne I.

No indications of climatic change are found in this mountain area at this time. Because of the lack of ^{14}C-dates in this part of the

HALNE I, Eidfjord, Hordaland. 1135 m s.m.

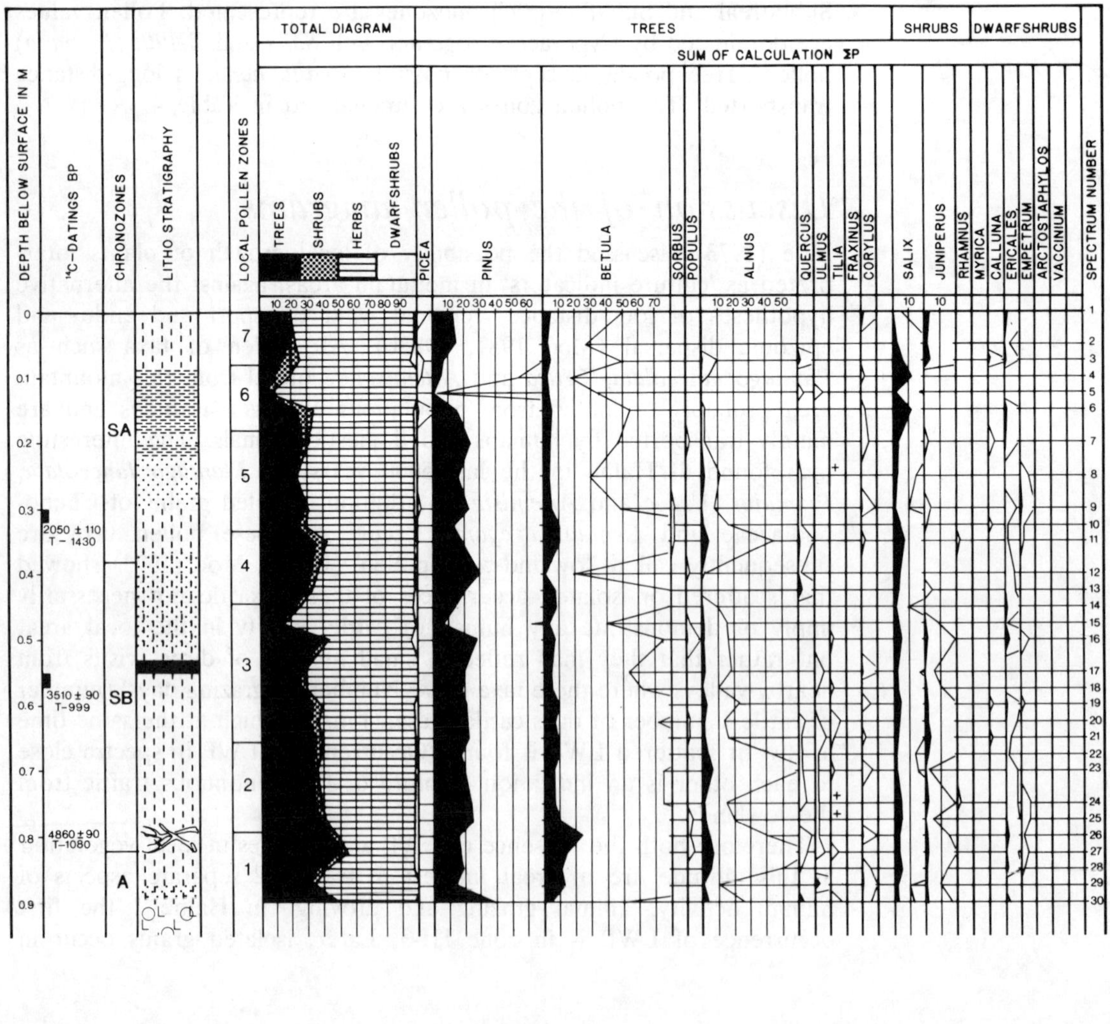

diagram, the ages of the different events are only approximate.

To some extent the local vegetation changes in Halne II (Fig. 4) are less dramatic compared with Halne I. However, some changes are found. In HII-2 the amounts of *Salix* and tall-herb pollen are low, and a small rise of Ericales is found. The peak of *Salix* in HII-3, together with some Apiaceae, may indicate a temporary reduction of grazing. However, the amount of grazing in HII-2 and HII-4 was probably small.

The 2 sites are compared in Figure 5. The pollen curves of the most important taxa used for discussing human impact on the local vegetation are directly compared by plotting them against the ^{14}C-timescale. Because the sites are so close (500 m), it is expected that local grazing effects would be similar in both. Also the transport of LWP must have been nearly equal at the sites, because the same

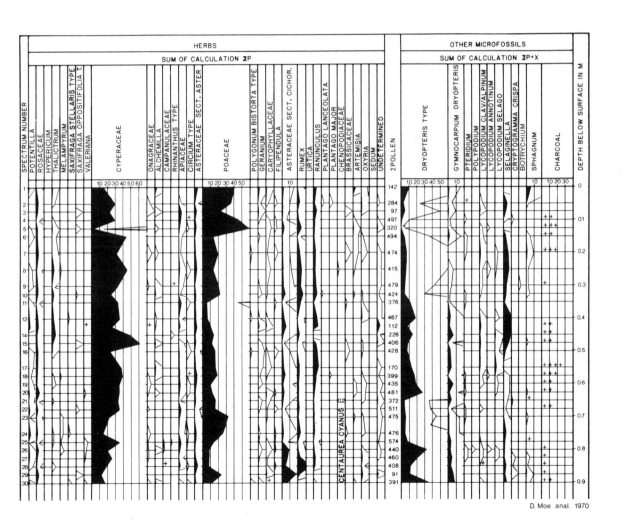

D. Moe anal. 1970

track passed by both. In addition, wind-transported *Picea* pollen is expected to start at about the same time. The dating of Halne II is based on this and on correlations with pollen-zone boundaries in Halne I. To the right, columns for each pollen site show the interpretation of (1) assumed grazing effect caused by domestic animals on the local vegetation, and (2) the amount of LWP reflecting its direct or indirect transport from mountain valleys or lowlands beyond the mountain plateau by domestic animal movements (see legend for Fig. 5).

Fig. 4. Pollen diagram from Halne II.

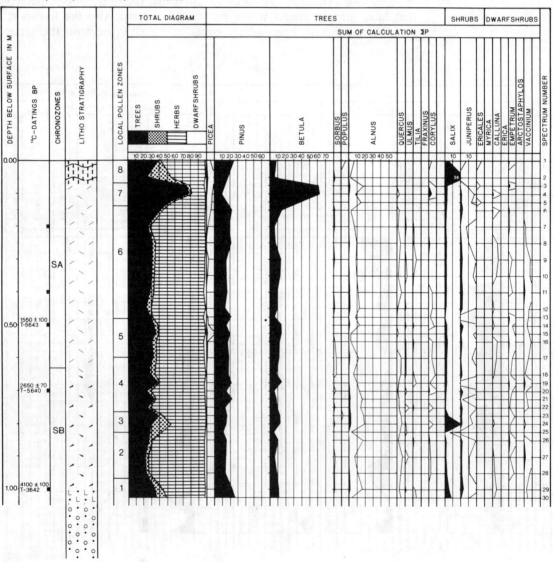

HALNE II, Hol, Buskerud, 1134 m s.m.

Grazing seems to have been continuous for a long period from 4900 to 2800 B.P. Between 2800 and 2200 B.P., there was little or no grazing effect. Grazing increased again after 2200 B.P., and this impact was maintained more or less continuously to the present.

Continuous records of LWP are found from 4100 to 3500 B.P., and from 2200 B.P., and a more or less continuous record to the present. Using alternative II chronology for Halne II (Fig. 6), the oldest stage starts at 3500 B.P. and ends at 2900 B.P. in addition to a short episode close to 2600 B.P.

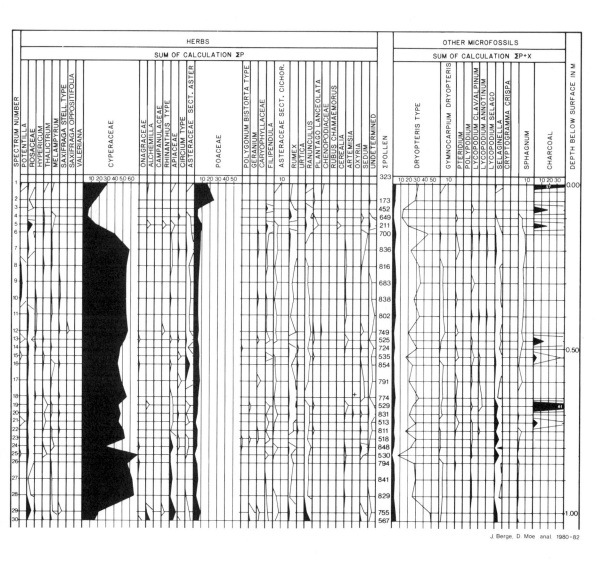

J. Berge, D. Moe anal, 1980-82

Zone no.	Depth (m)	Estimated age (years B.P.)		Local pollen assemblage zone
		Alternative I	Alternative II	
HII-8	Surface - 0.07	present - 350	present - 350	*Salix*, Cyperaceae, Poaceae
HII-7	0.07 - 0.14	350 - 750	350 - 700	*Betula (B. nana)*
HII-6	0.14 - 0.47	750 - 1750	700 - 1500	Cyperaceae (I)
HII-5	0.47 - 0.59	1750 - 2250	1500 - 2000	Cyperaceae, Asteraceae Sect. Asteroidae
HII-4	0.59 - 0.76	2250 - 3350	2000 - 2900	Cyperaceae (II)
HII-3	0.76 - 0.83	3350 - 3750	2900 - 3300	Cyperaceae, *Salix*
HII-2	0.83 - 0.97	3750 - 4500	3300 - 4000	Cyperaceae (III)
HII-1	0.97 - 1.03	4500	4000	*Salix*, Apiaceae, Ranunculaceae

Table 4. Pollen zone characteristics for the Halne II profile (Fig. 4).

Discussion and Conclusions

In Figure 6 the complete archaeological and palynological data are brought together. This compilation gives a consistent picture of the main use of the small area of Halne.

The oldest finds date from 7520 ± 150 B.P. However, the main information starts at 5000-4900 B.P. with both archaeological and palynological evidence, the time when the oldest dated evidence of Neolithic influence is found on the Hardangervidda (Indrelid & Moe 1982).

The archaeological finds at Halne show a change in types of flakes and tools, for example the arrows, just after 4000 B.P. The archaeological data from the Early and Middle Neolithic Periods (*ca.* 5000-4200 B.P.) have been related to pastoral nomadism, probably by groups which spent the winter season in the fjord and coastal districts to the south or south-east of the Hardangervidda. Around 4000 B.P., there was a change. More permanent settlements seem to have existed in the mountain valleys surrounding the mountain plateau in addition to coastal areas in the inner part of the Hardangerfjord (Sørfjorden) (Indrelid 1986). The transition stage between Early and Late Neolithic Periods is characterized by a continuous presence of LWP, and it is suggested (Moe 1973) that droving increased and became well established between 4000 and 3500 B.P. So far there is no other information about places involved in the movement. Instead of local grazing at Halne, animals were moved between lower pastures in the east and west, crossing the mountains at Halne in the process.

The archaeological Bronze Age on the Hardangervidda ends *ca.* 2500 B.P., while local grazing at Halne ends at 2800 B.P. (Alternative II 2500 B.P.). The locally delayed start of the Early Neolithic at Halne, and the abandonment of the site some years before the end of

Fig. 5. Combined pollen diagram of selected taxa from Halne I (——) and Halne II

(------) plotted against age. (Datings of levels in Halne I are based on the ¹⁴C-curve in Fig. 2; datings of levels in Halne II are explained in text). On the right, phases of human activity are indicated: solid lines = heavy local grazing and animal movement (indicated by the sum of *Artemisia, Plantago lanceolata, P. major, Centaurea cyanus* ≥ 0.5%); dotted lines = light local grazing and less animal movement (indicated by continuous curve of the above taxa but at < 0.5%).

the Bronze Age may reflect the fact that the area was probably not ideal pasture for cattle, sheep or goats.

An interesting gap in the palynological evidence for human activity (Fig. 6) occurs between 2800 B.P. (Alternative II 2500 B.P.) and 2200 B.P. There are several reasons for a such break in the use of this mountain area, such as climatic change, changes in the social economy, plagues, etc.

Although a general climatic amelioration is found in northern Norway between *ca.* 3400 B.P. and 2700-2600 B.P. (Moe 1983a), no similar changes have, so far, been found in southern Norway (Moe 1978).

Another explanation might be a fundamental change in settlement system, from different kinds of nomadism and extensive use of land to the permanent farm. Permanent farms are known in South Norway from 0 A.D., but we do not know how far back in time this economic

HALNE I, II.

D. Moe 1986

and social form can be traced. It is interesting to note records of LWP dated to *ca.* 2500-2600 B.P. from Hadlemyrane (1000 m) and Vøringsfossen (670 m) in a mountain valley to the west of the Hardangervidda (Moe 1978; Indrelid & Moe 1982). From the Sogn sub-alpine area to the north-west, evidence for settlement at the same time comes from ^{14}C-dated pollen diagrams and pits in the sub-alpine region, associated with summer pasturing activity (Gustafson 1980; Prescott 1986; Kvamme this volume; Magnus 1986). The renewed expansion at 2200 B.P. at Halne may reflect expansion at lower altitudes, and the need for pasture away from arable areas during the summer.

From 2200 B.P. to the present, before railway and roads crossed the mountains, there was a more or less continuous use of the area

HALNE

D. MOE 1986

for summer grazing, as a route for transhumance with domestic animals and as a base for hunting. An exception is a short break tentatively dated *ca.* 700 B.P. Unfortunately, no ^{14}C-datings are available from this stage. The Black Death severely reduced the population in the surrounding valleys, and the gap found at the end of Medieval time is probably a result of this.

Summary

(1) An area on the northern shore of Halnefjord on Hardangervidda has been studied archaeologically and palynologically.

(2) At this altitude, above the tree-line for 5000 years, significant quantities of pollen of lowland anthropochores have been found, distinct periods of grazing by domestic animals are recorded and numerous open archaeological sites with house constructions, flakes, tools, fire-cracked rocks, charcoal, horse-nails, etc. are found. People lived here during the summer.

(3) Results show an almost continuous summer-use of the area from at least 4800 B.P. to 2800 (or 2500) B.P., and from 2200 B.P. to the present.

(4) There is evidence that the area was used in three main ways: hunting and fishing; movement of domestic animals (droving) from lower altitudes; and semi-permanent occupation with grazing by domestic animals. House remains are found associated, to some extent, with all these activities.

(5) Halne is a hunting area today, and hunting equipment has been found dating from 7500 B.P. to present.

(6) Greater droving activity occurred between 4000 B.P. and 3500 (3000) B.P., around 2800 B.P., and from 2200 B.P. more or less continuously to today. These suggested periods of droving are not always concurrent with periods of intensive grazing.

(7) From 4000 B.P. to 2800 (2500) B.P. grazing is associated with quantities of fire-cracked rocks and two kinds of arrows. From 2200 B.P. to today, grazing activity is linked to house remains and pits. The earlier period may reflect a regular seasonal occupation resembling 'driftlæger' (base-camp sites for shepherds), as known from the last century rather than 'seter'-activity with dairy produce.

(8) Since 1250 A.D. discontinuities in the record may result from the Great Plague, but no precise datings are available.

Acknowledgements

The authors are most grateful to Hilary Birks and to the referee for linguistic revision and for improvements to the manuscript.

◄

Fig. 6. Comparison diagram of known archaeological sites (with locality number) dated by typology, ^{14}C, and thermoluminescence, and estimated dated human and domestic animal influence on the vegetation in the area (Halne I, Halne II).
Archaeological sites:
Full line = ^{14}C-datings;
broken line = typological or thermoluminescence datings (TL).
Other abbreviations:
H = house-grounds;
C = charcoal;
P = pits with charcoal;
Ft = flakes and tools;
Fc = fire-cracked rocks.
Palynological sites: see legend of Fig. 5. For the geographical position of the sites see Fig. 1.

Nomenclature

Plant nomenclature follows J. Lid (1963) *Norsk og svensk flora* (3rd edition), Det norske Samlaget, Oslo.

Vegetation and Habitation History of the Callanish Area, Isle of Lewis, Scotland

S. J. P. Bohncke

Introduction

During peat-cutting by local crofters on a peninsula near Callanish, Isle of Lewis, Outer Hebrides, parts of a sub-peat 'field-wall' system became visible. The peninsula, which lacks a local name, extends south from Callanish (6⁰45'W, 58⁰12'N), and is low-lying, reaching an elevation of only a few metres elevation (Fig. 1). It is west of a small inlet, Tob nan Leobag, an arm of Loch Ceann Hulavig, which is, in turn, part of Loch Roag.

Apart from a fringe of improved land at the south end, the peninsula is peat-covered and most of it has been extensively cut for domestic fuel. Past peat-cutting has removed 1 m or more of peat over most of the area. Recently, cutting was abandoned and then resumed a few years ago on a smaller scale. It has removed the peat to within a few cm of the underlying glacial till. During this cutting, linear stone features were partially uncovered and are gradually being revealed further each year. Parts of these features, now shown to be the remains of ancient field and enclosure walls, are visible in the present-day intertidal zone of a small bay on the west side of the Tob nan Leobag inlet. Peat erosion and submergence at high tide of an exposed stretch of field wall presupposes a considerable degree of coastal submergence. The peninsula thus appears to form the western edge of a now submerged depression. Although recognized as a general trend in the Outer Hebrides (cf. Ritchie 1966) datings for stages of this transgression are imperfectly known.

The proximity of the site to the major prehistoric monuments of Callanish (Fig. 1) enhanced the scientific importance of the peninsula for more detailed survey and excavation and, above all, for palaeoecological study. Survey and trial excavations of the visible archaeological features on the peninsula were carried out in 1979 in advance of further peat-cutting, under the direction of T. Cowie (Central Excavation Unit, Scottish Development Department (Ancient Monuments Branch)).

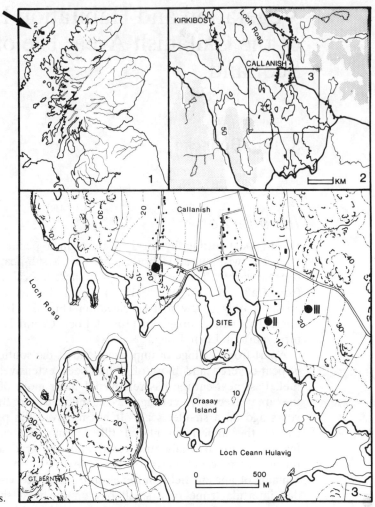

Fig. 1. Map showing
(1) Scotland and the position
of the Outer Hebrides;
(2) the Callanish area and its
present settlements;
(3) the locations of the peat
columns in relation to the
nearby monuments:
I = the standing stones of
Callanish;
II and III = small stone circles.

In some cases the walls appeared to have been constructed after peat growth was well established. It was hoped that a pollen diagram from such an area might show human influences on vegetation which might be linked, for example, with prehistoric activities that culminated in the erection of the Megalithic monuments (Fig. 1), including a Neolithic chambered tomb at the stone circle (Henshall 1972; Burl 1976). Lewis and Harris have been settled since Neolithic times (Burl 1976). A birch-forest clearance has been dated to 2460 ± 79 B.P. (Simpson in Burleigh *et al*. 1973).

The presence of archaeological features on the peninsula was first reported to the Ancient Monuments Branch of the Scottish Development Department in 1977. That autumn, J. W. Barber collected a 1 m deep column, Callanish-1 (CN-1) (Figs. 2, 3), from a peat face next to

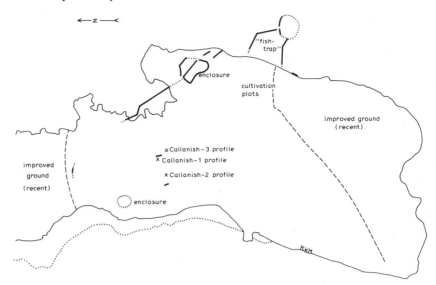

Fig. 2. Enlarged sketch of the peninsula (SITE in Fig. 1(3)) showing the principal features and the position of the Callanish samples. The thick black lines indicate the position of the 'field-wall' remains uncovered during the excavations.

an exposure of a field wall, and a 1.5 m deep column (CN-2), next to the footpath to the peat cuttings, where the peat surface seemed to be preserved intact. In autumn 1978, T. Cowie collected Callanish-3 (CN-3), from peat containing birch wood, initially thought to be an artificial deposit.

Three transects of peat borings showed that the central part of the peninsula forms a reasonably distinct basin. The CN-1 and CN-3 sections are therefore likely to contain older peat layers than CN-2 which comes from a relatively higher area (Fig. 3). The coring results also showed that sub-peat features, partially revealed in the excavations, and interpreted as cultivation plots (T. Cowie unpublished) are located on the higher margins of the peninsula.

Methods

The profiles

The peat columns were collected by hammering metal boxes (0.5 m x 0.1 m) into the cleaned peat face. The boxes were then cut out with a long knife and spade, and sealed in polythene. The stratigraphy of CN-1 was described in the field. The stratigraphy of CN-2 and CN-3 was described from the boxes in the laboratory (Fig. 3, Table 1).

The peat of CN-1 is undisturbed up to an undulating indistinct division between 67 and 75 cm above the till. This division is almost certainly the result of peat-cutting. The top sods of the peat-bank that were cut in one year were replaced on the bank that was cut in the preceding year. The vegetation on the sods, under the prevailing

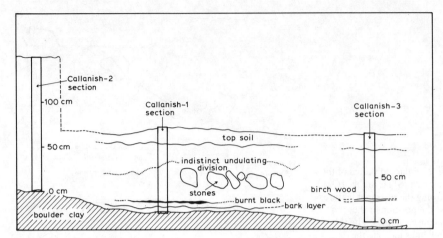

Fig. 3. Schematic impression
of the relative position of the
analysed sections at Callanish.
The indistinct undulating
divisions represent the depth
to which peat-cutting had
proceeded in the past. The
linear band of stones is the
possible remains of a
'field-wall'.

wet conditions, keeps on growing in the new position, and the root
system penetrates the underlying layer, resulting in the appearance
of continuous peat growth. The transition between 79-83 cm in CN-3
probably represents the same feature. Hence, in both CN-1 and CN-3
the top part was not analysed.

In the laboratory, pollen samples were taken at 5 cm intervals, and
prepared according to Fægri & Iversen (1975), using HF treatment
where necessary to remove silica. Additional samples at 1 cm intervals
were later taken from CN-3.

Radiocarbon dates

Thirteen ^{14}C-dates from peat and 2 from wood are listed in Table 2,
in ^{14}C-years B.P. Radiocarbon age against depth for CN-3 is plotted
in Figure 4. The age of the upper and lower limit of each pollen
(sub)zone was estimated by linear interpolation between pairs of dates,
using mean age and mean depth for each ^{14}C-date.

The pollen diagrams

The pollen sum includes total tree and herb pollen, but excludes pollen
of aquatics and spores. A summary diagram contains the following
groups: (1) regional trees, (2) local trees, (3) blanket-mire species, (4)
indifferent herbs, (5) Poaceae, and (6) anthropogenic herbs. The
criteria used to separate Cerealia-type pollen from wild grasses follow
Beug (1961) except for grain size. A minimum grain size of 42 μm
was used.

Because CN-3 contains the oldest peat, it will be discussed first,
followed by the next oldest CN-2, and lastly by CN-1.

Zonation of the pollen diagrams

The pollen diagrams are subdivided into local zones (see Figs. 5a, 5b, 5c), mainly based on reflections of human activity in the landscape. They are prefixed by the site designation CaN and are numbered from the oldest to the youngest. The levels at which I think the sections overlap are indicated in the zone column on Figure 5.

Inferred vegetational history of the Callanish area

Zone CaN-1 (60-54.5 cm in CN-3; up to ca. 8400 B.P.)

The extremely high *Betula* pollen percentages during this zone contrast with findings from other areas in the Outer Hebrides (e.g. Birks & Madsen 1979; Blackburn 1946; Ritchie 1966; Walker 1984b) or comparable areas such as the Shetlands (Jóhansen 1975, 1985) or Orkneys (Moar 1969; Keatinge & Dickson 1979). However, finds of *Betula* and *Salix* wood to the north of Callanish and dated to 7980 ± 55 B.P. and 9140 ± 65 B.P. respectively (Wilkins 1984), prove the presence of these trees in the Outer Hebrides at these times. The absence of birch wood in the peat of CN-3 suggests that birch was not actually growing on the site. It may have occupied sheltered valleys presently

Callanish-1 column (CN-1)

47 -	0 cm	rich brown peat
48 -	47 cm	black burnt layer
56 -	48 cm	rich brown peat
57 -	56 cm	layer with birch-bark remains
63 -	57 cm	black, greasy peat, strongly humified
65 -	63 cm	basal peat with rock fragments

Callanish-2 column (CN-2)

	0 cm	present vegetation cover
20 -	0 cm	light brown peat with recent root infiltrations
45 -	20 cm	rich brown peat, gradually becoming lighter towards the top. *Eriophorum* remains at 40 cm
61 -	45 cm	fibrous brown peat with *Eriophorum* remains
69 -	61 cm	*Eriophorum* peat
85 -	69 cm	dark brown, humified peat with *Eriophorum* remains
134 -	85 cm	fibrous brown peat with *Eriophorum* remains
139 -	134 cm	fibrous dark brown peat
148 -	139 cm	black greasy peat, strongly humified
150 -	148 cm	basal part with rock fragments

Callanish-3 column (CN-3)

24 -	0 cm	brown, less strongly decomposed peat
33 -	24 cm	brown, fibrous peat
35 -	33 cm	birch wood layer
60 -	35 cm	strongly decomposed, amorphous, black peat

Table 1. Sediment descriptions.

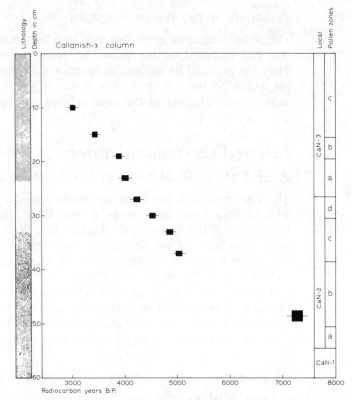

Fig. 4. Plot of ^{14}C-age against depth for the Callanish-3 column. The thick horizontal bars represent one standard deviation of the age determination, the thin horizontal lines represent two standard deviations.

submerged. *Salix, Sorbus* and *Corylus* were associated with *Betula* in suitable sites, and under the light canopy, plants such as *Melampyrum, Lonicera periclymenum* and *Empetrum* and ferns such as *Pteridium aquilinum* and *Dryopteris* spp. occurred. The peat-forming vegetation at CN-3 contained *Sphagnum* and perhaps *Salix*.

Zone CaN-2 (54.5-26.5 cm in CN-3; ca. 8400-ca. 4200 B.P.)

Subzone CaN-2a (54.5-50.5 cm; ca. 8400-ca. 7650 B.P.)

The sharp decline in *Betula* pollen may be part of a regional event around 7900 B.P. (Wilkins 1984) induced by a wetter climate, which, besides killing the trees, allowed wood to be preserved. Although the *Salix* pollen-increase immediately after the *Betula*-decline could be caused by increased wetness, other changes at the same time cannot be explained in this way. Increases in *Calluna*, Poaceae and *Potentilla*-type suggest the expansion of heather moor and grassland. A rise in charcoal particles suggests an increase in fire incidence. A grain of *Chamaenerion*-type at 55 cm in this context probably derives from *Epilobium angustifolium*, a plant associated with natural disturbance such as wind-thrown forest, forest fires and dieback in grassland following drought and various types of human disturbance such as forest felling and woodland clearance, especially if fire is used (Myerscough 1980; Stockmarr 1975). In general, deciduous forests are less vulnerable to forest fires induced by lightning than coniferous forests (Rackham 1980; Rowley-Conwy 1981). These changes, taken together with the fact that the decline in birch forest occurs over several levels, could mean that Mesolithic people were burning the forest (Edwards & Ralston 1985). The peak of *Salix* pollen suggests that willow, growing in wetter areas, was not affected, but was even favoured near the sampling site. The increase in open woodland towards the end of this subzone, as indicated by increases in *Melampyrum* and *Pteridium* (Stockmarr 1975) and heather moor and grassland with *Potentilla* may have attracted deer and other mammals, and in return these conditions may have supported a hunter-gatherer community.

The low pollen values of *Ulmus*, *Quercus* and *Pinus* throughout the Holocene are probably derived by long-distance transport. Pine was locally present on central Lewis *ca.* 4800 B.P. (Wilkins 1984).

Subzone CaN-2b (50.5-38.5 cm; ca. 7650-ca. 5320 B.P.)

Betula pollen percentages rise at the subzone CaN-2a/b boundary, but do not reach their former level before declining again. The regenerating birch wood is assumed to have repressed *Salix*, but it was, in turn, partially replaced by *Salix* and *Populus* in subzone CaN-2b. *Betula* values do not fall below 10%, suggesting that *Betula* was still locally present. The abundance of charcoal, together with the fluctuating *Betula* curve, probably reflects continuing human activity on the

Laboratory reference no.	Depth in cm below the top of the column		Radiocarbon age years BP
GU-1234	49-48 cm	CN-3	7270 ± 100
GU-1993	37 cm	CN-3	5035 ± 60
GU-1150	35-33 cm	CN-3	5180 ± 90
GU-1151	35-33 cm	CN-3	5180 ± 90
GU-1992	33 cm	CN-3	4860 ± 60
GU-1991	30 cm	CN-3	4525 ± 60
GU-1990	27 cm	CN-3	4255 ± 65
GU-1989	23 cm	CN-3	4005 ± 60
GU-1988	19 cm	CN-3	3890 ± 55
GU-1987	15 cm	CN-3	3430 ± 55
GU-1986	10 cm	CN-3	3010 ± 50
GU-1095	65-63 cm	CN-1	4810 ± 60
GU-1171	50-48 cm	CN-1	3220 ± 65
GU-1170	20-18 cm	CN-1	2355 ± 65
GU-1205	31-29 cm	CN-2	225 ± 65

Table 2. Radiocarbon dates.

peninsula. The *Corylus* curve is unaffected, suggesting that the inland birch-hazel woods remained intact.

Locally, a wet grassland developed, supporting *Narthecium ossifragum*, *Sphagnum*, *Potentilla*, Cyperaceae, *Rumex* and *Succisa pratensis*. *Calluna* expanded greatly on drier soils cleared of forest and on wetter moorland.

The interpolated date for the start of the continuous *Alnus* pollen curve at 42 cm is 6000 B.P. *Alnus* was never locally present, and its pollen probably comes from elsewhere in Scotland, where it was spreading from Skye (6000-6500 B.P., Birks 1977) northwards to Orkney (5700 B.P., Keatinge & Dickson 1979). *Ulmus*, although not native on the island, shows a clear decline at the top of the subzone, which probably reflects the regional elm-decline on the mainland.

Subzone CaN-2c (38.5-30.5 cm; ca. 5320-ca. 4580 B.P.)

At 37 cm a *Betula* pollen decline is dated to 5035 ± 60 B.P. Birch branches from nearby peat at this level have been dated to 5180 ± 70 B.P. and 5110 ± 90 B.P. The wood dates are a little older than from the peat, not surprisingly since the wood was growing some time

Fig. 5a. Pollen percentage diagram from Callanish, column Callanish-3. The scale at the base of the diagram is percentages of total pollen, excluding aquatics, for black silhouettes; unshaded silhouettes are exaggerated 10x scale. In the summary diagram
1 = regional trees;
2 = local trees;
3 = blanket-mire species;
4 = indifferent herbs;
5 = Poaceae;
6 = anthropogenic herbs.

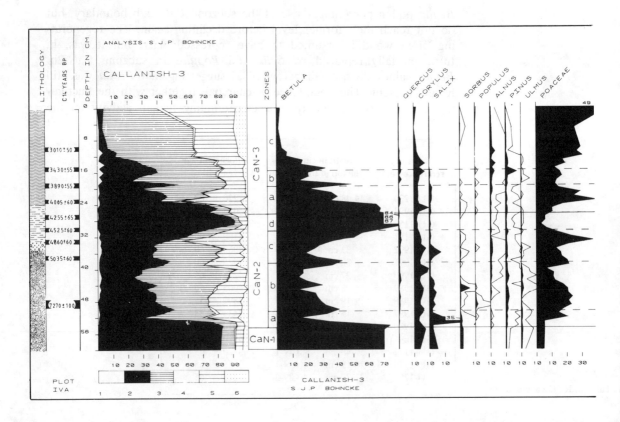

before it was buried in the peat. On the evidence available, the birch-wood layer can be dated to a period between 5180 ± 70 B.P. and 4860 ± 60 B.P.

A similar *Betula*-decline on the Orkneys after a mainland elm-decline was assumed by Keatinge & Dickson (1979) to be caused by an increase in onshore winds which reduced shrub vegetation at about 5000 B.P. At Callanish, this *Betula*-decline is followed by an increase in *Salix* and Ericaceae undiff. pollen, which may have been a response to increased wetness that killed the birch trees growing on peat and allowed their remains to be preserved by continuing peat growth.

Subzone CaN-2c is distinguished from the previous subzone by the occurrence of Cerealia-type pollen and low frequencies of *Plantago lanceolata*, reflecting local Neolithic activity. The rarity of *P. lanceolata* pollen during the Early Neolithic is also recorded in Northern Ireland, where Pilcher *et al.* (1971) explain it as the effect of a farming stage with an emphasis on arable farming (cf. Edwards this volume). The charcoal curve reaches enormous values. At first *Calluna* invaded the areas cleared of birch, but later it was replaced by grasses. This may be a result of agriculture, or exploitation of the *Calluna* heaths for grazing by regular burning, eventually converting them to grass heaths.

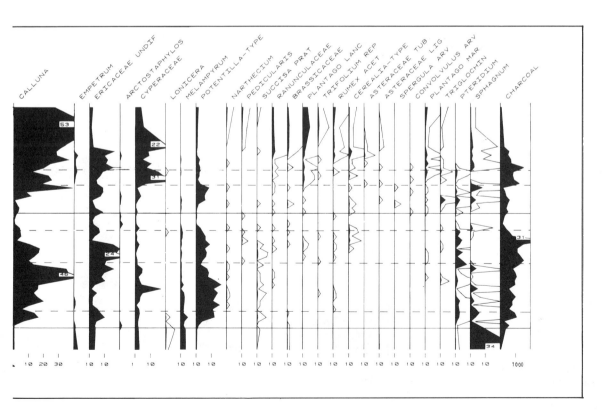

The increase of *Pteridium* spores may be a response to grazing of the remaining birch woods (Behre 1981), or it may reflect the spread of *Pteridium* on to degraded heather moor and grassland.

From 25 cm onwards *Betula* percentages gradually rise, reaching a maximum in subzone CaN-2d. The brown fibrous peat possibly derives from strongly decayed birch-wood, suggesting that birch colonized the peat, possibly as a result of drying out.

Subzone CaN-2d (30.5-26.5 cm; ca. 4580-ca. 4190 B.P.)

This subzone is a phase of uninterrupted birch regeneration due, presumably, to the absence of prehistoric people in the close vicinity as suggested by the absence of *Plantago lanceolata*, Cerealia-type pollen and a significant decline in the charcoal curve. *Lonicera periclymenum* reappears in the understorey.

Zone CaN-3 (26.5-0 cm in CN-3; 65-24.5 cm in CN-1; ca. 4190-ca. 2200 B.P.)

Subzone CaN-3a (26.5-19.5 cm in CN-3; ca. 4190-ca. 3900 B.P.)

The decline in birch together with the reappearance of Cerealia-type and *Plantago lanceolata* pollen and a distinct increase in the charcoal

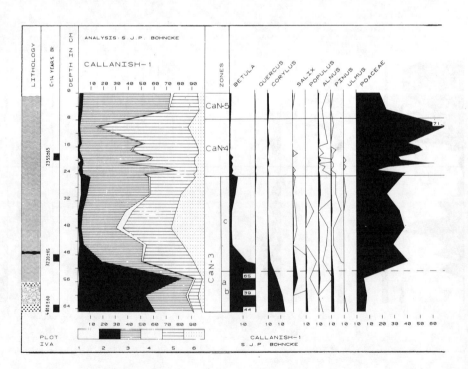

Fig. 5b. Pollen percentage diagram from Callanish, column Callanish-1.

curve suggests a second Neolithic clearance. Again, as in subzone CaN-2c, very few herbs that are usually associated with prehistoric agriculture occur in this zone. During the second half of this zone cereal pollen becomes less frequent, and grasses and *Potentilla* (?*P. erecta*) became more abundant. These herbs probably reflect a change from an economy with an emphasis on arable farming to an economy mainly based on stock grazing.

The end of this subzone seems to reflect abandonment of the area by prehistoric people followed by a regeneration of natural vegetation. At first, a species-rich Callunetum spread, succeeded, in some places, by birch woodland.

Subzone CaN-3b (19.5-15.5 cm in CN-3; ca. 3900-ca. 3400 B.P.)

The *Betula* regeneration is suddenly interrupted. Simultaneously there is a strong increase in the charcoal curve, and Cerealia-type pollen appear, together, this time, with a slightly more diverse herb vegetation. Again the combined changes are ascribed to human interference on the landscape to provide arable and pasture. This local clearance may relate to the Late Neolithic or Early Bronze Age.

The peak of *Pinus* pollen in one sample corresponds in time to the date of 3910 ± 70 B.P. for pine from south-east Lewis (Wilkins 1984). *Pinus* shows a widespread decline in north-west Scotland *ca.*

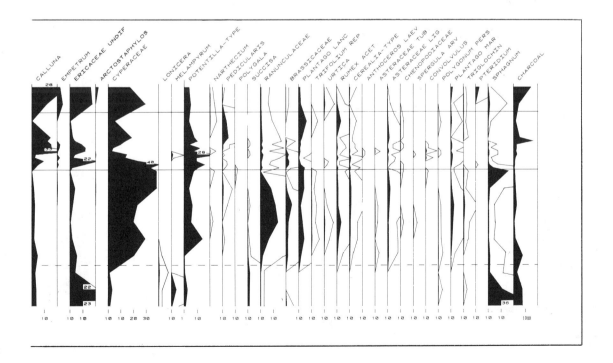

4500-4000 B.P., possibly due to increased climatic oceanicity leading to waterlogging of soils and spread of blanket bog (Birks 1975). This increased oceanicity may be reflected at Callanish by the expansion of *Plantago maritima* responding to increased sea-spray (Walker 1984b). Sea-spray is common in the Outer Hebrides so that *P. maritima*, besides occurring on coastal rocks and in salt marshes, occurs in grass-lands and heaths (H. J. B. Birks personal communication). However, the increase in *P. maritima* may also reflect the maximum of the marine transgression, when all low-lying areas around the peninsula were inundated.

Subzone CaN-3c (15.5-0 cm in CN-3; 52.5-24.5 cm in CN-1; ca. 3490-ca. 2520 B.P.)

The 2 previous clearance phases had only affected birch growing locally on the peninsula and in the valleys now inundated around Callanish. Now a regional clearance is recorded, also affecting *Corylus*.

From 15 cm (CN-3) onwards there is a strong increase in Poaceae and *Plantago lanceolata* becomes abundant. Together with *Trifolium repens, Urtica, Rumex, Potentilla*-type (probably *P. erecta*), Brassicaceae

Fig. 5c. Pollen percentage diagram from Callanish, column Callanish-2.

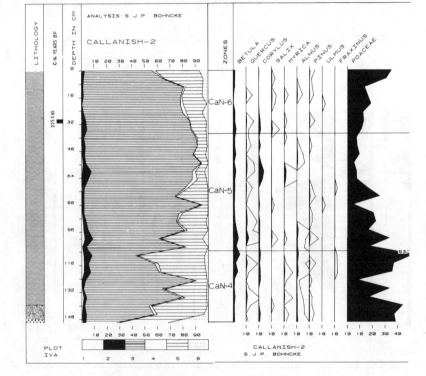

LEGEND:

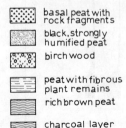

basal peat with rock fragments

black, strongly humified peat

birch wood

peat with fibrous plant remains

rich brown peat

charcoal layer

and Ranunculaceae, they provide strong evidence for the formation of pasture land. According to Iversen (1973) *Trifolium repens* is a good indicator of rough pasture, and *Potentilla erecta* is often associated with high grazing pressure (Behre 1981). Besides these herbs indicating grazing, Cerealia-type and herbs often associated with agriculture, such as Chenopodiaceae, *Spergula arvensis*, *Convolvulus arvensis* and *Polygonum persicaria*-type, appear in this zone. It is clear that stock breeding was not the sole means of subsistence of the prehistoric community. Agriculture was probably practised on higher ground on the edges of the peninsula (Fig. 2) and probably also on the surrounding hill slopes. However, the archaeological evidence for prehistoric field systems has been largely obscured by the peat cover and the recent improvement of the higher ground.

Towards the end of this zone there is a temporary decrease in many herbs associated with tillage and grazing. Instead the curves of Cyperaceae and *Sphagnum* accompanied by *Narthecium* and *Polygala* increase in CN-1. Taken together they indicate that local conditions became wetter. A decline in human population may also account for these changes. One effect of the decline in human impact is the regeneration of birch, but apparently only in small stands.

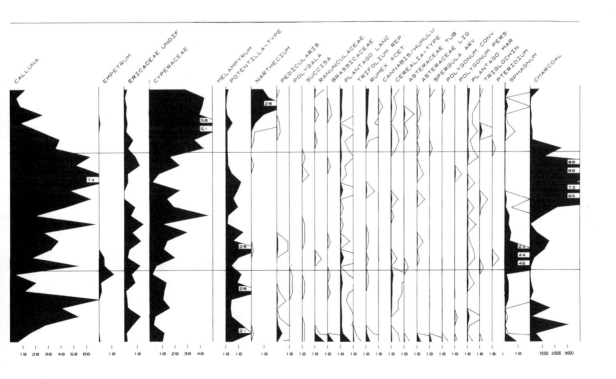

Zone CaN-4 (24.5-7.5 cm in CN-1; 150-106.5 cm in CN-2; ca. 2520-ca. 2030 B.P.)

From the start of this zone, tillage is resumed, as shown by the increase in Cerealia, *Convolvulus arvensis*, *Spergula arvensis*, Asteraceae and, surprisingly, spores of the liverwort *Anthoceros laevis* L. (CN-2). *A. laevis* is xerophilous and often occurs in stubble fields after harvest (Koelbloed & Kroeze 1965). Following this phase with an emphasis on tillage, *Plantago lanceolata* and *Rumex* spread. Together with Poaceae, they form part of the communities that invade fallow fields. Probably by this time rotational cultivation was practised.

Zone CaN-5 (7.5-0 cm in CN-1; 106.5-37.5 cm in CN-2; ca. 2030-ca. 650 B.P.(1300 A.D.))

During this zone there is an enormous increase in *Calluna*. Although *Calluna* pollen was present previously in considerable percentages (CN-3, 20-0 cm) the correlation between the different columns suggests that they must have been of local origin. The high values during and after CaN-4 indicate expansion of heather moor. Pollen of *Pedicularis*, *Succisa pratensis* and *Polygala* are present, reflecting locally wet conditions on the peninsula, which must have led to the formation of bog and wet moorland with *Sphagnum*.

The sparse occurrence of Cerealia-type pollen indicates that there were still some cultivation plots. However, few herbs associated with arable land occur. Brassicaceae, *Rumex* and *Plantago lanceolata* are more likely to have formed part of pasture-land vegetation. The combination of pollen types might suggest a different working of the land as the result of the increased oceanicity of the climate, and also the introduction of the iron plough (cf. Behre 1981). Perhaps the run-rig system (Dodgshon this volume) came into use. With reference to run-rig on Lewis, MacDonald (1978) writes "...the arable land, although held in run rig, was also regarded as part of the common grazing except between seed time and harvest". Also the presence of extensive moorland in the Callanish area was best suited to rearing livestock. The increase of charcoal may show that people started to manage the heather moors around Callanish to provide cattle fodder "the suitability of heather for management in this way depends on its capacity for vegetative regeneration, that is to say the production of young shoots from the old stem bases which are not usually killed by the passage of fire" (Gimingham 1975). Regular burning of heather produces even-aged stands which have a high productivity of edible new shoots.

Zone CaN-6 (37.5-0 cm in CN-2; ca. 1300 A.D.-recent)

At 35 cm in CN-2, the charcoal curve falls dramatically and the presence of pollen of Cerealia-type, Caryophyllaceae, *Spergula arvensis* and *Rumex*, indicate the start of a new agricultural phase. A date of 225 ± 65 B.P. (Fig. 5c) does not seem likely. "The ^{14}C activity of this sample is compatible with 20th century ^{14}C level due to the burning of fossil fuels since industrialization", (M. J. Stenhouse personal communication).

From the literature it appears that there was an increase in the population after 1266 A.D., when Lewis was ceded to Scotland under the terms of the Treaty of Perth. "With the cession of the Hebrides to Scotland numerous immigrants arrived from the mainland" (Mac-Donald 1978). As a result, agricultural activity might be expected to have increased on the island, and could have included the area around Callanish. According to MacDonald (1978), crop cultivation was of secondary importance to stock rearing, since pasture and heather moorland were much more widespread.

The record of *Humulus/Cannabis* pollen at 10 cm probably originates from *Cannabis sativa*. MacDonald (1978) writes: "Before the introduction of the potato in 1756, the stable crops were barley (or bere), *Hordeum* spec. and small black oats, *Avena strigosa*. Towards the end of the eighteenth century flax, *Linum usitatissimum*, and hemp, *Cannabis sativa*, were grown for a while, but the increasing number of cotton spinning mills on the mainland lowered the demand for island spun material and these two crops were eventually abandoned".

Other characteristics of this zone include the declines in *Calluna* and Ericaceae undiff. percentages from 25 cm onwards, and the enormous increase in *Narthecium ossifragum* to a maximumm of 28% at 10 cm. Heather suffered from land improvement (there are still lazy beds on and around the peninsula) and peat-cutting. The increase of Poaceae at the end of CaN-6 probably reflects abandonment of the lazy beds (Dodgshon this volume) and their use today as pasture. Nowadays, heather is scarce on the peninsula. In wet places such as peat-cutting pits, *Erica tetralix* and *Narthecium* are present. The continuous curve of *Plantago maritima* reflects proximity of the sea, and perhaps a renewed relative rise in the sea-level.

Discussion and conclusions

The presence of birch forest during the early Holocene in the Callanish area, as evident from the Callanish-3 profile, contrasts with the findings of other authors (e.g. Birks & Madsen 1979). The mountains on the island of Great Bernera probably protected the presently partly flooded low-lying area west of Callanish from westerly gales. The predominant westerly winds are probably the reason why correlation between the pollen diagrams from Little Loch Roag (which is only 11 km from Callanish) and Callanish is virtually impossible. More of these sheltered valleys must have been present in the early Holocene and although the overall impression of Lewis will have been of a barren landscape, pockets of woodland, mainly consisting of birch were present.

This woodland would have attracted prehistoric people. A possible clearance phase in zone CaN-2a between *ca.* 8400 and 7650 B.P. could belong to a period of more or less contemporary Mesolithic influences in the Inner Hebrides, such as on Jura (Mercer 1969-1970) and Oronsay (Mellars 1971). No Mesolithic sites have so far been discovered in the Outer Hebrides, but the earlier Holocene coastline has nearly all been removed by geomorphological changes so that coastal sites, such as shell middens, may not have survived. Inland, the extensive peat cover makes archaeological fieldwork impracticable.

It is difficult to correlate the building of the Callanish stone circle with any of the human-influence phases deduced from the pollen diagrams. The phase with broad-scale clearance, starting at 15 cm in CN-3, however, perhaps shows the influence of a community large and organized enough to build a monument such as the stone circle. The start of zone CaN-3c can be dated to around 3490 B.P., conventionally corresponding to the Bronze Age. The timber gained by the clearance might have been useful in the construction of such stone circles.

Before the elm-decline, there is little trace of maritime influence in the pollen diagram. But immediately after the birch-decline (26 cm in CN-3; 45 cm in CN-1) a continuous *Plantago maritima* curve starts. Except for the work of Ritchie (1966), little is known about sea-level changes in the Outer Hebrides. The start of the influence of a transgressing sea in the Callanish area, as far as can be deduced from the *Plantago maritima* and *Triglochin* curves, can be dated to between 4190 B.P. and 3900 B.P. but may have been closer to 5000 B.P., taking into account the low frequencies towards the end of zone CaN-2b. It should be borne in mind that, in addition to the clearances during zone CaN-3, there was a great loss of birch due to the inundation of former valleys where Loch Roag and Loch Ceann Hulavig are presently situated. In CN-2 the *Plantago maritima* curve becomes discontinuous, and phases with and without *P. maritima* may or may not correlate with transgression and regression phases. The increase in *P. maritima* towards the top of CN-2 reflects the present situation around the peninsula. Coastal erosion of the peat cover due to the relative rise in sea-level is still going on around Callanish and along much of the western coast of the Outer Hebrides.

Summary

(1) Pollen diagrams from three overlapping peat columns from the Callanish area, Isle of Lewis, show that from *ca.* 9000 B.P. until the Late Neolithic (*ca.* 4000 B.P.), the landscape has not always been as

treeless as traditionally supposed. Woodland occupied sheltered valleys protected from westerly gales. Around Callanish, it consisted of *Betula, Corylus, Salix, Sorbus* and probably *Populus,* with an understorey containing *Lonicera, Melampyrum* and ferns.

(2) These woodland patches may have attracted prehistoric people as early as *ca.* 8400-7650 B.P. Local Neolithic clearances are recorded at *ca.* 5035 ± 60 B.P. and 4225 ± 65 B.P.

(3) More extensive clearances, also affecting *Corylus*, occurred during the Early Bronze Age. Influences attributable to later prehistoric communities are also present in the pollen diagrams, but after *ca.* 500 A.D. the area seems to have been sparsely inhabited. Renewed human influences in the upper part of the Callanish profile are probably Medieval, perhaps dating to the late 13th-14th century, following the cession of Lewis to Scotland, under the terms of the Treaty of Perth of 1266.

Acknowledgements

T. Cowie is acknowledged for his invaluable assistance with archaeological matters and for his permission to use unpublished information. I am also grateful to S. E. Durno for information concerning the vegetational history of the Outer Hebrides, to M. J. Stenhouse for determining the ^{14}C-ages, and to H. J. B. Birks for his critical reading of an earlier version of the manuscript. The Scottish Development Department (Ancient Monument Branch) supported this research and I am especially grateful to Patrick Ashmore (S. D. D.) who provided me with the results of ^{14}C-determinations. R. de Vries is acknowledged for typing the manuscript.

Nomenclature

Plant nomenclature follows A. R. Clapham, T. G. Tutin & E. F. Warburg (1962) *Flora of the British Isles* (2nd edition), Cambridge University Press, Cambridge, except for Asteraceae (= Compositae), Brassicaceae (= Cruciferae) and Poaceae (= Gramineae).

Conclusions

H. J. B. Birks

The chapters in this volume represent the range of approaches currently used in cultural landscape research in Europe. Although there is a strong and important ecological basis to European Quaternary pollen analysis, there is surprisingly little in common between the questions about the cultural landscape being addressed by pollen analysts and the problems being studied by cultural landscape ecologists. In these conclusions I discuss why this is and I suggest how, in the future, ecological and palaeoecological studies could be brought nearer together. I also discuss the problems of hypothesis testing in cultural landscape research and suggest possible future research directions that might permit more rigorous hypothesis-testing and help to unify palaeoecological studies with contemporary landscape ecology.

The basic dichotomy between ecological and palaeoecological studies arises from differences in scale (Birks 1986a). Ecologists largely work within a time scale of 1-100 years and a spatial scale of 10^0-10^6 m². The scales that palaeoecologists study are forced on them by the inherent pollen-trapping characteristics of the sites studied pollen analytically. The pollen-source area of many bogs and lakes is in the order of 10^9 m². Therefore the spatial scales of contemporary ecology and Quaternary palaeoecology hardly overlap, except for pollen profiles from very small hollows (10-30 m diameter) (e.g. Andersen this volume) and soil profiles (e.g. Odgaard this volume) with pollen-source areas of 2-3 x 10^3 m². Similarly because sediments commonly cover 100-10 000 years, there is little overlap in the temporal scales studied. Pollen-stratigraphical data covering 10 000 years rarely have the temporal precision (10-50 years per sample) or sampling density (1 sample per 100-500 years) of ecological data (1-5 years per sample, 1 sample per 1-10 years).

One important development in strengthening links between palaeo-ecology and ecology is very fine temporal precision and close sampling in pollen-stratigraphical studies, for example from annually laminated sediments (e.g. Tolonen 1978, 1981) and from very narrow, contiguous samples (e.g. Sturludottir & Turner 1985). Detection of important biotic changes over time scales of 50 and even 5-10 years from palaeo-ecological data are possible, as shown by current palaeolimnological research on effects of acid precipitation on lakes (e.g. Flower et al. 1987).

Another important development is the study of fine-resolution pollen profiles from a network of sites of different sizes and hence different pollen-source areas within a cultural landscape (e.g. Odgaard this volume; Behre & Kučan 1986) and from pairs of sites of comparable size and morphometry in contrasting landscapes (e.g. Jacobson 1979).

The challenge is how to interpret the observed geographical palynological information from many sites in terms of pollen-representation theory (e.g. Janssen 1984; Prentice 1985; Prentice *et al.* 1987), known landscape and archaeological features (e.g. Digerfeldt & Welinder 1985), and models of spatial processes (e.g. Burrough 1986). It is only through the detailed study of a network of sites that the spatial precision and density of pollen-analytical data will become relevant to the spatial scales of interest to landscape ecologists.

Cultural landscape ecology and palaeoecology are primarily descriptive and narrative (*sensu* Ball 1975) in their underlying philosophies, as illustrated by many of the chapters. Emphasis is almost entirely placed on the detection or reconstruction, description and classification of patterns (descriptive approach) or on the presentation of plausible but largely untested inductively-based explanations for the observations (narrative approach). There is, ideally, a need to develop an analytical approach, in which falsifiable hypotheses are proposed, evaluated, tested and rejected. This approach involves not only description but also explanation and is considered an essential component of any mature science (Ball 1975). Such an approach is an attractive and desirable goal for future palaeoecological research (e.g. Edwards 1983; Birks 1985). Is it possible in cultural landscape research?

It is often difficult, if not impossible in some instances, to present valid and accurate palaeoecological interpretations as realistic falsifiable hypotheses. Heck & McCoy (1979) discuss the very real problems of formulating and testing hypotheses in historical biogeography. Much of their discussion is directly relevant to palaeoecology (see Birks & Birks 1980). The approach they recommend is to accumulate all available data without any *a priori* assumptions about underlying causes, and then select from multiple causal explanations the one that is most consistent with the available evidence. This is Chamberlin's (1890) method of multiple working hypotheses. If alternative explanations are supported equally, the criterion of parsimony should be applied and the explanation selected that involves the fewest assumptions. Iversen's (1941) classic landnam paper exemplifies perfectly the Heck & McCoy approach. In a Popperian sense the approach is not strictly scientific because no potentially falsifiable hypotheses are being tested.

In a historical science such as palaeoecology it is inevitable that some observations may be so localized in time or space that they reflect isolated, unique, so-called singular phenomena resulting from, for example, idiosyncratic human behavioral ecology or chance events. Any causal hypotheses to explain singular observations may simply not be testable with data from other sites, as historical events may not be reproducible in time or space. Although Popper (1980) emphasized that explanations of historical features in terms of falsifiable hypotheses should not be abandoned solely because they may be unique, in practice Heck & McCoy's (1979) approach is often the most rigorous and certainly the only realistic one in much of palaeoecology. However, as Simberloff (1983) warns, when testable hypotheses are really not possible we must be extremely cautious about the generality of any palaeoecological conclusions. We must acknowledge that much of what has been regarded as accepted or proven causal interpretations may, in reality, be local descriptions, coherent, attractive inductively-derived narratives, or plausible but untested and untestable scenarios. Within cultural landscape palaeoecology, this warning seems particularly important in discussions of causes of expansion and regression phases, the extent of pre-Neolithic and Mesolithic impacts on vegetation, the magnitude of land-use changes in marginal areas and, of course, the keenly debated cause(s) of the *Ulmus*-decline in north-west Europe. The method of multiple working hypotheses (Chamberlin 1890) remains the most important and practical approach in palaeoecology.

Hypothesis testing in descriptive or historical sciences may often become possible if the observed patterns or reconstructions can, in some way, be quantified. There has been a surprising dearth of attempts at quantification in cultural landscape palaeoecology. An important first step is to study modern pollen in relation to contemporary land-use (e.g. Berglund *et al.* 1986; Vuorela 1973; Hicks 1985a, 1985c) and then to quantify pollen/vegetation/land-use relationships by, for example, ter Braak's (1986, 1987) canonical correspondence analysis (Birks & Berglund unpublished). Fine-precision studies of recent (50-300 years) pollen stratigraphical changes in relation to known agrarian history, crop-production data, grazing figures, etc. (e.g. Núnèz & Vuorela 1979; Joosten 1985) or landscape development (e.g. Solomon & Kroener 1971) can form the basis for simple quantitative calibrations between modern and historical pollen and land-use. These could then be used to reconstruct in quantitative terms past land-use changes and agricultural intensity from prehistoric pollen assemblages, as is done in quantitative palaeolimnology (e.g. Binford *et al.* 1987; Flower *et al.* 1987). Although such land-use reconstructions would inevitably be rather crude, they would aid in linking quantitative ecological, process-orientated studies (e.g. Olsson this volume; Emanuelsson this volume) with detailed palaeoecological reconstructions (e.g. Gaillard & Berglund this volume). They would also provide hypotheses that could, in some circumstances, be falsified by independent archaeological or ecological evidence.

Major changes are occurring in many European landscapes today (e.g. Ihse this volume; Barr *et al.* 1986). Such changes can be documented not only by aerial photography, cartography and ecological studies but also by fine-scale pollen analysis of recent sediments (as in acid-rain palaeolimnological studies). Landscape/pollen relationships could be established (cf. Solomon & Kroener 1971), again helping to link ecological and palaeoecological studies at the cultural landscape scale.

Although this book is almost entirely European in its geographical coverage, the relevance of the cultural landscape concept is, of course, global. Recent work in North America (e.g. McAndrews 1976; Delcourt *et al.* 1986) illustrates the impact of prehistoric people on American vegetation. Studies of cultural landscape ecology and palaeoecology will continue to be a major part of ecological and palaeoecological research in Europe, and they will probably assume growing importance in other areas too, as ecologists and palaeoecologists increasingly focus on patterns and processes at the landscape scale. A major challenge for the future is the interaction of temporally precise and spatially defined palaeoecological data into landscape ecology as a means of testing and validating current concepts in landscape ecology and management (e.g. Forman & Godron 1986; Risser *et al.* 1984; Romme & Knight 1982). Hopefully, by presenting a variety of different approaches within one cover, this book emphasizes the need for interdisciplinary integration and can make a contribution towards integration of research on past, present and future cultural landscapes.

Part Three

Abstracts

Abstracts

The abstracts describe papers or posters presented at the Cultural Landscape Symposium, in addition to those in the first two parts of this book.

Soil Deterioration after the Removal of the Rain Forest and its Replacement by Tree and Field Crops

J. O. Adejuwon and O. Ekanade

The tropical rain-forest at maturity together with its soil, micro- and macro-climate are in a state of dynamic equilibrium. Data collected in the cocoa-producing areas of south-western Nigeria show that when this forest is removed and replaced by field or tree crops, the balance between vegetation and soil breaks down and instability sets in. Subsequently, new soil-vegetation balances are established under tree-crop and field-crop complexes, respectively. The soil quality in the new balance shows considerable deterioration.

The percentage deterioration of various soil attributes for field and tree crops, respectively include bulk density (28%; 31%) total porosity (12%; 18%); pH (16.2%; 13.2%); organic matter (29%; 26%); nitrate-nitrogen (35%; 27%); available P (19%; 19%); Ca (24%; 41%); Na (50%; 50%); K (27%; 43%); Mg (26%; 47%); cation exchange capacity (36%; 34%); base saturation (5.4%; 8.7%).

Soil deterioration under the shifting cultivation system of agriculture has already been extensively investigated. The present study, however, represents the first consideration of such deterioration using observations on peasant farmers' cocoa plots.

Methodological Aspects of the Recording of Habitation Phases in Pollen Diagrams

K.-E. Behre

In the Pleistocene area of north-west Germany an interdisciplinary project between archaeology and palaeo-ethnobotany has been continued for 15 years. Pollen investigations closely associated with excavations have been concentrated on the Siedlungskammer Flügeln, a limited habitation area, surrounded by large raised bogs. From an area of about 25 km², 10 pollen diagrams have been analysed, partly from the surrounding raised mires, but mainly from small kettle-hole mires that are scattered over the Siedlungskammer. These densely spaced pollen diagrams reveal the occupation phases since the Neolithic and reflect the varying intensity of human activities in the different parts of the Siedlungskammer.

This study focuses on two main questions:

(1) how certain is the assumption of continuous habitation recorded by pollen diagrams in a limited area and what are the problems in recognizing short interruptions in the habitation history, and

(2) what is the relative reliability of pollen diagrams taken at different distances from known settlements and field areas for elucidating former habitation phases by the reflection of anthropogenic indicators?

The combination of the archaeological evidence and the network of numerous pollen diagrams provides an opportunity to study these problems and to obtain an estimate of former pollen deposition patterns.

Palaeoenvironment and Human Impact in North-Eastern Bulgaria

E. D. Bozilova

The area of north-eastern Bulgaria along the Black Sea has been populated since ancient times. Palaeoecological investigations of the coastal lakes in the last 10 years resulted in the collection of valuable palynological, ethnobotanical and archaeological information about the activity and the impact of humans on the natural landscape since Neolithic time.

During the second half of the 7th millenium B.C. in the Balkan peninsula a new centre of early farming originated, connected with the Neolithic in Asia Minor.

In the vicinity of Lake Durankulak and Lake Shabla the inhabitants influenced the natural vegetation cover by clearing deciduous trees during the Eneolithic time (4800-4000 B.C.) for building material as well as for obtaining open land for farming and stock breeding. The cultivation of *Triticum monococcum, T. dicoccum* and *Hordeum* was widely practised. As a result of these activities, xerothermic herbaceous plant communities remained dominant.

The pollen-analytical and palaeo-ethnobotanical data from the area of Lake Varna showed human interference on the deciduous forest cover after the 5th millenium B.C. when the number of Eneolithic settlements increased. The pollen diagrams indicate the presence of characteristic anthropophytes and crops which had been cultivated. The charred wood identified from the cultural layers of the Bronze Age shows that *Quercus, Fraxinus, Corylus, Alnus* and *Cornus* were used for different purposes.

This long-term and continuous anthropogenic activity has resulted in considerable changes in the natural vegetation cover leading to the formation of the recent cultural landscape.

Medieval Land-use in Berlin

A. Brande

Palynological research of Medieval settlement sites and their surroundings shows clear differences of intensity, type, and areas of land-use within 3 main periods.

After a relatively high density of settlements during the Roman Imperial Age (23 archaeological sites in West Berlin) with a predominant livestock economy, the early Medieval Migration Period (2 sites) is a time of woodland regeneration. In the Slav Period (19 sites) the Burgwall at Spandau developed into a fortified urban settlement. There, at about 800 A.D., clearance of the surrounding *Quercus* and *Alnus* woodland and cutting of *Calluna* took place, but during the whole period it was more or less restricted to an area of less than 8 km in diameter, including some villages. So already by *ca.* 980 A.D., *Pinus*, as the most frequent tree in this pine-oak landscape, was used for the Burgwall timber instead of *Quercus*. Besides the development of meadows on former *Alnus* stands, *Secale, Panicum, Hordeum,*

Avena and *Triticum* were cultivated (carbonized remains) on drier sites. Field areas and/or *Secale* cultivation increased after *ca*. 1100 A.D.

The German colonization period after 1157 (2 sites) began with a typical landnam phase in woodlands mainly on the ground moraine plateaux. There erosion, colluvial processes and water runoff improved the hydrological budget of kettle-hole mires and ponds, while in the valley area lower parts of potential meadow land paludified as a consequence of mill barrages. Thus in both landscape units a well-marked stratigraphical transition was formed due to these hydrological changes. Since Late Medieval times *Secale* was grown on a large scale, and also, for the first time, *Cannabis*. *Fagopyrum* was present, and *Juglans* and *Vitis* plantations also date to this period. In some parts of the landscape, villages were already abandoned by the end of the Middle Ages, before the 30 years war. Other parts remained woodland throughout the Medieval centuries and in modern times were changed into forests rich in pine that still persist today.

On Estonian Cultural Landscapes

J. Eilart and L. Saarse

Estonia offers a wide variety of landscapes among which arable land comprises 25%, grasslands 8%, forests 30%, and peatlands, water bodies, settlements, roads etc. the remaining 37%. Due to its small area and rather dense population, the Estonian landscape is highly affected by human activities, which started at *ca*. 9500 B.P. when the first hunters and fishermen appeared on the Pärnu River bank at Pulli. During the Late Neolithic period primitive-cattle breeding and tillage were started which became widespread in the Bronze Age. In the second half of the first millennium, besides farming on long-fallow land and burnt-over clearings, permanent fields began to be cultivated.

The feudal relations, foreign conquest and subdivision of Estonia between three powers were the main reasons why agriculture, that flourished at the end of prehistoric times, remained on the same level for centuries. In the late 19th century land-use was broadened and expansion started. After the Second World War, manual work was replaced by powerful machines, and the small farmsteads by modern rural settlements. Estonia became a region of intensive agriculture with problems of landscape management being very urgent. It was emphasized that landscape management should be based on ecological principles with special attention to the natural uniqueness and historical development of the countryside.

During the last 35 years, *ca*. 700 000 ha of agricultural land and 300 000 ha of forests have been drained. When this expansion programme ends 75-80% of the total territory of the republic will have undergone land improvement. Present-day human activity seeks a compromise between high productivity and nature conservation. 3200 km^2, (7.1% of the whole territory) of the ESSR are under protection.

Anthropogenic Indicators in Pollen Diagrams from South-East England

J. Empett, A. E. Evans, E. Long, P. D. Moore,
S. E. Moseley and I. Perry

The main problem in the investigation of vegetation and land-use history in southern England using pollen analysis is the shortage of appropriate sites. Very few lake and peatland sites of any antiquity and size are available in the south-east so here the major sources of pollen evidence are small mires, mor humus profiles and small water bodies such as hammer ponds and moats. Each of these presents specific problems in the interpretation of pollen indicators of past human activity.

(1) Small peat deposits are strongly influenced by local hydrological features, but these may prove of value in themselves for deducing land-use history.

(2) Mor humus profiles of woodland sites show interesting responses when the canopy is opened by a selective process, such as elm removal by disease. An example is presented from Kent.

(3) Hammer ponds, dating from Medieval times, have rapid sedimentation rates, and contain evidence of former flooded surfaces and of erosion events.

(4) Moats, which once surrounded fortified dwellings, present particular difficulties in the interpretation of infill and erosion processes. The profiles vary considerably within a single moat site.

The combination of all sources of evidence is beginning to supply information about recent woodland history and human effects on the development of the woodland landscape.

Transhumance and Archaeology

A. Fleming

This is a 'state of the art' contribution, in which I discuss the evidence which has been put forward for transhumance in prehistoric and early historic times in the British Isles, and the extent to which transhumance can be demonstrated and investigated through archaeological evidence and environmental considerations.

On the Reconstruction of Past Cultural Landscape by Pollen and Diatom Analytical Data of Lacustrine Deposits in Lithuania

M. Kabailiené

Research on sedimentation processes, delivery and settling of pollen and diatoms in different lakes in Lithuania has shown that the most convenient way to study human impact on past cultural landscapes is to use the sediments of small lakes fed by surface water from small catchments under intense agriculture. Sediments in such lakes are rich in pollen of synantropic species including cereals. It should be noted that the largest quantities of cereal pollen and other herbaceous plants are deposited not at the shore but in the deep zone of the lake (error probability <4%). The diatom flora of such lakes is dominated by species characteristic of eutrophication of human origin. *Stephanodiscus*, as well as *Fragilaria crotonensis, Cyclotella meneghiniana, C. operculata*, etc. occur most frequently in Lithuania. There are increased quantities of halophils, alkalophils, alkalibionts and α-mesosaprobs. High accumulation rates of both diatoms (up to 16 500 frustules m^{-2} s^{-1}) and sediments (up to 7.65×10^{-5} g m^{-2} s^{-1}) are characteristic of such lakes in Lithuania, being considerably higher than those for diatoms and sediments in lakes located within forests and fed mainly by groundwater (603-4420 frustules m^{-2} s^{-1} and 1.42-2.04×10^{-5} g m^{-2} s^{-1}), respectively.

According to pollen and diatom analysis data, the development of cultural landscapes in Lithuania began in the second half of the Atlantic, and the beginning of agriculture is dated to the second half of the Subboreal.

The Development of Blanket Mires in Western Norway

P. E. Kaland

Ombrotrophic mires are characteristic landscape elements in the heath area of western Norway. The classification of this type of mire has, however, been insufficiently studied, but the majority of this type is tentatively classified as Atlantic mires.

Only some of the ombrotrophic mires are equivalent to the extensive blanket mires of Great Britain. In western Norway blanket mires occur at the extreme coast and around the alpine tree-limit in the interior.

Studies on mire development in the heath area north of Bergen have shown that the peat-accumulation rate increased considerably after human-induced deforestation. They also indicate that the blanket mires represent the terminal stage of ombrotrophic mire development, and that the Atlantic mires represent different intermediate succession stages. Well-developed blanket mires have only been observed in areas with a very early deforestation. At the coast the results therefore support the hypothesis proposed by P. D. Moore of human-induced formation of blanket mires.

In contrast, preliminary studies of blanket bogs around the alpine forest limit indicate that the heavy precipitation of this area is the primary reason for ombrotrophic mire-formation.

My preliminary conclusion is that the anthropogenic factor is only important for blanket bog formation in areas with relatively low precipitation.

Reconstruction of the Cultural Landscape of a Loess Region in the Early Neolithic

A. J. Kalis and J. Lüning

The first human societies that - at least in part - depended on agriculture and cattle-breeding came into existence in the central and north-west European lowlands about 7500 years ago. These people, archaeologically known as the carriers of the 'Linnearbandkeramik' culture (Linear Pottery Culture), were the first who successfully attacked the primeval

Tilia forests, partly transforming them into probably the first north-west European cultural landscape. They founded an economic and social system dependent on agriculture and cattle-breeding that lasted more or less unaltered for more than 4 centuries. The reconstruction of this intriguing society and its natural environment is one of the main objects of study by the Seminar für Vor- und Frühgeschichte, University of Frankfurt.

Traces of settlement of the Linearbandkeramik culture have been found exclusively in loess soils, which were, and still are, among the best soils for agriculture. One particular loess region, the Aldenhovener Platte near Aachen (Rhineland, Germany), has been studied in great detail. There large-scale open-cast lignite mining is destroying extensive areas including all archaeological remains. Large excavations and archaeological field surveys were carried out between 1971 and 1982 in order to record the traces of prehistoric settlements before they were lost forever. The results of these archaeological investigations in an area of 85 km², combined with the results of supplementary geological and palaeobotanical studies, enable the reconstruction of the Early Neolithic landscape of the Aldenhovener Platte.

First Farmers in Central Europe - Some Botanical Aspects of the Earliest Neolithic Settlements in Western Germany

A. Kreuz

Investigations of the period at the beginning of the Neolithic are of particular interest because of the first strong impact of humans on the environment. During that time also a completely new economic and cultural organization of society originated. About 10 settlements belonging to the earliest phase of the Early Neolithic Linear Pottery Culture (Linearbandkeramik) and located between the Hungarian border and central western Germany, were excavated as part of a long-term research project of the Seminar für Vor- and Frühgeschichte, University of Frankfurt. To obtain representative data on the ecological, economical and social background of the Linearbandkeramik culture, many sites in different types of landscape have to be considered.

One of the most important aspects of this project is the reconstruction of the environment for these settlements in connection with economic practices, e.g. crop processing, woodland economy, etc., by means of archaeobotany (palynology, macrofossil analysis). Archaeo-botanical methods for obtaining macrofossils from carbonized crops, weeds, wood, etc. have been developed. The irregular distribution of the plant remains observed in the excavation area can be explained as a consequence of different centres of activity in the area surrounding the houses.

Experiments in Processing Grain by Neolithic Methods

J. Lüning

From 1979-1985 the Institut für Ur- und Frühgeschichte, University of Cologne conducted experiments on various aspects of Neolithic agriculture. In a forest clearing near Cologne the closed cycle of growing and harvesting emmer, einkorn, spelt and barley was system-atically observed for 7 consecutive years, during which all necessary steps from sowing to harvesting the grain were repeated annually. Special attention was given to harvesting techniques utilizing different types of sickles. In food preparation, emphasis was placed on separat-ing the corn from the glumes by the use of mortars. The studies were completed with a series of baking experiments using both underground and ground-level Neolithic ovens.

Studies of Remaining Species-Rich Hay Meadows in Norway

A. Norderhaug

Species-rich hay meadows, characteristic elements of the cultural landscape for centuries, are now rapidly disappearing.

In 1985, a pilot project, focusing on initial studies of species-rich hay meadows in Norway was organized by Økoforsk. It was conducted between Agder in the south and Finnmark in the northernmost part of Norway. Its purposes were to:

(1) clarify various aspects of meadow terminology,

(2) summarize available background information relating to the cultural history of meadows in Norway and Sweden, and

(3) review the status of remaining hay meadows in Norway

From this preliminary study, the following conclusions can be drawn:

(1) Most of the hay meadows with continuous use over a long period of time have already disappeared, or have undergone significant changes.

(2) Accordingly, the remaining hay meadows have a considerable value for science, conservation and recreation.

(3) The hay meadow ecosystem is not well understood, and their plant communities are among the least known in Scandinavia.

It is concluded that more extensive work is urgently needed for the study, conservation and proper management of the remaining hay meadows in Norway.

Statistical Analysis of Anthropogenic Indicators in Pollen Diagrams from North Poland

Magdalena Ralska-Jasiewiczowa

The aim of this work was to characterize and compare the records of cultural phases of different age (Late Neolithic, Late Bronze Age and Roman) in selected pollen diagrams from north Poland using statistical methods.

Data from 4 sites with the most detailed record of man-made changes were used. The sites are located in contrasting ecological situations, and have different cultural histories.

The anthropogenic indicators recorded in the pollen diagrams were divided into 6 groups representing roughly "various farming contexts" (*sensu* Behre 1981). The sums of pollen indicators within each group, for each of the 3 cultural phases in all 4 diagrams were used as basic data. The statistical characteristics, including the arithmetic means, standard deviation and coefficient of variation were calculated for each phase. The sample assemblages within all groups and phases were compared using Wilcoxon's test. The sample assemblages were classified using cluster analysis on original and standardized data, and the results were presented as dendrograms and plexus diagrams. The same procedures were applied to analyse the variation of samples within the phases in all sites.

In sites located in contrasting ecological contexts (very fertile soils, very poor soils) the individuality of sites dominated strongly over possible similarities in pollen records of human phases representing similar cultures. The Late Bronze Age phase was mostly characterized by the highest values of general apophytes, and the Roman phase by the highest values of poor, dry meadow and agriculture indicators. In sites on very fertile soils with early, intensive Neolithic settlements, the highest values of fresh meadow indicators were recorded in the Late Neolithic phase. The sample assemblages of all 3 phases from the area of poor soils reveal similarities to the Late Neolithic record from other sites. However, to make the conclusions reliable, more sites should be included.

Mapping the Vegetation of the Past - An Example from South Sweden

G. Regnéll

The substance of this abstract is already published as Malmer, N. & Regnéll, G. 1986. Mapping present and past vegetation. In *Handbook of Holocene Palaeoecology and Palaeohydrology* (ed. B. E. Berglund), pp. 203-218, John Wiley & Sons Ltd., Chichester.

The vegetational mantle is one of the chief factors making up the environment for prehistoric people and wildlife. Prehistoric vegetation is usually described verbally mainly on the basis of pollen diagrams. However, if the main types of vegetation can be determined for a given period and their distribution mapped, the past environment for humans can be more clearly demonstrated. The maps provide a sounder basis for framing hypotheses that can be tested in collaboration with archaeologists and other scientists.

The work must be based on a knowledge of plant communities in general, and specifically of the composition of the prehistoric flora as well as the fauna, the ecological demands of the species, and the prevailing climate and soil conditions during the period in question. Also, the time dimension must be taken into account, since succession has been a characteristic of many prehistoric plant communities.

I have prepared a map showing the reconstructed vegetation in an area of *ca*. 5 km² in the vicinity of present-day Ystad on the south coast of Sweden during the late Middle Neolithic and the Late Neolithic. The reconstruction is based on (1) a map of the present-day vegetation, also showing the topography of the area; (2) a geological map, showing the subsoil at 0.5 m; (3) Skånska rekognosceringskartan (1812-20) - a military reconnaissance map giving valuable information on the hydro-logical conditions before the period of extensive draining during the 19th century; and (4) a pollen diagram from the nearby lake Bergsjöholmssjön.

The reconstruction shows a landscape with considerable areas of surface water and wet woodland comprising mainly *Alnus glutinosa* and *Fraxinus excelsior*. A few rather small dry sites were presumably dominated by *Quercus* with *Tilia*. Extensive intermediate sites are believed to have been occupied by woodland dominated by *Ulmus glabra* in association with *Fraxinus*, and *Ulmus* woodland in places where the canopy had been opened up by man.

Areas judged as suitable for settlement have been indicated, but the actual location of settlements is a task for the archaeologist.

The Cultural Landscape During 6000 Years - Regional Vegetational History at Lake Krageholm, Southern Sweden

J. Regnéll

Human impact on the vegetation since the beginning of the Early Neolithic (3800 B.C.) is reconstructed in a palaeoecological study of a sediment core from Lake Krageholm in southernmost Sweden.

Major changes in pollen content from both non-trees and trees are interpreted as expansion periods of human impact, alternating with periods of stagnation or regression. Fine-scale variations in the amount of pollen from anthropogenic indicators including cereals, *Plantago*, *Rumex* and *Artemisia*, as well as changes in the pollen content of tree genera, are mainly regarded as reflecting changes in land-use. The underlying causes of these changes are considered with contributions from other sciences, including plant ecology, archaeology, history and social geography.

Neolithic Cultivation at Weier, Switzerland - Some New Evidence

D. E. Robinson

The Neolithic village at Weier was built on the dried-out shore of an islet at a time of climatically-induced low water-level. A wooden palisade encircled houses, byres and barns and a wooden causeway linked the village to the terrace bordering the lake. Dendrochronological and ^{14}C-dating show that there were 3 separate periods of occupation, beginning *ca.* 3100 B.C. and spanning 2 centuries. On the terrace there was evidence of small arable plots contemporary with the village. Material from these had been washed down and incorporated into the lake sediments. Analysis revealed that cereal crops (mostly *Triticum aestivum* s.l., but also *T. dicoccum, T. monococcum* and both naked and hulled forms of *Hordeum vulgare*), flax and opium poppy were cultivated. The presence of large numbers of puparia of house fly (*Musca domestica*) was interpreted by Troels-Smith as showing that the plots had been manured.

Recently new samples of the washed-in material were examined. In addition to confirming the previous findings, an exciting new piece of evidence was revealed in the form of compacted layers of uncarbonized cereal debris and dicotyledonous leaf fragments, with eggs of the intestinal parasite *Trichuris* adhering to them. The cereal debris is coarse, comprising mainly testa fragments of *Triticum* and rachis internodes and spikelets referable to *T. aestivum*. The most likely source of this material is dung used in manuring the plots. Its coarse nature is consistent with animal rather than human faeces, which in turn suggests that animals were being fed on grain or grain-rich chaff, probably as a supplement to their winter fodder. The cultivation of the plots must therefore have been such that there was a surplus of grain which could be used in this way. The introduction of feeding with cereals may explain why the winter carrying capacity of the village for animals apparently increased over the successive periods of occupation.

Wetlands in the Cultural Landscape of the North-Eastern United States

J. A. Schmit

Most of the north-eastern states enacted laws in the late 1960's and early 1970's to protect coastal wetlands from filling for development. Permit programmes restricted new fill in privately owned wetlands and coastal waterways to those uses which require a waterfront location. State efforts were reinforced by passage of the federal Clean Water Act in 1972. Implementation of the federal act has been slow. Fill was regulated first in coastal areas and along large rivers where the federal government traditionally had been responsible for navigation and commerce. By the mid 1980's the federal restrictions became evident in freshwater, inland wetlands.

Along the Atlantic coast the rate at which wetlands were being converted to intensive use after 1945 has been cut dramatically. An estimated 15 000 ha of New Jersey's tidal wetlands were lost between 1950 and 1970. Since 1970 the loss has been less than 500 ha; more than 90 000 ha remain. Similar results have been achieved in other north-eastern states with extensive tidal wetlands.

Inland, the protection of freshwater wetlands has lagged by at least 10 years. New Jersey is the most densely populated of the United States and the New Jersey legislature is currently considering a new state law to protect inland wetlands. At work are both the continuing insistence of conservationists on resource protection through regulatory programmes and a desire to reestablish state-level preeminence in wetland decision-making through delegation of federal permits. Already the 40% of the 500 000 ha Pinelands National Reserve in New Jersey that are wetlands receive stringent regulatory protection, including regulatory control over a 100 m wide upland buffer surrounding the wetlands. There are extensive wetlands in the unglaciated coastal plain outside the Pinelands Reserve as well as in the mostly glaciated piedmont and Appalachian highlands. Here major regulatory battles are being waged over large projects that need federal and state approvals for wetland fills. Similar permit court fights are underway over Massachusetts development projects.

To avoid delays and costs of the permit process, as well as the great expense of creating new wetland to offset those filled, designers are clustering residential and commercial developments into uplands, even in Pennsylvania where wetlands are confined to narrow stream valleys. Assessors are being forced to recognize the reduced value of wetlands for future development when calculating local property taxes.

Stream courses, and their woods and marshes, are being preserved in suburban areas as never before in the north-eastern United States. Architectural solutions are being developed to avoid wetlands as a major goal in site design. In consequence the look of the wild and rural landscape will be preserved in the suburban megalopolis of the late 20th century to a much greater extent than it was during the preceding century.

Palaeovegetation Maps and Early Cultural Landscape from 8000 B.P. to 1200 B.P. in South-West Finland

Mirjami Tolonen

This study is part of a research project that attempts to describe the history of human activity in the Paimionjoki River Valley, south-west Finland by means of archaeological and palaeoecological methods. The area lies on the Baltic coast where the present-day land uplift is 4.5 mm yr^{-1}.

The palaeovegetational maps were constructed for the periods *ca.* 8000 B.P., 5000 B.P., 4000 B.P., 2600 B.P. and 1200 B.P.; ± 500 years. The maps are based on the reconstructed distribution of dry land and different soil types at particular times, knowledge about the primary soil types at these times, knowledge about the primary successions on the coastal area, other known ecological factors, and 7 ^{14}C-dated pollen diagrams from peat bogs in the area (250 km^2). Five of these pollen diagrams are from a 'semi-quantitatively' studied central 'control area'. Extrapolations were done for marginal areas.

The vegetation type was primarily controlled by land uplift and associated ecological and soil factors. Nevertheless, the long-term post-glacial succession clearly followed the general pattern of forest history described for northern Europe by Iversen. The natural changes included, *in an ecological sense*, the protocratic and mesocratic stages.

The first slight signs of human impact come in the earlier part of the telocratic stage (*ca.* 3000-1500 B.P.). This influence on the natural forest did not cause changes as great as did climatic and pedogenic factors. Scattered openings in the forest cover by clearances accompanied by a regression of *Picea abies* are discernible on the 1200 B.P. map.

Protected Natural Sites in Bulgaria

S. B. Tonkov

The protected natural sites in Bulgaria are rare places in the landscape with valuable scientific, national economic and historical interest. The Law for the Protection of Nature defines the following categories of protected natural sites: National Parks, Reserves, Sites of Natural Beauty, Protected Localities and Historical Sites.

Of the 7 National Parks the most popular and famous are Pirin, Ropotamo and Vitosha. Scientific research is conducted and encouraged in the National Parks. Tourist activity is unrestricted except in the zones that have the status of Reserves.

More than 60 localities characterized by animal and plant life valuable for science or threatened by reduction or even extinction are declared as Reserves. Some of them have obtained the status of Biosphere UNESCO Reserves and can be visited only with special permission on clearly marked roads and paths.

Protected Localities and Sites of Natural Beauty include interesting rock formations, gorges, sand dunes, soil pyramids, waterfalls, lakes, marshes, old trees, etc., all of which are distinguished by their picturesque beauty. The Historical Sites are localities where historical events of national or local importance have occurred.

Every year the number of protected natural sites is increasing so that these unique parts of the Bulgarian landscape can be preserved for future generations.

Man as a Modifier of Mire Landscapes in North-East Finland

Y. Vasari

Open, treeless aapa-mires are a characteristic feature of the landscape in north-east Finland. In the 1950's about one third of all these mires were open, either oligo-mesotrophic *Sphagnum* mires or meso-eutrophic brown-moss mires. As a whole, mires then covered about one third of the total land area.

Through the centuries, and up to the present day, human influence upon mires in north-east Finland has been profound. In the early 18th century exploitation of mires began on a vast scale in order to obtain enough winter-fodder for cattle. Hay was collected from open or semi-open mires and alluvial meadows. In many cases trees were cleared from mires and river shores and sophisticated methods were applied in order to improve hay production. Methods by which otherwise poor mires could be made more productive were damming and building of catchworks over meadows. Depending upon the productivity of natural meadows, hay could be collected every year, every second year, or once in 3 years.

The great influence that this hay-making practice had upon the mires became clearly visible in the 1960's, after the end of this traditional practice. Since then, very many previously open mires are becoming forested again. It is now evident that the abundance of open mires was essentially due to human influence.

Since the Second World War widespread drainage of mires for agricultural purposes has taken place. Hardly any of the larger mire complexes have remained untouched. Deplorably, it was not long before an extensive migration away from the marginal agricultural areas began and many of the new fields have been abandoned. More recently, drainage for afforestation has become common. These modern ways of peat-land exploitation, together with the abandonment of natural meadows, have completely changed the mire landscapes in north-east Finland.

The Cultural History of Rouveen

J. A. J. Vervloet

Commissioned by the Government Service for Land and Water Use, the Netherlands Soil Survey Institute has made an inventory of the remnants of the occupation history in the cultural landscape. It has summarized the results in the map "Ontginning en Bewoning" (Reclamation and Occupation). On the basis of these two activities, the historical background of the cultural landscape is sketched. When plans for land consolidation are outlined, the remnants of historical structures will be taken into account. Different aspects are shown on the map, including an overall reconstruction of soils at the time of occupation and reclamations, and several important linear elements; natural watercourses, lines of occupation, boundaries, dykes, artificial watercourses, old roads, and the main lines of division. Some other elements are also shown such as old churchyards, settlements, decoys, sluices and locks.

The inventory of the Rouveen-Staphorst area is discussed. A peat-swamp reclamation with elongated holdings was originally situated behind a river bank, but gradually extended into the bog by the lengthening of the strip-holdings of the reclaimers.

The Agricultural History of Southern Finland as Reflected in Pollen Diagrams

I. Vuorela

The agricultural history of southern Finland is examined in the light of summarized pollen data from 27 previously published diagrams.

The development from the earliest, isolated, slash-and-burn cultivation clearings in dense virgin forest to modern field cultivation is discussed on the basis of anthropogenic pollen data and an attempt is made to define the most typical phenomena encountered among this pollen group at the absolute, empiric and rational Cerealia limits.

Special attention is paid to the question of pollen dispersal under the changing circumstances of forest structure during the period of traditional agricultural practice. Evidently, the pollen evidence, which commences *ca.* 4000 B.P. and gradually increases, does not primarily indicate the agricultural activity which produced it but rather the opening up of the landscape and improved pollen dispersal within the forest which was eventually completely destroyed in the 18-19th centuries.

After the termination not only of this traditional agricultural practice but also of the particular crop combination, pollen evidence usually decreases, in spite of the ever expanding permanent fields. The main reason for this seems to lie in the forest edges which again effectively prevent pollen dispersal from the cultivated areas.

References

Aaby, B. (1976). Cyclic climatic variations in climate over the past 5,500 yr reflected in raised bogs. *Nature, London,* **263,** 281-4.

Aaby, B. (1983). Forest development, soil genesis and human activity illustrated by pollen and hypha analysis of two neighbouring podzols in Draved Forest, Denmark. *Danmarks geologiske Undersøgelse,* Series II, **114,** 1-114.

Aaby, B. (1986a). Mennesket og naturen på Abkæregnen gennem 6000 år. Resultater af et forskningsprosjekt. *Sønderjysk Månedsskrift,* **62,** 277-90.

Aaby, B. (1986b). Trees as anthropogenic indicators in regional pollen diagrams from eastern Denmark. *Anthropogenic Indicators in Pollen Diagrams* (Ed. by K.-E. Behre). A. A. Balkema, Rotterdam.

Aaby, B. & Berglund, B. E. (1986). Characterization of peat and lake deposits. *Handbook of Holocene Palaeoecology and Palaeohydrology* (Ed. by B. E. Berglund). J. Wiley & Sons, Chichester.

Aaby, B. & Tauber, H. (1975). Rates of peat formation in relation to degree of humification and local environment, as shown by studies of a raised bog in Denmark. *Boreas,* **4,** 1-17.

Aalen, F. H. A. (1983). Perspectives on the Irish landscape in prehistory and history. *Landscape Archaeology in Ireland* (Ed. by T. Reeves-Smyth and F. Hamond). British Archaeological Reports, **116,** Oxford.

Ádám, L. & Marosi, S. (1975). *A Kisalföld és a Nyugat-magyarországi peremvidék (The Little Plain and the West-Hungarian Borderland).* Akadémiai Kiadó, Budapest.

Ahti, T. & Hämet-Ahti, L. (1971). Hemerophilous flora of the Kuusamo district, northeast Finland, and the adjacent part of Karelia, and its origin. *Annales Botanici Fennici,* **8,** 1-91.

Ahti, T., Hämet-Ahti, L. & Jalas, J. (1968). Vegetation zones and their sections in northwestern Europe. *Annales Botanici Fennici,* **5,** 169-211.

Aitchison, J. (1986). *The Statistical Analysis of Compositional Data.* Chapman & Hall, London & New York.

Åkerblom, F. (1891). *Historiska anteckningar om Sveriges nötkreatursafvel.* N. J. Gumperts, Göteborg.

Albertus Magnus (1260). *De vegetabilibus libri VII. Liber V, tractatus I, capitulum 7.*

Alcock, M. R. (1982). *Yorkshire grasslands: a botanical survey of hay meadows within the Yorkshire Dales National Park.* England Field Unit, Report **10,** Nature Conservancy Council, Banbury.

Alestalo, J. (1979). Land uplift and development of the littoral and aeolian morphology on Hailuoto, Finland. *Palaeohydrology of the Temperate Zone* (Ed. by Y. Vasari, M. Saarnisto and M. Seppälä). Acta Universitatis Ouluensis, Series A **82,** 109-20.

Allen, D. (1979). Excavations at Hafod y Nant Criafolen, Brenig Valley, Clwyd 1973-4. *Post-Medieval Archaeology,* **13,** 1-59.

Alm, G. & Nordberg, M.-L. (1985). Remote sensing and computer cartography in studies of landscape development. Description on case studies (in Swedish, English abstract). *Statens Naturvårdsverk - PM,* **3050,** 72 pp.

Ammann, B. (1986). Litho- and palynostratigraphy at Lobsigensee: Evidences of trophic changes during the Holocene. Studies in the Late-Quaternary of Lobsigensee, 13. *Hydrobiologia,* **143,** 301-7.

Ammann, B., Andrée, M., Chaix, L., Eicher, U., Elias, S. A., Hofmann, W., Oeschger, H., Siegenthaler, U., Tobolski, K., Wilkinson, B. & Züllig, H. (1985). Lobsigensee - late-glacial and Holocene environments of a lake on the Central Swiss Plateau. An attempt at a palaeoecological synthesis. *Dissertationes Botanicae,* **87,** 165-70.

Ammann, B., Chaix, L., Eicher, U., Elias, S. A., Gaillard, M.-J., Hofmann, W., Siegenthaler, U., Tobolski, K. & Wilkinson, B. (1983). Vegetation, insects, molluscs and stable isotopes from Late-Würm deposits at Lobsigensee (Swiss

Plateau). Studies in the Late Quaternary of Lobsigensee, **7.** *Revue de Paléobiologie,* **2,** 221-7.

Ammann-Moser, B. (1975). Vegetationskundliche und pollenanalytische Untersuchungen auf dem Heideneweg im Bielersee. *Beiträge zur geobotanischen Landesaufnahme der Schweiz,* **56,** 74 pp.

Ammerman, A. J. & Cavalli-Sforza, L. L. (1984). *The Neolithic Transition and the Genetics of Populations in Europe.* Princeton University Press, Princeton.

Andersen, A. (1954). Two standard pollen diagrams from South Jutland. *Danmarks geologiske Undersøgelse,* Series II, **80,** 188-209.

Andersen, S. Th. (1970). The relative pollen productivity of North European trees, and correction factors for tree pollen spectra. *Danmarks geologiske Undersøgelse,* Series II, **96,** 1-99.

Andersen, S. Th. (1973). The differential pollen productivity of trees and its significance for the interpretation of a pollen diagram from a forested region. *Quaternary Plant Ecology* (Ed. by H. J. B. Birks and R. G. West). Blackwell Scientific Publications, Oxford.

Andersen, S. Th. (1974). Wind conditions and pollen deposition in a mixed deciduous forest. II. Seasonal and annual pollen deposition 1967-1972. *Grana,* **14,** 64-77.

Andersen, S. Th. (1975). The Eemian freshwater deposit at Egersund, South Jylland, and the Eemian landscape development in Denmark. *Danmarks geologiske Undersøgelse Årbog,* **1974,** 49-70.

Andersen, S. Th. (1978). Local and regional vegetation development in eastern Denmark in the Holocene. *Danmarks geologiske Undersøgelse Årbog,* **1976,** 5-27.

Andersen, S. Th. (1979a). Brown earth and podzol: soil genesis illuminated by microfossil analysis. *Boreas,* **8,** 59-73.

Andersen, S. Th. (1979b). Identification of wild grasses and cereal pollen. *Danmarks geologiske Undersøgelse Årbog,* **1978,** 69-72.

Andersen, S. Th. (1980). The relative pollen productivity of the common forest trees in the early Holocene in Denmark. *Danmarks geologiske Undersøgelse Årbog,* **1979,** 5-19.

Andersen, S. Th. (1984). Forests of Løvenholm, Djursland, Denmark, at present and in the past. *Det Kongelige Danske Videnskabernes Selskab Biologiske Skrifter,* **24** (1), 208 pp.

Andersen, S. Th. (1985). Natur- og kulturlandskabet i Næsbyholm Storskov siden istiden. *Fortidsminder. Antikvariske Studier,* **7,** 85-107.

Andersen, S. Th., Aaby, B. & Odgaard, B. V. (1983). Environment and Man. Current Studies in Vegetational History at the Geological Survey of Denmark. *Journal of Danish Archaeology,* **2,** 184-96.

Annexed Estates (1973). *Reports on the Annexed Estates 1755-1769* (Ed. by V. Wills). Her Majesty's Stationery Office, Edinburgh.

APS (1814-75). *Acts of the Parliaments of Scotland.* 12 volumes. T. Thomson, London.

Argyll Estate (1964). *Argyll Estate Instructions (Mull, Morvern, Tiree) 1771-1805.* (Ed. by E. Cregeen). Scottish History Society, Series 4, **1,** Edinburgh.

Aune, E. I. (1973). Forest vegetation in Hemne, Sør-Trøndelag, western central Norway. *Det Kongelige norske Videnskabers Selskab Miscellanea,* **12,** 87 pp.

Austad, I. (1985a). Vegetasjon i kulturlandskapet. Bjørkehager og einerbakker. *Sogn og Fjordane distriktshøgskule. Skrifter* **1985** (1), 36 pp.

Austad, I. (1985b). Vegetasjon i kulturlandskapet. Lauvingstrær. *Sogn og Fjordane distriktshøgskule. Skrifter* **1985** (2), 43 pp.

Austad, I. & Hauge, L. (1988). Galdane i Lærdal kommune. Metode opplegg for istandsetting og skjøtsel av kultur-landskapet. *Økoforsk rapport* (in preparation).

Austad, I., Lea, B. O. & Skogen, A. (1985). Kulturpåvirkete edellauvskoger. Utprøving av et metodeopplegg for istandsetting og skjøtsel. *Økoforsk rapport,* **1985** (1), 56 pp.

Baker, C. A., Moxey, P. A. & Oxford, P. M. (1978). Woodland continuity and change in Epping Forest. *Field Studies,* **4,** 645-69.

Baker, H. (1937). Alluvial meadows: a comparative study of grazed and mown meadows. *Journal of Ecology,* **25,** 408-20.

Bakker, M. & van Smeerdijk, D. G. (1982). A palaeoecological study of a late Holocene section from 'Het Ilperveld', western Netherlands. *Review of Palaeobotany and Palynology,* **36,** 95-163.

Ball, I. R. (1975). Nature and formulation of biogeographical hypotheses. *Systematic Zoology,* **24,** 407-30.

Barber, K. E. (1976). History of vegetation. *Methods in Plant Ecology* (Ed. by S. B. Chapman). Blackwell Scientific Publications, Oxford.

Barr, C., Benefield, C., Bunce, B., Ridsdale, H. & Whittaker, M. (1986). *Landscape changes in Britain.* Institute of Terrestrial Ecology, Abbots Ripton.

Beckett, S. C. (1979). Pollen influx in peat deposits: values from raised bogs in the Somerset levels, south-western England. *New Phytologist,* **83,** 839-47.

Behre, K.-E. (1981). The interpretation of anthropogenic indicators in pollen diagrams. *Pollen et Spores,* **23,** 225-45.

Behre, K.-E. (1986). *Anthropogenic Indicators in Pollen Diagrams.* A. A. Balkema, Rotterdam.

Behre, K.-E., Brandt, K., Kučan, D., Schmid, P. & Zimmermann, W. H. (1982). *Mit dem spaten in die vergangenheit. - 5000 Jahre Siedlung und Wirtschaft im Elbe-Weser-Dreieck.* Landkreis, Cuxhaven in cooperation with Niedersächsisches Landesinstitut für Marschen- und Wurtenforschung, Wilhelmshaven.

Behre, K.-E. & Kučan, D. (1986). Die Reflekton archäologisch bekannter Siedlungen in Pollendiagrammen verschiedener Entfernung - Beispiele aus der Siedlungskammer Flögeln, Nordwestdeutschland. *Anthropogenic Indicators in Pollen Diagrams* (Ed. by K.-E. Behre). A. A. Balkema, Rotterdam.

Bendixen, B. E. (1892). Fornlevninger i Hardanger. *Foreningen til Norske Fortidsmindesmerkers Bevaring. Aarsberetning,* **1891,** 13-59.

Bendixen, B. E. (1893). Undersøgelser og Udgravninger i Eidfjord. *Foreningen til Norske Fortidsmindesmerkers Bevaring. Aarsberetning,* **1892,** 14-32.

Bengtsson, G. & Kristiansson, S. (1958). *Gödsling och kalkning.* (6th edition) LTs förlag, Stockholm.

Bengtsson, S., Larsson, H. & Petersson, S. (1973). Vombs ängar, vegetation, fågelliv, markanvändning samt synpunkter på restaurering och skötsel. *Miljövårdsprogrammet vid Lunds Universitet och Tekniska Högskola.* Länsstyrelsen i Malmöhus län, 55 pp.

Bergendorff, C. & Emanuelsson, U. (1982). Skottskogen - en för summad del av vårt kulturlandskap. *Svensk Botanisk Tidskrift,* **76,** 91-100.

Berglund, B. E. (1969). Vegetation and human influence in South Scandinavia during prehistoric time. *Oikos Supplement,* **12,** 9-28.

Berglund, B. E. (1985a). Early agriculture in Scandinavia: research problems related to pollen-analytical studies. *Norwegian Archaeological Review,* **18,** 77-105.

Berglund, B. E. (1985b). Det sydsvenska kulturlandskapets förändringar under 6000 år - en presentation av Ystadprosjektet. *Kulturlandskapet - dess framväxt och förändring.* Symposium September 1984 (Ed. by G. Regnéll). Lund University, Lund.

Berglund, B. E. (1986). The cultural landscape in a long-term perspective. Methods and theories behind the research on land-use and landscape dynamics. *Striae,* **24,** 79-87.

Berglund, B. E., Emanuelsson, U., Persson, S. & Persson, T. (1986). Pollen vegetation relationships in grazed and mowed plant communities of south Sweden. *Anthropogenic Indicators in Pollen Diagrams* (Ed. by K.-E. Behre). A. A. Balkema, Rotterdam.

Berglund, B. E. & Ralska-Jasiewiczowa, M. (1986). Pollen analysis and pollen diagrams. *Handbook of Holocene Palaeoecology and Palaeohydrology* (Ed. by B. E. Berglund). J. Wiley & Sons, Chichester.

Berglund, B. E. & Stjernquist, B. (1981). Ystadsprojektet - det sydsvenska kulturlandskapets förändringar under 6000 år. Luleälvssymposiet 1-3 juni 1981. *Skrifter från Luleälvsprojektet,* **1,** 161-85, Umeå.

Bertelsen, R. (1985). *Lofoten og Vesterålens historie. Fra den eldste tida til ca. 1500 e.Kr.* Kommunene i Lofoten og Vesterålen.

Besteman, J. C. (1974). Carolingian Medemblik. *Berichten van de Rijksdienst voor het Oudheidkundig Bodemonderzoek*, **24**, 43-106.

Besteman, J. C. & Guiran, A. J. (1983). Het middeleeuws- archeologisch onderzoek in Assendelft, een vroege veenontginning in middeleeuws Kennemerland. *Westerheem*, **32**, 144-76.

Beug, H.-J. (1961). *Leitfaden der Pollenbestimmung für Mitteleuropa und angrenzende Gebiete*. Gustav Fischer Verlag, Stuttgart.

Binford, M. W., Brenner, M., Whitmore, T. J., Higuera-Gundy, A., Deevey, E. S. & Leyden, B. (1987). Ecosystems, paleoecology and human disturbance in subtropical and tropical America. *Quaternary Science Reviews*, **6**, 115-28.

Birch, S. P., Hughes, J. C. & Huntley, B. (1988). Fertiliser trials on hay meadow vegetation, with particular reference to Palace Leas, Cockle Park, Morpeth, Northumberland, Northern England (in preparation).

Birks, H. H. (1975). Studies in the vegetational history of Scotland IV. Pine stumps in Scottish blanket peats. *Philosophical Transactions of the Royal Society of London B*, **270**, 181-226.

Birks, H. J. B. (1977). The Flandrian forest history of Scotland: a preliminary synthesis. *British Quaternary Studies. Recent Advances* (Ed. by F. W. Shotton). Clarendon Press, Oxford.

Birks, H. J. B. (1982). Mid-Flandrian forest history of Roudsea Wood National Nature Reserve, Cumbria. *New Phytologist*, **90**, 339-54.

Birks, H. J. B. (1985). Recent and possible future mathematical developments in quantitative palaeoecology. *Palaeogeography, Palaeoclimatology, Palaeoecology*, **50**, 107-47.

Birks, H. J. B. (1986a). Late-Quaternary biotic changes in terrestrial and lacustrine environments, with particular reference to north-west Europe. *Handbook of Holocene Palaeoecology and Palaeohydrology* (Ed. by B. E. Berglund). J. Wiley & Sons, Chichester.

Birks, H. J. B. (1986b). Numerical zonation, comparison and correlation of Quaternary pollen-stratigraphical data. *Handbook of Holocene Palaeoecology and Palaeohydrology* (Ed. by B. E. Berglund). J. Wiley & Sons, Chichester.

Birks, H. J. B. (1986c). The Cultural Landscape - Past, Present and Future Excursion Guide. *Botanisk institutt, Universitetet i Bergen Rapport* **42**, 150 pp.

Birks, H. J. B. & Birks, H. H. (1980). *Quaternary Palaeoecology*. Edward Arnold, London.

Birks, H. J. B., Deacon, J. & Peglar, S. M. (1975). Pollen maps for the British Isles 5000 years ago. *Proceedings of the Royal Society of London B*, **189**, 87-105.

Birks, H. J. B. & Madsen, B. J. (1979). Flandrian vegetational history of Little Loch Roag, Isle of Lewis, Scotland. *Journal of Ecology*, **67**, 825-842.

Birks, H. J. B. & Moe, D. (1986). Comments on early agriculture in Scandinavia. *Norwegian Archaeological Review*, **19**, 39-43.

Bjørgo, T. (1986). Mountain archaeology. Preliminary results from Nyset-Steggje. *Norwegian Archaeological Review*, **19**, 122-7.

Black Book (1855). *The Black Book of Taymouth* (Ed. by C. Innes). Bannatyne Club, Edinburgh.

Blackburn, K. B. (1946). On a peat on the Island of Barra, Outer Hebrides. Data for the study of post-glacial history. X. *New Phytologist*, **45**, 44-9.

Blehr, O. (1973). Traditional reindeer hunting and social change in the local communities surrounding Hardangervidda. *Norwegian Archaeological Review*, **6**, 102-12.

Böcher, T. & Jørgensen, C. A. (1972). Jyske dværgbuskheder. Eksperimentelle undersøgelser af forskellige kulturindgrebs indflydelse på vegetationen. *Det Kongelige Danske Videnskabernes Selskab Biologiske Skrifter*, **19** (5), 55 pp.

Bøe, J. (1942). Til høgfjellets forhistorie. *Bergens Museums Skrifter*, **21**, 1-96.

Bøe, J. (1951). Da Hardangervidda ble oppdaget. *Drammens og Oplands Turistforenings Årbok*, **1951**, 23-34.

Bognár-Kutzián, I. (1972). *The Early Copper Age Tisza-polgár Culture in the Carpathian basin*. Akadémiai Kiadó, Budapest.

Bondorff, K. A. (1939). *Forelaesninger over landbrukets jorddyrkning. II. Gødningslaeren*. Den Kongelige Veterinaer- och Landbohøjskole, Copenhagen.

Bonham-Carter, G. F., Gradstein, F. M. & D'Iorio, M. A. (1986). Distribution of

Cenozoic foraminifera from the northwestern Atlantic margin analyzed by correspondence analysis. *Computers and Geosciences*, **12**, 621-35.

Bos, J. M. (1983). Veldnamen in verband met de nederzettingsgeschiedenis van Waterland (N.H.). *Naamkunde*, **15**, 120-8.

Bos, J. M. (1985a). Stadse fratsen in een Hallands veengebied. Archeologie van stad en land. *Westerheem*, **34**, 110-23.

Bos, J. M. (1985b). Archeologische streekbeschrijving; een handleiding. *Archeologische Werkgemeenschap Nederland monografieën*, **4** (1).

Bos, J. M. (1986). Ransdorp in Waterland. De ruimtelijke ontwikkeling van een veennederzetting. *Historisch-Geografish Tijdschrift*, **4**, 1-5.

Bos, J. M. (1988). *A 14th century industrial complex at Monnickendam, and the preceding events*. Cingula (in press).

Boserup, E. (1973). *Jordbruksutveckling och befolkningstillväxt*. Gleerups, Lund.

Bowen, E. G. & Gresham, C. A. (1967). *History of Merioneth*, **1**. Dolgellau.

Bowen, H. C. & Fowler, P.J. (1978). *Early Land Allotment*. British Archaeological Reports, **48**, Oxford.

Bradley, R. (1984). *The Social Foundations of Prehistoric Britain*. Longman, London.

Bradshaw, R. H. W. (1981a). Modern pollen-representation factors for woods in southeast England. *Journal of Ecology*, **69**, 45-70.

Bradshaw, R. H. W. (1981b). Quantitative reconstruction of local woodland vegetation using pollen analysis from a small basin in Norfolk, England. *Journal of Ecology*, **69**, 941-55.

Bradshaw, R. H. W. & Browne, P. (1987). Changing patterns in the postglacial distribution of *Pinus sylvestris* in Ireland. *Journal of Biogeography*, **14**, 237-48.

Brand, J. (1701). *A Brief Description of Orkney, Zetland, Pightland-Firth and Caithness*. Edinburgh.

Brandt, R. W. (1983). De archeologie van de Zaanstreek. *Westerheem*, **32**, 120-37.

Brelin, B., Den Braver, E. & Rudin, Ø. (1979). Näringsvärde i några olika typer av lövsly. *Fårskötsel*, **59**, 6-7.

Briggs, C. S. (1985). Problems of the early agricultural landscape in upland Wales, as illustrated by an example from the Brecon Beacons. *Upland Settlement in Britain. The Second Millennium BC and After* (Ed. by D. Spratt and C. Burgess). British Archaeological Reports, **143**, Oxford.

Bringéus, N.-A. (1964). *Tradition och förändring i 1800-talets skånska lanthushållning*. Kristianstads läns hushållningssällskap 1814-1964, Kristiansand.

Brink, N. (1983). Gödselanvändningens miljöproblem. *Ekohydrologi*, **14**, 15-9.

Brown, A. H. F. & Oosterhuis, L. (1981). The role of buried seed in coppice woods. *Biological Conservation*, **21**, 19-38.

Brun, I. (1962). *The air temperature in Norway 1931-60*. Det norske Meteorologiske Institutt, Oslo.

Bryce, T. H. (1904). On the cairns and tumuli of the Island of Bute. *Proceedings of the Society of Antiquaries of Scotland*, **38**, 17-81.

Bryhni, I. (1977). Jotundekket og dets underlag i Sogn. *Norges Geologiske Undersøkelse, Rapport* **1560**/28.

Buckland, P. C. & Edwards, K. J. (1984). The longevity of pastoral episodes in pollen diagrams - the role of post-occupation grazing. *Journal of Biogeography*, **11**, 243-49.

Burenhult, G. (1982). *Arkeologi i Sverige 1. Fångstfolk och herdar*. Wiken, Höganäs.

Burl, A. (1976). *The Stone Circles of the British Isles*. Yale University Press, New Haven.

Burleigh, R., Evans, J. G. & Simpson, D. A. G. (1973). Radiocarbon dates for Northton, Outer Hebrides. *Antiquity*, **47**, 61-4.

Burrough, P. A. (1986). *Principles of Geographical Information Systems for Land Resources Assessment*. Clarendon Press, Oxford.

Burt's Letters (1754). *Burt's Letters from the North of Scotland*. 1876 edition, 2 volumes, Edinburgh.

Campbell, A. (1802). *A Journey from Edinburgh through Parts of North Britain*. Edinburgh.

Campbell, Å. (1927). *Skånska bygder under förra hälften av 1700-talet. Etnografisk studie över den skånska allmogens äldre odlingar, hägnader och byggnader*. A.-B. Lundequistska Bokhandeln, Uppsala.

Campbell, Å. (1928). "Risbygden" i Skåne. *Västsvenska Hembygdsstudier*, **1928**, 92-118.

Carr, A. D. (1982). *Medieval Anglesey.* Llangefni.

Case, H. (1969). Neolithic explanations. *Antiquity,* **43,** 176-86.

Casseldine, C. J. (1985). Surface pollen studies across Bankhead Moss, Fife, Scotland. *Journal of Biogeography,* **8,** 7-25.

Caulfield, S. (1978). Neolithic fields: the Irish evidence. *Early Land Allotment* (Ed. by H. C. Bowen and P. J. Fowler). British Archaeological Reports, **48,** Oxford.

CCR (1886). *Crofters Commission Reports.* Her Majesty's Stationery Office, London.

Chamberlin, T. C. (1890). The method of multiple working hypotheses. *Science,* **15,** 92-6.

Chambers, F. M. & Price, S.-M. (1985). Palaeoecology of *Alnus* (alder): early post-glacial rise in a valley mire, North-West Wales. *New Phytologist,* **101,** 333-44.

Childe, V. G. (1958). *The Prehistory of European Society.* Penguin Ltd., London.

Chorley, G. P. H. (1981). The agricultural revolution in northern Europe, 1750-1880: nitrogen, legumes, and crop productivity. *The Economic History Review,* **34,** 71-93.

Christensen, P. G. (1981). *Status over hedeplejemetoder.* Fredningsstyrelsen, Copenhagen.

Christiansen, S. (1978). Infield-outfield systems - characteristics and development in different climatic environments. *Geografisk Tidskrift,* **77,** 1-5.

Clark, R. M. (1975). A calibration curve for radiocarbon dates. *Antiquity,* **49,** 251-266.

Cloudsley-Thompson, J. L. (1977). *Man and the Biology of Arid Zones.* Edward Arnold, London.

Clymo, R. S. & Mackay, D. (1987). Upwash and downwash of pollen and spores in the unsaturated surface layer of *Sphagnum*-dominated peat. *New Phytologist,* **105,** 175-83.

Coles, J. (1976). Forest farmers: some archaeological, historical and experimental evidence. *Acculturation and Continuity in Atlantic Europe* (Ed. by S. J. De Laet). IV Atlantic Colloquium, Brugge.

Coles, J. M. & Orme, B. J. (1977). Neolithic hurdles from Walton Heath, Somerset. *Somerset Levels Papers,* **3,** 6-29.

Coles, J. M. & Orme, B. J. (1983). *Homo sapiens* or *Castor fiber? Antiquity,* **57,** 95-102.

Connolly, G. (1930). The vegetation of southern Connemara. *Proceedings of the Royal Irish Academy B,* **39,** 203-31.

Cooney, G. (1987). An unrecorded Wedge-tomb at Scrahallia, Cashel, Connemara, Co. Galway. *Journal of the Galway Archaeological and Historical Society,* **40** (in press).

Coxon, P. (1987). A post-glacial diagram from Clare Island, Co Mayo. *Irish Naturalists' Journal,* **22,** 219-23.

Crabtree, K. (1982). Evidence for the Burren's forest cover. *Archaeological Aspects of Woodland Ecology* (Ed. by S. Limbrey and M. Bell). British Archaeological Reports, S146, Oxford.

Craig, A. J. (1978). Pollen percentage and influx analyses in south-east Ireland: a contribution to the ecological history of the Late-glacial period. *Journal of Ecology,* **66,** 297-324.

Crowther, R. E. & Patch, D. (1981). How much wood for the stove? *Forestry Commission Research Note,* 55/80/SILS.

Dahl, S. (1942). Torna och Bara. Studier i Skånes bebyggelse och näringsgeografi före 1860. *Meddelanden från Geografiska Institution, Lunds Universitet. Avhandlingar,* **VI,** 247 pp.

Dahl, S. (1978). Two Scanian types of two-field system. *Geographica Polonica,* **38,** 37-40.

Daniel, E. (1986). Beskrivning till jordartskartorna Tomelilla SO/Simrishamn SV och Ystad NO/rnahusen NV. *Sveriges Geologiska Undersökning.* Serie Ae, 65-66. Uppsala.

Dansgaard, W. (1975). Indlandsisen. *Grønland* (Ed. by P. Koch). Gyldendal, Copenhagen.

Dargie, T. C. D. (1986). Species richness and distortion in reciprocal averaging and detrended correspondence analysis. *Vegetatio,* **65,** 95-8.

Darling, F. (1955). *West Highland Survey.* Clarendon Press, Oxford.

Davies, E. (1979). *Hendre* and *hafod* in Caernarvonshire. *Transactions of the Caernarvonshire Historical Society,* **40,** 17-46.

de Cock, J. K. (1975). Historische geografie van Waterland. *Holland,* **7,** 329-49.

de Valera, R. & Ó Nualláin, S. (1972). *Survey of the Megalithic Tombs of Ireland.* Vol. 3. Stationery Office, Dublin.

Delcourt, P. A., Delcourt, H. R., Cridlebaugh, P. A. & Chapman, J. (1986). Holocene ethnobotanical and paleoecological record of human impact on vegetation in the Little Tennessee River Valley, Tennessee. *Quaternary Research*, **25**, 330-49.

Dennell, R. (1983). *European Economic Prehistory - a New Approach*. Pergamon Press, London.

Dennell, R. (1985). The hunter-gatherer/agricultural frontier in prehistoric temperate Europe. *The Archaeology of Frontiers and Boundaries* (Ed. by S. W. Green and S. Perlman). Academic Press, New York.

Digerfeldt, G. (1972). The Post-glacial development of Lake Trummen. Regional vegetation history, water level changes and palaeolimnology. *Folia Limnologica Scandinavica*, **16**, 1-104.

Digerfeldt, G. (1974). The Post-glacial development of the Ranviken bay in Lake Immeln. I. The history of the regional vegetation and II. The water-level changes. *Geologiska Föreningens i Stockholm Forhandlingar*, **96**, 3-32.

Digerfeldt, G. (1975). Post-glacial water-level changes in Lake Växjösjön, central southern Sweden. *Geologiska Föreningens i Stockholm Förhandlingar*, **97**, 167-73.

Digerfeldt, G. & Welinder, S. (1985). An example of the establishment of the Bronze Age cultural landscape in SW Scandinavia. *Norwegian Archaeological Review*, **18**, 106-14.

Dimbleby, G. V. (1985). *The Palynology of Archaeological Sites*. Academic Press, London.

Dodgshon, R. A. (1981). *Land and Society in Early Scotland*. Clarendon Press, Oxford.

Dodgshon, R. A. (1983). Medieval Rural Scotland. *An Historical Geography of Scotland* (Ed. by G. Whittington and I. D. Whyte). Academic Press, London.

Dodgshon, R. A. (1987). West Highland Chiefdoms, 1500-1750: a study in redistributive exchange. *Scotland and Ireland: A Study in Comparative Development* (Ed. by R. Mitchison and P. Roebuck). J. Donald, Edinburgh (in press).

Dresser, P. Q. (1985). University College, Cardiff radiocarbon dates I. *Radiocarbon*, **27**, 338-85.

Driver, C. (1985). Charcoal: another use for neglected woodland. *Quarterly Journal of Forestry*, **79**, 29-32.

Drury, S. M. (1984). The use of wild plants as famine foods in eighteenth century Scotland and Ireland. *Plant-Lore Studies* (Ed. by R. Vickery). Folklore Society, London.

Dubois, A. D. & Ferguson, D. K. (1985). The climatic history of pine in the Cairngorms based on radiocarbon dates and stable isotope analysis, with an account of the events leading up to its colonization. *Review of Palaeobotany and Palynology*, **46**, 55-80.

Duke of Argyll, George Douglas (1883). *Crofts and Farms in the Hebrides: An Account of the Management of an Island Estate for 130 Years*. Privately printed, Edinburgh.

Dupont, L. M. (1985). *Temperature and Rainfall Variation in a Raised Bog Ecosystem*. Thesis, University of Amsterdam.

Dupont, L. M. & Brenninkmeijer, C. A. M. (1984). Palaeobotanic and isotopic analysis of late Subboreal and early Subatlantic peat from Engbertsdijksveen VII, The Netherlands. *Review of Palaeobotany and Palynology*, **41**, 241-71.

Dymond, D. P. (1985). *The Norfolk Landscape*. Hodder & Stoughton, London.

Ecsedi, I. (1914). *A Hortobágy puszta és élete (The Hortobágy(Puszta and its Life)*. Debrecen, Hungary.

Edwards, K. J. (1979). Palynological and temporal inference in the context of prehistory, with special reference to the evidence from lake and peat deposits. *Journal of Archaeological Science*, **6**, 255-70.

Edwards, K. J. (1982). Man, space and the woodland edge - speculations on the detection and interpretation of human impact in pollen profiles. *Archaeological Aspects of Woodland Ecology* (Ed. by S. Limbrey and M. Bell). British Archaeological Reports, S146, Oxford.

Edwards, K. J. (1983). Quaternary palynology: consideration of a discipline. *Progress in Physical Geography*, **7**, 113-25.

Edwards, K. J. (1985a). The anthropogenic factor in vegetational history. *The Quaternary History of Ireland* (Ed. by K. J. Edwards and W. P. Warren). Academic Press, London.

Edwards, K. J. (1985b). Radiocarbon dating. *The Quaternary History of Ireland* (Ed. by

K. J. Edwards and W. P. Warren). Academic Press, London.

Edwards, K. J. (1988). Meso-neolithic vegetational impacts in Scotland and beyond: palynological considerations. *The Mesolithic in Europe: Proceedings of the Third International Symposium, Edinburgh* (Ed. by C. Bonsall). J. Donald, Edinburgh (in press).

Edwards, K. J. & Hirons, K. R. (1984). Cereal pollen grains in pre-elm decline deposits: implications for the earliest agriculture in Britain and Ireland. *Journal of Archaeological Science*, **11**, 71-80.

Edwards, K. J., McIntosh, C. J. & Robinson, D. E. (1986). Optimising the detection of cereal-type pollen grains in pre-elm decline deposits. *Circaea*, **4**, 11-3.

Edwards, K. J. & Ralston, I. (1984). Post-glacial hunter-gatherers and vegetational history in Scotland. *Proceedings of the Society of Antiquaries of Scotland*, **114**, 15-34.

Ekholm, K. (1977). External exchange and the transformation of central African chiefdoms. *The Evolution of Social Systems* (Ed. by J. Friedman and M. W. Rowlands). Duckworth, London.

El-Daoushy, F. (1986). The value of ^{210}Pb in dating Scandinavian aquatic and peat deposits. *Radiocarbon*, **28**, 1031-40.

Ellenberg, H. (1978). *Vegetation Mitteleuropas mit den Alpen in ökologischer Sicht.* (2nd edition) Ulmer, Stuttgart.

Ellenberg, H. & Klötzli, F. (1972). Waldgesellschaften und Waldstandorte der Schweiz. *Mitteilungen Schweizerische Anstalt für das forstliche Versuchswesen*, **46**, 587-930.

Emanuelsson, U. (1985). *Bygg landskapsvård på kunnskap om markens historia. Läplantering.* Lunds Universitet, Lund.

Emanuelsson, U., Bergendorff, C., Carlsson, B., Lewan, N. & Nordell, O. (1985). *Det skånska kulturlandskapet.* Bokförlaget Signum, Lund.

Ervasti, S. (1978). *Kuusamon historia 1.* Koillissanomat Oy, Kuusamo.

Eskeröd, A. (1973). *Jordbruk under femtusen år. Redskapen och maskinerna.* LTs förlag, Borås.

Etherington, J. R. (1975). *Environment and Plant Ecology.* J. Wiley & Sons, London.

Fægri, K. (1944). On the introduction of agriculture in western Norway. *Geologiska Föreningens i Stockholm Förhandlingar*, **66**, 449-62.

Fægri, K. (1954a). Den uberørte natur. *Nordisk sommeruniversitet, 1952*, 58-9.

Fægri, K. (1954b). On age and origin of the beech forest (*Fagus silvatica* L.) at Lygrefjorden, near Bergen (Norway). *Danmarks geologiske Undersøgelse*, Series II, **80**, 230-49.

Fægri, K. (1960). Maps of distribution of Norwegian vascular plants. I. Coast plants. *Universitetet i Bergen Skrifter*, **26**, 134 pp.

Fægri, K. (1962). Hvorfor. *Den norske turistforenings årbok 1962*, 48-53.

Fægri, K. (1981). Some pages of the history of pollen analysis. *Striae*, **14**, 42-7.

Fægri, K., Hartvedt, G. H. & Nyquist, F. P. (1981). *Fjordheimen.* Grøndahl & Søn, Oslo.

Fægri, K. & Iversen, J. (1975). *Textbook of Pollen Analysis.* (3rd edition revised by K. Fægri) Blackwell Scientific Publications, Oxford.

Fasteland, A. (1971a). Eit framentarisk blad av Hardangerviddas historie. *Nicolay*, **10**, 3-7.

Fasteland, A. (1971b). *Utnyttinga av den sentrale og nordlege delen av Hardangervidda i ikkje-steinbrukande tid. Ein studie med utgangspunkt i fangstbuanlegga. Forskningshistorik og eit første analyseforsøk.* Thesis, University of Bergen.

Fenton, A. (1976). *Scottish Country Life.* J. Donald, Edinburgh.

Fenton, A. (1986). *The Shape of the Past 2: Essays in Scottish Ethnology.* J. Donald, Edinburgh.

Fett, P. (1961). *Førhistoriske Minne i Fjordane, Stryn Prestegjeld.* Historisk museum rapport, Universitetet i Bergen, Bergen.

Firbas, F. (1934). Über die Bestimmung der Walddichte und der Vegetation waldloser Gebiete mit Hilfe der Pollenanalyse. *Planta*, **22**, 109-44.

Firbas, F. (1937). Der pollenanalytische Nachweis des Getreidebaus. *Zeitschrift für Botanik*, **31**, 447-78.

Fitter, A. H. & Hay, R. K. M. (1983). *Environmental Physiology of Plants.* Academic Press, London.

Flenley, J. R. (1981). Some recent and possible future developments in palynological techniques (abstract). *Quaternary Newsletter*, **35**, 33.

References 495

Flinn, M. W. (1979). *Scottish Population History from the 17th Century to the 1930s.* Cambridge University Press, Cambridge.

Flower, R. J., Battarbee, R. W. & Appleby, P. G. (1987). The recent palaeolimnology of acid lakes in Galloway: diatom analysis, pH trends, and the rôle of afforestation. *Journal of Ecology,* **75,** 797-824.

Fogelfors, H. (1979). Floraförändringar i jordbrukslandskapet. Åkermark. *Svenska Lantbruks Universitetet rapport* **5** , 65 pp.

Forestry Commission (1984). *Broadleaves in Britain: a Consultative Paper.* Forestry Commission, Edinburgh.

Forestry Commission (1985a). *The Policy for Broadleaved Woodlands.* Forestry Commission, Edinburgh.

Forestry Commission. (1985b). *Guidelines for the Management of Broadleaved Woodland.* Forestry Commission, Edinburgh.

Forman, R. T. T. & Godron, M. (1986). *Landscape Ecology.* J. Wiley & Sons, New York.

Forster, J. A. & Morris, D. (1977). Native pinewood conservation in north-east Scotland. *Native Pinewoods of Scotland* (Ed. by R. G. H. Bunce and J. N. R. Jeffers). Institute of Terrestrial Ecology, Cambridge.

Fox, C. (1932). *The Personality of Britain.* National Museum of Wales, Cardiff.

Fream, W. (1888). On the flora of water-meadows, with notes on the species. *Journal of the Linnean Society (Botany),* **24,** 454-64.

Fredskild, B. (1973). Studies in the vegetational history of Greenland. Palaeobotanical investigations of some Holocene lake and bog deposits. *Meddelelser om Grønland,* **198** (4), 1-245.

Fredskild, B. (1978). Palaeobotanical investigations of some peat deposits of Norse age at Qagssiarssuk, South Greenland. *Meddelelser om Grønland,* **204** (5), 1-41.

Fredskild, B. (1983). The Holocene vegetational development of the Godthåbsfjord area, West Greenland. *Meddelelser om Grønland, Geoscience,* **10,** 1-28.

Freidman, J. (1982). Continuity and catastrophe in social evolution. *Theory and Explanation in Archaeology* (Ed. by C. Renfrew, M. W. Rowlands and B. A. Segraves). Academic Press, London.

Fremstad, E. & Moe, B. (1982). Botaniske undersøkelser i Vetlefjordsvassdraget, Sogn og Fjordane. *Botanisk institutt, Universitetet i Bergen Rapport* **25,** 72 pp.

Fries, M. (1958). Vegetationsutveckling och odlingshistoria i Varnhemstrakten. En pollenanalytisk undersökning i Västergötland. *Acta Phytogeographica Suecica,* **39,** 1-64.

Gaál, L. (1966). *A magyar állattenyésztés múltja (Livestock Breeding in Hungary in the Past).* Akadémiai Kiadó, Budapest.

Gábori, M. (1964). A késöi paleolitikum Magyarországon (Late Palaeolithic in Hungary). *Régészeti Tanulmányok* **III.** Akadémiai Kiadó, Budapest.

Gaffney, V. (1960). *The Lordship of Strathavon.* Third Spalding Club, Aberdeen.

Gaillard, M.-J. (1984). A palaeohydrological study of Krageholmssjön (Scania, South Sweden). Regional vegetation history and water level changes. *LUNDQUA Report,* **25,** 1-40.

Gaillard, M.-J. (1985a). Palaeohydrology, palaeoclimate and cultural landscape - a palaeohydrological study in the context of the Ystad project. *Kulturlandskapet - dess framväxt och förändring.* Symposium September 1984 (Ed. by G. Regnéll), Lund University, Lund.

Gaillard, M.-J. (1985b). Postglacial palaeoclimatic changes in Scandinavia and central Europe. A tentative correlation based on studies of lake level fluctuations. *Ecologia Mediterranea,* **11,** 159-75.

Garthwaite, P. F. (1977). Management and marketing of hardwoods in south-east England. *Quarterly Journal of Forestry,* **71,** 67-77, 144, 150.

Gauch, H. G. (1982). *Multivariate Analysis in Community Ecology.* Cambridge University Press, Cambridge.

Germundsson, T. (1987). Population, land-holding and the landscape. *En arbetesrapport från Kulturlandskapet under 6000 år.* Institute of Human Geography, Lund University, Lund.

Gimingham, C. H. (1972). *Ecology of Heathlands.* Chapman & Hall, London.

Gimingham, C. H. (1975). *An Introduction to Heathland Ecology.* Oliver & Boyd, Edinburgh.

Girling, M. A. & Greig, J. (1985). A first fossil record for *Scolytus scolytus* (F.) (elm bark beetle): its occurrence in elm decline deposits from London and the implications for Neolithic elm disease. *Journal of Archaeological Science,* **12,** 347-51.

Gissel, S., Jutikkala, E., Osterberg, E., Sandnes, J. & Teitsson, B. (1981). *Desertion and Land Colonization in the Nordic Countries c. 1300-1600.* Almquist & Wiksell, Stockholm.

Gjerdåker, B. (1951). *Introduksjon til feltgranskingar på ein gamaldags vestlandsgard (Havrå i Haus).* Avhandling til magistergrad i folkelivsgransking. 385 pp. (Unpublished manuscript).

Góczán, L., Lóczy, D., Molnár, K. & Tósza, I. (1983). Application of remote sensing in monitoring and predicting changes in land use and ecological conditions. *Földrajzi Közlemények, Budapest,* **31,** 295-308.

Godwin, H. (1975). *The History of the British Flora.* (2nd edition) Cambridge University Press, Cambridge.

Göransson, G. (1986). *Populationsekologi* (Ed. by T. Alerstam *et al.*). Liber Tryck AB, Stockholm.

Göransson, H. (1986). Man and the forests of nemoral broad-leaved trees during the Stone Age. *Striae,* **24,** 143-52.

Göransson, H. (1987). On arguing in a circle, on common sense, on the smashing of paradigms, on thistles among flowers, and on other things. *Norwegian Archaeological Review,* **20,** 43-5.

Gordon, A. D. (1982). Numerical methods in Quaternary palaeoecology. V. Simultaneous graphical representation of the levels and taxa in a pollen diagram. *Review of Palaeobotany and Palynology,* **37,** 155-83.

Gower, J. C. (1984). Multivariate analysis: ordination, multidimensional scaling and allied topics. *Handbook of Applicable Mathematics VI: Statistics Part B* (Ed. by E. Lloyd). J. Wiley & Sons, Chichester.

Graff, O. & Makeschin, F. (1980). Crop yield of rye grass influenced by excretions of three earthworm species. *Pedobiologia,* **20,** 176-80.

Graham, J. M. (1970). Rural society in Connacht 1600-1640. *Irish Geographical Studies* (Ed. by N. Stephens and R. E. Glasscock). Queen's University, Belfast.

Grahn, B. & Hansson, A. (1961). *Handledning om gödselmedel och kalk.* (2nd edition) Gödsel- och kalkindustriernas Samarbetsorganisation, Stockholm.

Gram, K., Jørgensen, C. A. & Køie, M. (1944). De jyske egekrat og deres flora. *Det Kongelige Danske Videnskabernes Selskab Biologiske Skrifter,* **3** (3), 1-210.

Gräslund, B. (1980). Climatic fluctuations in the early Subboreal period. A preliminary discussion. *Striae,* **14,** 13-22.

Gray, M. (1955). Economic welfare and money income in the Highlands 1750-1850. *Scottish Journal of Political Economy,* **II,** 47-63.

Gray, M. (1957). *The Highland Economy 1750-1850.* Oliver & Boyd, Edinburgh.

Gray, T. R. G. & Williams, S. T. (1971). *Soil Microorganisms.* Longman, London & New York.

Greenacre, M. J. (1984). *Theory and Applications of Correspondence Analysis.* Academic Press, London.

Gregersen, H. V. (1974). Forsvundne kirker langs med Hærvejen gennem Nordslesvig. *Sønderjysk Månedsskrift 50 årgang,* 237-47.

Gregersen, H. V. (1977). *Egnen omkring Vojens. En Sønderjysk kommunes historie indtil 1864.* J. P. Schmidt's bogtrykkeri, Vojens.

Groenman-van Waateringe, W. (1979). The origin of crop weed communities composed of summer annuals. *Vegetatio,* **41,** 57-9.

Groenman-van Waateringe, W. (1983). The early agricultural utilization of the Irish landscape: the last word on the elm decline? *Landscape Archaeology in Ireland* (Ed. by T. Reeves-Smyth and F. Hamond). *British Archaeological Reports,* **116,** Oxford.

Grohne, U. (1957). Die Bedeutung des Phasenkontrastverfahrens für die Pollenanalyse, dargelegt am Beispiel der Gramineen-pollen vom Getreidetyp. *Photographie und Forschung,* **7,** 237-48.

Gross, H. (1931). Das problem der nacheiszeitlichen Klima- und Florenentwicklung in Nord- und Mitteleuropa. *Beihefte Botanischen Zentralblatt,* **47,** 1-110.

Gustafson, A. (1982). Växtnäringsförluster från åkermark i Sverige. *Ekohydrologi*, 11, 19-27.

Gustafson, A. (1985). Växtnäringsläckage och motåtgärder. *Ekohydrologi*, 20, 44-59.

Gustafson, L. (1980). Kullgroper i fjellstrøk. *Arkeo*, 1, 18-23.

Gustafson, L. & Indrelid, S. (1972). *Registreringer i Halne/Hein-området sommeren 1972*. Historisk museum, Universitetet i Bergen. (Unpublished manuscript).

Guyan, W. U. (1981). Zur Viehhaltung im Steinzeitdorf Thayngen-Weier II. *Archaeologie der Schweiz*, 4, 112-9.

Hæggström, C.-A. (1983). Vegetation and soil of the wooded meadows in Nåtö, Åland. *Acta Botanica Fennica*, 120, 1-66.

Haffey, D. (1983). *A Classification and Evaluation of Traditional Hay Meadows in the Northumberland National Park*. Northumberland National Park and Countryside Committee, Northumberland County Council.

Hafsten, U. (1965). Vegetational history and land occupation in Valldalen in the sub-alpine region of central South Norway traced by pollen analysis and radiocarbon measurements. *Årbok Universitetet i Bergen. Matematisk-naturvitenskapelig Serie*, 1965 (3), 26 pp.

Håkansson, S. (1986). University of Lund radiocarbon dates XIX. *Radiocarbon*, 28, 1111-32.

Halila, A. (1954). *Pohjois-Pohjanmaan ja Lapin historia 1721-1775*. Kaleva, Oulu.

Hallberg, G. (1975). *Skånes ortnamn. Serie A. Bebyggelsenamn. III.* Ljunits härad, Lund.

Hämet-Ahti, L., Suominen, I., Ulvinen, T., Uotila, P. & Vuokko, S. (1984). *Retkeilykasvio*. Suomen Luonnonsuojelun Tuki Oy, Helsinki.

Hannerberg, D. (1971). *Svenskt agrarsamhälle under 1200 år. - Gård och åker. Sörd och boskap.* Universitetsförlaget, Stockholm.

Hansen, J. & Kyllingsbaek, A. (1983). Kvaelstof og planteproduktion. *Statens Planteavisforsøg, Report S*, 1669. Copenhagen.

Hansen, K. (1976). Ecological studies in Danish heath vegetation. *Dansk Botanisk Arkiv*, 31 (2), 1-118.

Harris, M. (1975). *Culture, People, Nature - An Introduction to General Anthropology*. Crowell, New York.

Heck, K. L. & McCoy, E. D. (1979). Biogeography of seagrasses: evidence from associated organisms. *Proceedings 1st Symposium Marine Biogeography and Evolution in the Southern Hemisphere*, 1, 109-27.

Hedeager, L. & Kristiansen, K. (1985). *Archaeologi i Leksikon*. Politikens Forlag A/S, Stockholm.

Hegg, O. (1980). Die heutige Pflanzenwelt der Region Biel. *Jahrbuch der Geographischen Gesellschaft von Bern*, 53, 43-70.

Heidinga, H. A. (1984). Indications of severe drought during the 10th century AD from an inland dune area in the Central Netherlands. *Geologie en Mijnbouw*, 63, 241-8.

Heikinheimo, O. (1915). Kaskiviljelyn vaikutus Suomen metsiin. *Acta Forestalia Fennica*, 4 (2), 1-264.

Hemp, W. J. & Gresham, C. A. (1944). Hut circles in North-West Wales. *Antiquity*, 18, 183-96.

Henshall, A. S. (1972). *The Chambered Tombs of Scotland II*. Edinburgh University Press, Edinburgh.

Hesjedal, O. (1974). Vegetasjonskartlegging av potensielle magasinområder for Dagali kraftverk. *Rapport 1974-03, Institutt for jordregistrering, Ås*, 36 pp.

Hicks, S. (1975). Variations in pollen frequency in a bog at Kangerjoki, N. E. Finland during the Flandrian. *Commentationes Biologicae*, 80, 1-28.

Hicks, S. (1985a). Pollen values and field size: an experimental example from Hailuoto. *Iskos*, 5, 101-3.

Hicks, S. (1985b). Modern pollen deposition records from Kuusamo, Finland. 1. Seasonal and annual variation. *Grana*, 24, 167-84.

Hicks, S. (1985c). Problems and possibilities in correlating historical/archaeological and pollen analytical evidence in a northern boreal environment: an example from Kuusamo, Finland. *Fennoscandia archaeologica*, 11, 5-84.

Hicks, S. (1986). Modern pollen deposition records from Kuusamo, Finland. II. The establishment of pollen:vegetation analogues. *Grana*, 25, 183-204.

Hicks, S. & Hyvärinen, V.-P. (1986). Sampling modern pollen deposition by means of "Tauber traps": some considerations. *Pollen et Spores,* **28,** 219-42.

Highlands 1750 (1898). *The Highlands of Scotland in 1750* (introduction by A. Lang). Edinburgh.

Hill, M. O. (1974). Correspondence analysis: a neglected multivariate method. *Applied Statistics,* **23,** 340-54.

Hill, M. O. (1979). *TWINSPAN - A FORTRAN program for arranging multivariate data in an ordered two-way table by classification of the individuals and attributes.* Section of Ecology and Systematics, Cornell University, Ithaca, New York.

Hill, M. O. (1982). Correspondence analysis. *Encyclopedia of Statistical Sciences* 2 (Ed. by S. Kotz. and N. L. Johnson). J. Wiley & Sons, Chichester.

Hill, M. O. & Gauch, H. G. (1980). Detrended correspondence analysis: an improved ordination technique. *Vegetatio,* **42,** 47-58.

Hirons, K. R. & Edwards, K. J. (1986). Events at and around the first and second *Ulmus* declines: palaeoecological investigations in Co. Tyrone, Northern Ireland. *New Phytologist,* **104,** 131-53.

Hjelmqvist, H. (1955). Die älteste Geschichte der Kulturpflanzen in Schweden. *Opera Botanica,* **1,** 1-186.

Hjelmqvist, H. (1963). Frön och frukter från det äldsta Lund. *Acta Archaeologica Lundensia,* **2,** 233-70.

Hjelmqvist, H. (1969). Getreideabdrücke in den bronzezeitlichen Funden aus Hötofta. *Acta Archaeologica Lundensia,* **8,** 208-16.

Hjelmqvist, H. (1979). Beiträge zur Kenntnis der prähistorischen Nutzpflanzen in Schweden. *Opera Botanica,* **47,** 1-57.

Hjelmroos, M. (1985a). Vegetational history of Fårarps Mosse, South Scania, in the Early Subboreal. *Acta Archaeologica,* **54,** 45-50.

Hjelmroos, M. (1985b). Mänsklig miljöpåverkan runt Fårarps Mosse, Skåne. Preliminäre resultat från en paleoekologisk undersökning. *Kulturlandskapet - dess framväxt och förändring.* Symposium September 1984 (Ed. by G. Regnéll). Lund University, Lund.

Høeg, O. A. (1974). *Planter og tradisjon. Floraen i levende tale og tradisjon i Norge 1925-1973.* Universitetsforlaget, Oslo.

Høeg, O. A. (1976). *Planter og tradisjon. Floraen i levende tale og tradisjon i Norge 1925-1973.* (3rd edition) Universitetsforlaget, Oslo.

Hogg, A. H. A. (1960). Garn Boduan and Tre'r Ceiri, excavations at two Caernarvonshire hill-forts. *Archaeological Journal,* **117,** 1-39.

Holzner, W. (1978). Weed species and weed communities. *Vegetatio,* **38,** 13-20.

Hornstein, F. v. (1951). *Wald und Mensch.* Ravensburg.

Hougen, B. (1947). *Fra Sæter til Gård.* Norsk Arkeologisk Selskap, Oslo.

Hultgren, T., Johansen, O. S. & Lie, R. W. (1984). Stiurhelleren i Rana. Dokumentasjon av korn, husdyr og sild i yngre steinalder. *Viking,* **48,** 83-102.

Hulthén, B. & Welinder, S. (1981). *A Stone Age Economy.* Theses and papers in European prehistory, **11,** Institute of Archaeology, University of Stockholm.

Huntley, B. & Birks, H. J. B. (1983). *An Atlas of Past and Present Pollen Maps for Europe: 0-13000 Years Ago.* Cambridge University Press, Cambridge.

Hurlbert, S. H. (1971). The nonconcept of species diversity: a critique and alternative parameters. *Ecology,* **52,** 577-86.

Husdjursskötsel (1919). Avel af svin, får, getter, fjäderfän och bin. *Skrifter utgivna av de skånska hushållningssällskapan med anledning av deras hundraårsjubileum år 1914.* C. W. K. Gleerups forlag, Lund.

Ihse, M. (1978). Flygbildstolkning av vegetation i syd- och mellan-Sverige - en metodstudie för översiktlig karteringer. *Statens Naturvårdsverk-PM,* 165 pp.

Ihse, M. (1984). Försvinnande biotoper i jordbrukslandskapet. Jämförande studier i flygbilder från 1940-talet till nutid i Ystadsområdet. *Kulturlandskapet - dess framväxt och förändring.* Symposium September 1984 (Ed by G. Regnéll). Lund University, Lund.

Ihse, M. (1985). Skåne - kulturlandskap i förvandling. *Kulturminnesvård* **5/85,** 3-11.

Ihse, M. & Lewan, N. (1986). Odlingslandskapets förändringar på Svenstorp, studerade i flygbilder från 1940-talet och framåt. *Ale, historisk tidskrift för Skåneland,* **1986** (2), 1-17.

Ihse, M. & Nordberg, M.-L. (1984). Landsbygdens förvandling - studerad med flygbilder och datateknik. *Svenska Sällskapet för Antropologi och Geografi, årsbok Ymer,* **1984,** 53-72.

Indrelid, S. (1986). *Fangstfolk og bønder i fjellet. Bidrag til Hardangerviddas historie 8500-2500 år før nåtid.* Thesis, University of Bergen.

Indrelid, S. & Moe, D. (1982). Februk på Hardangervidda i yngre steinalder. *Viking,* **46,** 36-71.

Iversen, J. (1934). Moorgeologische Untersuchungen auf Grönland. *Meddelelser fra Dansk Geologisk Forening,* **8,** 341-58.

Iversen, J. (1941). Landnam i Danmarks Stenalder. Land occupation in Denmark's Stone Age. *Danmarks geologiske Undersøgelse,* Series II, **66,** 1-65.

Iversen, J. (1960). Problems of the early Postglacial forest development in Denmark. *Danmarks geologiske Undersøgelse,* Series IV, **4** (3), 1-32.

Iversen, J. (1969). Retrogressive development of a forest ecosystem demonstrated by pollen diagrams from fossil mor. *Oikos Supplement,* **12,** 35-49.

Iversen, J. (1973). The development of Denmark's nature since the last glacial. *Danmarks geologiske Undersøgelse,* Series V, **7-C,** 126 pp.

Ivimey-Cook, R. M. & Proctor, M. C. F. (1966). The plant communities of the Burren, County Clare. *Proceedings of the Royal Irish Academy B,* **64,** 211-301.

Jacobi, R. M. (1980). The early Holocene settlement of Wales. *Culture and Environment in Prehistoric Wales* (Ed. by J. A. Taylor). British Archaeological Reports, **76,** Oxford.

Jacobi, R. M., Tallis, J. H. & Mellars, P. A. (1976). The southern Pennine Mesolithic and the ecological record. *Journal of Archaeological Science,* **3,** 307-20.

Jacobsen, N. K. & Jakobsen, B. H. (1986). C[14] datering af en fossil overfladehorisont ved Igaliku Kujalleq, Sydgrønland, set i relation til nordboernes landnam. *Geografisk Tidsskrift,* **86,** 74-7.

Jacobson, G. L. (1979). The palaeoecology of white pine (*Pinus strobus*) in Minnesota. *Journal of Ecology,* **67,** 697-726.

Jacobson, G. L. & Bradshaw, R. H. W. (1981). The selection of sites for paleovegetational studies. *Quaternary Research,* **16,** 80-96.

Jacobson, G. L. & Grimm, E. C. (1986). A numerical analysis of Holocene forest and prairie vegetation in central Minnesota. *Ecology,* **67,** 958-66.

Jakobsen, B. H. (1986). *Jordbundsgeografiske undersøgelser i Sydvestgrønland.* Arbejdsgruppen vedrørende miljø og fåreavl, rapport 8, Ministeriet for Grønland, Copenhagen.

Janssen, C. R. (1959). *Alnus* as a disturbing factor in pollen diagrams. *Acta Botanica Neerlandica,* **8,** 55-8.

Janssen, C. R. (1984). Modern pollen assemblages and vegetation in the Myrtle Lake Peatland, Minnesota. *Ecological Monographs,* **54,** 213-52.

Jansson, S. L. (1956). Stallgödseln, dess egenskaper, vård och användning. *Handbok om växtnäring. III. Markvård.* Gödsel- och kalkindustriernas Samarbetsorganisation, Stockholm.

Járai-Komlódi, M. (1966). Adatok az Aföld negyedkori klímaés vegetációtörténetéhez (Contributions to the Quaternary climatic and vegetation history of the Great Hungarian Plain), I. *Botanikai Közlemények, Budapest,* **53,** 191-201.

Járai-Komlódi, M. (1969). Adatok az Alföld negyedkori klimaés vegetációtörténetéhez (Contributions to the Quaternary climatic and vegetation history of the Great Hungarian Plain), II. *Botanikai Közlemények, Budapest,* **56,** 43-55.

Jessen, C. (1867). *Alberti Magni ex ordine praedicatorum de vegetabilibus libri VII, historiae naturalis pars XVIII.* Georg Reimer, Berlin.

Jessen, K. (1933). Planterester fra den aeldre Jernalder i Thy. *Botanisk Tidsskrift,* **42,** 257-65.

Jessen, K. (1935). Archaeological dating in the history of North Jutland's vegetation. *Acta Archaeologica,* **5,** 183-214.

Jessen, K. (1949). Studies in late Quaternary deposits and flora-history of Ireland. *Proceedings of the Royal Irish Academy B,* **52,** 85-290.

Jessen, K. (1951). Old tidens korn dyrkning i Danmark. *Viking,* **15,** 15-21.

Johannessen, T. W. (1970). The climate of Scandinavia. *Climates of Northern and*

Western Europe (Ed. by C. C. Wallén). World Survey of Climatology, **5.** Elsevier, Amsterdam.

Johansen, A. B. (1973). The Hardangervidda project for interdisciplinary cultural research. A presentation. *Norwegian Archaeological Review,* **6,** 62-6.

Jóhansen, J. (1975). Pollen diagrams for the Shetland and Faroe Islands. *New Phytologist,* **74,** 369-89.

Jóhansen, J. (1985). Studies in the vegetational history of the Faroe and Shetland Islands. *Annales Societatis Scientiarum Faroensis Supplementum,* **11,** 1-117.

Johansen, O. S. (1982). Den eldste bosettinga i Borge og Valberg. *Borge og Valberg bygdebok, I* (Ed. by M. Krogtoft). Borge og Valberg bygdeboknemnd, Bodø.

Johansson, I. (1953). *Husdjursraserna.* LTs förlag, Stockholm.

Johnson, S. (1971). *A Journey to the Western Islands of Scotland* (Ed. by M. Lascelles). Yale University Press, New Haven.

Jonassen, H. (1950). Recent pollen sedimentation and Jutland heath diagrams. *Dansk Botanisk Arkiv,* **13** (7), 1-168.

Jonassen, H. (1957). Bidrag til Filsøegnens naturhistorie. *Meddelelser fra Dansk Geologisk Forening,* **13,** 192-205.

Jones, G. (1987). *The Norse Atlantic Saga.* Oxford University Press, Oxford.

Jones, G. R. J. (1961). Early territorial organisation in England and Wales. *Geografiska Annaler,* **43,** 174-81.

Jones, G. R. J. (1963). Early settlement in Arfon: the setting of Tre'r Ceiri. *Transactions of the Caernarvonshire Historical Society,* **24,** 1-20.

Jones, G. R. J. (1964). The distribution of bond settlements in north-west Wales. *Welsh Historical Review,* **2,** 19-36.

Jones, G. R. J. (1965). Agriculture in north-west Wales during the later Middle Ages. *Climatic Change with Special Reference to Wales and its Agriculture* (Ed. by J. A. Taylor). Department of Geography, University College of Wales, Aberystwyth.

Jones, G. R. J. (1972). Post-Roman Wales. *The Agrarian History of England and Wales, I* (Ed. by H. P. R. Finberg). Cambridge University Press, Cambridge.

Jones, G. R. J. (1973). Field systems of North Wales. *Studies of Field Systems in the British Isles* (Ed. by A. R. H. Baker and A. Butlin). Cambridge University Press, Cambridge.

Jones, G. R. J. (1979). Ancient British settlements in their organisational settings. *Paysages Ruraux Europeens: Travaux de la Conference Européene Permanente pour l'Étude du Paysage Rural* (Ed. by P. Flatres). Rennes University Press, Rennes.

Jones, M. (1986). Coppice wood management in the eighteenth century: an example from County Wicklow (Ireland). *Irish Forestry,* **43,** 15-31.

Jones, R. M. (1979). *The Vegetation of Upland Hay Meadows in the North of England with Experiments into the Causes of Diversity.* Thesis, University of Lancaster.

Jongman, R. H. G., ter Braak, C. J. F. & van Tongeren, O. F. R. (1987). *Data Analysis in Community and Landscape Ecology.* Pudoc, Wageningen.

Joosten, J. H. J. (1985). A 130 year micro- and macrofossil record from regeneration peat in former peasant peat pits in the Peel, The Netherlands: a palaeoecological study with agricultural and climatological implications. *Palaeogeography, Palaeoclimatology, Palaeoecology,* **49,** 277-312.

Jordbruksstatistisk årsbok (1985). *Sveriges officiella statistik.* Statistiska centralbyrån. Stockholm.

Kaakinen, E. & Saari, V. (1977). Lisätietoja Hailuodon kasvistoon. *Memoranda Societas Fauna et Flora Fennica,* **53,** 87-93.

Kääriäinen, E. (1953). On the recent uplift of the Earth's crust in Finland. *Fennia,* **77** (2), 106 pp.

Kaland, P. E. (1979). Landskapsutvikling og bosetningshistorie i Nordhordlands lyngheiområde. *På leiting etter den eldste garden* (Ed. by R. Fladby and J. Sandnes). Universitetsforlaget, Oslo.

Kaland, P. E. (1986). The origin and management of Norwegian coastal heaths as reflected by pollen analysis. *Anthropogenic Indicators in Pollen Diagrams* (Ed. by K.-E. Behre). A. A. Balkema, Rotterdam.

Kalicz, N. (1980). *Clay Gods. The Neolithic Period and the Copper Age in Hungary.* Corvina, Budapest.

Keatinge, T. H. & Dickson, J. H. (1979). Mid-Flandrian changes in vegetation on Mainland Orkney. *New Phytologist,* **82,** 585-612.

Keene, A. S. (1981). *Prehistoric Foraging in a Temperate Forest, a Linear Programming Model.* Academic Press, New York.

Kelly, R. S. (1982a). The Ardudwy survey: fieldwork in western Merioneth, 1979-81. *Journal of the Merioneth Historical and Records Society,* **9,** 121-62.

Kelly, R. S. (1982b). The excavation of a Medieval farmstead at Cefn Graeanog, Clynnog, Gwynedd. *Bulletin of the Board of Celtic Studies,* **19,** 859-908.

Kelly, R. S. (1983). Recent discoveries in the Morfa Dyffryn submerged forest. *Journal of the Merioneth Historical and Records Society,* **9,** 261-63.

Kielland-Lund, J. (1976). Beitets påvirkning på ulike skogsvegetasjonssamfunn. *Gjengroing av kulturmark* (Ed. by I. Solbu). Internordic symposium at NLH Ås, Norway, 1975.

Kindstrand, U. & Jungbeck, H. (1984). Markkarteringsdata för Malmöhus län. *Meddelande från Centralstyrelsen för Malmöhus läns försöks- och växtskyddsringar 1984,* **51,** 131-5.

Kirby, E. N. & O'Connell, M. (1982). Shannawoneen Wood, County Galway, Ireland: the woodland and saxicolous communities and the epiphytic flora. *Journal of Life Sciences, Royal Dublin Society,* **4,** 73-96.

Kirby, K. J. (1986). Forest and woodland evaluation. *Wildlife Conservation Evaluation* (Ed. by M. B. Usher). Chapman & Hall, London.

Kirby, K. J., Lister, J. & Eccles, C. (1988). The use of ancient woodland inventories to examine public access, landscape and nature conservation in woods in lowland England. *Proceedings, Recreation Ecology Research Group Meeting, Wye, 11-13.4.86* (in press).

Kirby, K. J., Peterken, G. F., Spencer, J. W. & Walker, G. J. (1984). Inventories of ancient and semi-natural woodland. *Focus on Conservation,* **6,** Nature Conservancy Council, Peterborough.

Kirch, P. V. (1980). Polynesian prehistory: cultural adaptation in island ecosystems. *American Scientist,* **68,** 39-48.

Kjekshus, H. (1977). *Ecology Control and Economic Development in East African History.* Heinemann, London.

Koelbloed, K. K. & Kroeze, J. M. (1965). Hauwmossen *(Anthoceros)* als cultuurbegeleiders. *Boor en Spade,* **14,** 104-9.

Köhler, E. & Lange, E. (1979). A contribution to distinguishing cereal from wild grass pollen grains by LM and SEM. *Grana,* **18,** 133-40.

Kortesalmi, J. J. (1975). *Kuusamon talonpoikaiselämä 1670-1970.* Kuusamen kunta, Helsinki.

Kossack, G. (1959). *Südbayern während der Hallstattzeit.* München.

Kossack, G. & Schmeidl, H. (1974/75). Vorneolithischer Getreidebau im Bayerischen Alpenvorland. *Jahresbericht der Bayerischen Bodendenkmalpflege,* **15/16,** 7-23.

Kovács, T. (1977). *The Bronze Age in Hungary.* Corvina, Budapest.

Kristiansen, K. (1984). *Settlement and Economy in Later Scandinavian Prehistory.* British Archaeological Reports, S211, Oxford.

Krogh, K. J. (1982). *Qallunaat-siaaqarfik Grønland. Erik den Rødes Grønland.* Nationalmuseets Forlag, Copenhagen.

Kujala, V. (1921). Havaintoja Kuusamon ja sen eteläpuolisten kuusimetsäalueiden metsäja suotyypeistä. *Acta Forestalia Fennica,* **18** (5), 1-65.

Küster, H. (1984). Botanische Untersuchungen zur Umweltgeschichte. *Universitas,* **39,** 739-48.

Küster, H. (1986). Werden und Wandel der Kulturlandschaft im Alpenvorland. *Germania,* **64** (2), 533-59.

Küttel, M. (1984). Veränderung von Diversität und Evenness der Tundra, aufgezeichnet in Pollendiagramm des Vuolep Allakasjaure. *Botanica Helvetica,* **94,** 279-83.

Kvamme, M. (1982). *En vegetasjonshistorisk undersøkelse av kulturlandskapets utvikling på Lurekalven, Lindås hd., Hordaland.* Thesis, University of Bergen.

Kvamme, M. & Randers, K. (1982). Breheimenundersøkelsene 1981. *Historisk museum, Universitetet i Bergen, Arkeologiske rapporter,* **3,** 140 pp.

Kyd, J. G. (1950). *Scottish Population Statistics.* Scottish History Society, Series 3, **44,** Edinburgh.

Laberg, H. (1938). *Lærdal og Borgund bygdabok. Bygd og ætter.* A. S. Lunde & Co.

Forlag, Bergen.

Lagerberg, T., Holmboe, J. & Nordhagen, R. (1958). *Våre ville planter* IV. J. G. Tanum, Oslo.

Lamb, H. H. (1982). *Climate History and the Modern World.* Methuen, London.

Lambert, G., Petrequin, P. & Richard, H. (1983). Périodicité de l'habitat lacustre néolithique et rythmes agricoles. *L'Anthropologie,* **87,** 393-411.

Lang, G. (1983). Spätquartäre See- und Moorentwicklung in der Schweiz. Stand und erste Ergebnisse eines Forschungsprogramms des Systematisch- Geobotanischen Instituts der Universität Bern. *Gesellschaft für Oekologie, Verhandlungen,* **11,** 383-90.

Lang, G. (1985). Palynologic and stratigraphic investigations of Swiss lake and mire deposits - A general view over a research program. *Dissertationes Botanicae,* **87,** 107-14.

Larsson, B. M. P. (1986). Jordbrukslandskapets vilda växtvärld - en hotad naturressurs. *Skaraborgsnatur,* **22,** 24-37.

Larsson, L. (1984). *Kulturlandskapet under neolitisk tid. Några aspekter på ekologiska förutsättningar och kulturella förhållanden - En arbetesrapport från Kulturlandskapet under 6000 år.* Institute of Human Geography, Lund University, Lund. 24 pp.

Larsson, M. (1985). The Early Neolithic funnel-beaker culture in south-west Scania, Sweden. *British Archaëological Reports,* S264, Oxford.

Leake, B. E., Tanner, P. W. G. & Senior, A. (1981). (Map of) *The Geology of Connemara.* University of Glasgow, Glasgow.

Lee, R. B. & de Vore, I. (1968). *Man the Hunter.* Aldine, Chicago.

Liese-Kleiber, H. (1985). Pollenanalysen in der Ufersiedlung Hornstaad-Hörnle I. Untersuchungen zur Sedimentation, Vegetation und Wirtschaft in einer neolith- ischen Station am Bodensee. *Materialhefte zur Vor- und Frühgeschichte in Baden-Würtemberg,* **6,** 1-149. Kommissionsverlag Konrad Theiss, Stuttgart.

Lindstrøm, J. & Lindstrøm, C. (1981). *Plassen Galdane, Lærdal i Sogn. Foreløpig registrering og restaureringsforslag m/ kostnadsoverslag.* (Unpublished stencil).

Linkola, K. (1917). Vanhan kulttuurin seurannaiskasveja maamme ruderaattija rikkaruohokasveistossa. *Terra,* **29,** 125-52.

Linkola, K. (1922). Niityt ja viljelysmaat. *Oma maa,* **3,** 1012-31.

Linné, C. von (1749). Skånska resa (Ed. by C.-O. von Sydow, 1975). Wahlström & Widstrand, Stockholm.

Livens, R. G. (1972). The Irish Sea element in the Welsh Mesolithic cultures. *Prehistoric Man in Wales and the West* (Ed. by F. M. Lynch and C. Burgess). Adams & Dart, Bath.

Lofs-Holmin, A. (1983). Influence of agricultural practices on earthworms (Lumbricidae). *Acta Agricultura,* **33,** 225-34.

Lunde, J. (1917). *Lauv som hjelpefôr.* Grøndahl & Søn Forlag, Kristiania.

Lutro, O. (1981). Berggrunnsgeologisk kart Borgund 1517 III 1:50 000. Førebels utgåve. *Norges Geologiske Undersøgelse.*

Lynch, F. (1984). Moel Goedog Circle I: a complex ring cairn near Harlech. *Archaeo- logia Cambrensis,* **133,** 8-50.

MacAodha, B. S. (1965). Clachán settlement in Iar-Connacht. *Irish Geography,* **5,** 20-8.

Macculloch, J. (1819). *A Description of the Western Islands of Scotland.* Edinburgh.

MacDonald, D. (1978). *Lewis. A History of the Island.* G. Wright, Edinburgh.

Macfarlane, A. (1978). *The Origins of English Individualism.* Oxford University Press, Oxford.

Macfarlane, W. (1905). *Geographical Collections Relating to Scotland.* Scottish History Society, Series 1, **53,** Edinburgh.

Mack, R. N., Rutter, N. W. & Valastro, S. (1979). Holocene vegetation history of the Okanogen Valley, Washington. *Quaternary Research,* **12,** 212-25.

MacKintosh, J. (1984). *Shetland Meadows Survey.* Scottish Field Unit, Report **14,** Nature Conservancy Council, Edinburgh.

MacKintosh, J. & Urquhart, U. (1984). *Survey of the Haymeadows of the Uists.* Scottish Field Unit, Report **18,** Nature Conservancy Council, Edinburgh.

Magnus, B. (1986). Iron Age exploitation of high mountain resources in Sogn. *Norwegian Archaeological Review,* **19,** 44-50.

Maguire, D., Ralph, N. & Fleming, A. (1983). Early land use on Dartmoor - palaeobotanical and pedological investigations on Holne Moor. *Integrating the Subsistence Economy* (Ed. by M. Jones). *British Archaeological Reports*, S181, Oxford.

Magyarország vízborította és árvízjárta területei az árvízmentesítő és lecsapoló munkálatok megkezdése elött (The inundated and waterlogged areas of Hungary before flood control and drainage measures). 1:60.000 (1938). M. Királyi Földmüvelésügi Minisztérium Vízrajzi Intézete, M. Királyi Honvéd Térképészeti Intézet (map).

Maier, R. A. (1985). Ein römerzeitlicher Brandopferplatz bei Schwangau und andere Zeugnisse einheimischer Religion in der Provinz Rätien. *Forschungen zur provinzialrömischen Archäologie in Bayerisch-Schwaben*, Special Volume, 231-56.

Malmgren, B. A. & Sigaroodi, M. M. (1985). Standardization of species counts - the usefulness of Hurlbert's diversity-index in paleontology. *Bulletin of the Geological Institutions of the University of Uppsala* N.S., **10**, 111-4.

Mangerud, J., Andersen, S. Th., Berglund, B. E. & Donner, J. J. (1974). Quaternary stratigraphy of Norden, a proposal for terminology and classification. *Boreas*, **3**, 109-28.

Marcotte, D. (1986). Book Review. *Journal of Mathematical Geology*, **18**, 513-5.

Markkola, J. & Merilä, E. (1982). Hailuodon niittykulttuuri ja sen romahdus. *Oulun Luonnonystävien Yhdistyksen Tiedotuksia*, **7** (1), 20-6.

Markkola, J. & Merilä, E. (1983). Hailuodon luonnonmaiden laiduntamisesta ja niittykulttuurista. *Hailuoto, Kuvauksia luonnosta ja kulttuurista*. Oulun seudun biologian ja maantieteen opettajat r.y., Oulu.

Marosi, S. & Szilárd, J. (1969). *A tiszai Alföld (The plain of the Tisza river)*. Akadémiai Kiadó, Budapest.

Martin, M. (1716). *A Description of the Western Islands of Scotland.* (2nd edition) Edinburgh.

Martin, P. S. & Wright, H. E. (1967). *Pleistocene Extinctions - The Search for a Cause.* Yale University Press, New Haven.

McAndrews, J. H. (1976). Fossil history of man's impact on the Canadian flora: an example from southern Ontario. *Canadian Botanical Association Bulletin Supplement*, **9**, 1-6.

McGovern, T. H., Bigelow, G. & Russell, D. (1985). Northern Islands, human error, and environmental degradation: a view of social and ecological change in the Medieval North Atlantic. Paper presented at the 1985 American Anthropological Association Meetings, Washington, D.C.

McIntosh, C. J. (1986). *Palaeoecological Investigations of Early Agriculture on the Isle of Arran and the Kintyre Peninsula.* Thesis, University of Birmingham.

McNally, A. & Doyle, G. J. (1984). A study of subfossil pine layers in a raised bog complex in the Irish midlands. I. Palaeowoodland extent and dynamics. II. Seral relationships and floristics. *Proceedings of the Royal Irish Academy B*, **84**, 57-70 & 71-81.

McVean, D. N. (1956). Ecology of *Alnus glutinosa* (L.) Gaertn. VI. Post-glacial history. *Journal of Ecology*, **44**, 331-3.

McVean, D. N. & Ratcliffe, D. A. (1962). *Plant Communities of the Scottish Highlands.* Her Majesty's Stationery Office, London.

Meldgaard, J. (1975). Historie (Nordboerne). *Grønland* (Ed. by P. Koch). Gyldendal, Copenhagen.

Mellars, P. A. (1971). Excavations of two Mesolithic shell middens on the Island of Oronsay (Inner Hebrides). *Nature, London*, **231**, 397-8.

Mellars, P. A. (1976). Fire ecology, animal populations and man: a study of some ecological relationships in prehistory. *Proceedings of the Prehistoric Society*, **42**, 15-45.

Mengel, K. & Kirkby, E. A. (1981). *Principles of Plant Nutrition.* (3rd edition) Worblaufen-Bern.

Merkt, J., Müller, H. & Streif, H. (1979). Stratigraphische Korrelierung spät-und postglazialer limnischer sedimente in Seebecken Südwestdeutschlands. *Schlussbericht Teil A, DFG-Forschungsvorhaben Str.*, **142/2**, 74 pp.

Mercer, J. (1969-70). Flint tools from the present tidal zone, Lussa Bay, Isle of Jura, Argyll. *Proceedings of the Society of Antiquaries of Scotland*, **102**, 1-30.

Middeldorp, A. A. (1982). Pollen concentration as a basis for indirect dating and quantifying net organic and fungal production in a peat bog ecosystem. *Review of Palaeobotany and Palynology*, **37**, 225-82.

Mikkelsen, V. M. (1949). Præstø Fjord. The development of the postglacial vegetation and a contribution to the history of the Baltic Sea. *Dansk Botanisk Arkiv*, **13** (5), 1-171.

Mikkelsen, V. M. (1986). Borup, man and vegetation. *Royal Danish Academy of Sciences and Letters' Commission for Research on the History of Agricultural Implements and Field Structures*, **4**, 1-42.

Miles, J. & Kinnaird J. W. (1981). Grazing with particular reference to birch, juniper and Scots pine in the Scottish Highlands. *Scottish Forestry*, **33**, 280-9.

Mitchell, F. (1986). *The Shell Guide to the Irish Landscape*. Country House, Dublin.

Moar, N. T. (1969). Two pollen diagrams from the Mainland, Orkney Islands. *New Phytologist*, **68**, 201-8.

Moe, D. (1973). Studies in the Holocene vegetation development on Hardangervidda, southern Norway. I. The occurrence and origin of pollen of plants favoured by man's activity. *Norwegian Archaeological Review*, **6**, 67-73.

Moe, D. (1978). *Studier over vegetasjonsutviklingen gjennom Holocen på Hardangervidda, Sør-Norge, II. Generell utvikling og tre-grensevariasjoner*. Part of thesis, University of Bergen.

Moe, D. (1979). Tregrensefluktuasjoner på Hardangervidda etter siste istid. *Fortiden i søkelyset* (Ed. by R. Nydal, S. Westin, U. Hafsten and S. Gulliksen). Strindheim, Trondheim.

Moe, D. (1981). Grønne sauer og hva ull kan inneholde. *Naturen*, **1**, 47-8.

Moe, D. (1983a). Studies in the vegetation history of Vestvågøy, Lofoten, North-Norway. A preboreal recession and the vegetational optimum. *Tromura, naturvitenskap*, **39**, 1-28.

Moe, D. (1983b). Palynology of sheep's faeces: relationship between pollen content, diet and local pollen rain. *Grana*, **22**, 105-13.

Møller, J. J. (1986). Holocene transgression maximum about 6000 years BP at Ramså, Vesterålen, North Norway. *Norsk geografisk Tidsskrift*, **40**, 77-84.

Molloy, K. & O'Connell, M. (1987). The nature of the vegetational changes at about 5000 B.P. with particular reference to the elm decline: fresh evidence from Connemara, western Ireland. *New Phytologist*, **106**, 203-20.

Monro, D. (1961). *Monro's Western Isles of Scotland* (Ed. by R. W. Munro). Oliver & Boyd, Edinburgh.

Moore, J. J. (1979). Zur Entstehung und Entwicklung der terrainbedeckendenmoore Westirlands. *Gesellschaftsentwicklung (Syndynamik)* (Ed. by R. Tüxen and W.-H. Sommer). Cramer, Vaduz.

Moore, P. D. (1973). The influence of prehistoric cultures upon the initiation and spread of blanket bog in upland Wales. *Nature, London*, **241**, 350-3.

Moore, P. D. (1980). Resolution limits of pollen analysis applied to archaeology. *MASCA Journal*, **1** (4), 118-20.

Moore, P. D. (1981). Neolithic land-use in mid-Wales. *Proceedings of the IVth International Palynological Conference (Lucknow)*, **3**, 279-90.

Moore, P. D. (1986). Hydrological changes in mires. *Handbook of Holocene Palaeoecology and Palaeohydrology* (Ed. by B. E. Berglund). J. Wiley & Sons, Chichester.

Moore, P. D. (1987). Burning issues in fire control. *Nature, London*, **325**, 486.

Moore, P. D. & Chater, E. H. (1969). Studies in the vegetational history of mid-Wales. I. The Post-glacial period in Cardiganshire. *New Phytologist*, **68**, 183-96.

Moore, P. D. & Webb, J. A. (1978). *An Illustrated Guide to Pollen Analysis*. Hodder & Stoughton, London.

Mukula, V. (1964). *Rikkaruohot ja niiden torjunta*. Kirjayhtymä, Helsinki.

Müller, H. (1962). Pollenanalytische Untersuchung eines Quartärprofiles durch die spät- und nach-eiszeitlichen Ablagerungen des Schleinsees (Süd westdeutschland). *Geologisches Jahrbuch*, **79**, 493-526.

Murray, A. (1740). *The True Interest of Great Britain, Ireland and Our Plantations*. Privately printed, London.

Myerscough, P. J. (1980). Biological Flora of the British Isles: *Epilobium angustifolium* L. *Journal of Ecology*, **68**, 1047-74.

Nagy, A. (1976). Erdök a Hortobágyon (Forests in the Hortobágy). *Hortobágy a nomád*

Pusztától a Nemzeti Parkig (The Hortobágy from the nomadic puszta to the National Park) (Ed. by G. Kovács and F. Salamon). Natura, Budapest.

Nannesson, L. (1914). Skånes nötkreatursskötsel från 1800-talets början till nuvarande tid. *Skrifter utgivna av de skånska hushållningssällskapen vid deras hundraårs-jubileum år 1914.* Berlings, Lund.

Nedkvitne, J. & Garmo, T. (1986). Conifer woodland as summer grazing for sheep. *Grazing Research at Northern Latitudes* (Ed. by O. Gudmundsson). Plenum Publishing Corporation, New York.

Negaard, H. (1911). Hardangerviddens ældste befolkning. Undersøkelser og fund. *Bergens Museums Årbok,* **1911** (4), 69 pp.

Nichols, H. (1967). Vegetational change, shoreline displacement and the human factor in the late Quaternary history of south-west Scotland. *Transactions of the Royal Society of Edinburgh,* **67,** 145-87.

Nielsen, P. (1984). Entomologiske undersøgelser i og omkring fåreholderområdet i Sydgrønland. Arbejdsgruppen vedrørende miljø og fåreavl, rapport 2, 28 pp. Ministeriet for Grønland, Copenhagen.

Nielsen, V. (1984). Prehistoric field boundaries in eastern Denmark. *Journal of Danish Archaeology,* **3,** 135-63.

Nielssen, A. R. (1977). Ødetida på Vestvågøya. Bosettingshistorien 1300-1600. Thesis, University of Tromsø.

Nilssen, E. (1983). Klima og vegetasjonshistoriske undersøkelser i Lofoten. Thesis, University of Tromsø.

Nilsson, J. (1970). Vegetationsförändringar i Steneryds lövängsområde vid dess restaurering. *Meddelanden 1 från Forskargruppen för skötsel av naturreservat.* Avdeling för ekologisk botanik, Lunds Universitet, Lund.

Nilsson, T. (1961). Ein neues Standardpollendiagramm aus Bjärsjöholmssjön in Schonen. *Lunds Universitets Årsskrift,* N.F.2, **56,** 1-34.

Nilsson, T. (1964). Standardpollendiagramme und C^{14}-Datierungen aus dem Ageröds Mosse im mittleren Schonen. *Lunds Universitets Årsskrift* N.F.2, **59,** 1-52.

Nordhagen, R. (1943). Sikilsdalen og Norges Fjellbeiter. *Bergens Museums Skrifter,* **22,** 607 pp.

Nordhagen, R. (1954). Om barkebrød og treslaget alm i kulturhistorisk belysning. *Danmarks geologiske Undersøgelse,* Series II, **80,** 262-308.

Noring, A. (1836). *Ett och annat rörande Boskaps- och Ladugårdsskötseln.* Berlings, Lund.

Noring, A. (1841-42). *Handbok i husdjursskötseln. Del 1-3.* Berlings, Lund.

NSA (1845). *New Statistical Account of Scotland.* 18 volumes, Edinburgh.

Núnèz, M. G. & Vuorela, I. (1979). A tentative evaluation of cultural pollen data in early agrarian development research. *Suomen Museo,* **1978,** 5-36.

O'Connell, M. (1980). The development history of Scragh Bog, Co. Westmeath and the vegetation history of its hinterland. *New Phytologist,* **85,** 301-19.

O'Connell, M. (1986). Reconstruction of local landscape development in the post-Atlantic based on palaeoecological investigations at Carrownaglogh prehistoric field system, County Mayo, Ireland. *Review of Palaeobotany and Palynology,* **49,** 117-76.

O'Connell, M. (1987). Early cereal-type pollen records from Connemara, western Ireland and their possible significance. *Pollen et Spores,* **29,** 207-224.

O'Connell, M., Mitchell, F. J. G., Readman, P. W., Doherty, T. J. & Murray, D. A. (1987). Palaeoecological investigations towards the reconstruction of the post-glacial environment at Lough Doo, County Mayo, Ireland. *Journal of Quaternary Science,* **2,** 149-64.

Odgaard, B. (1985). Kulturlandskabets historie i Vestjylland. Foreløbige resultater af nye pollenanalytiske undersøgelser. *Fortidsminder. Antikvariske Studier,* **7,** 48-59.

Odland, A. (1981). Pre- and subalpine tall herb and fern vegetation in Røldal, W Norway. *Nordic Journal of Botany,* **1,** 671-90.

Oeschger, H., Andrée, M., Moell, M., Riesen, T., Siegenthaler, U., Ammann, B., Tobolski, K., Bonani, B., Hofmann, H. J., Morenzoni, E., Nessi, M., Suter, M. & Wölfli, W. (1985). Radiocarbon chronology at Lobsigensee. Comparison of materials and methods. *Dissertationes Botanicae,* **87,** 135-39.

Oksbjerg, E. (1964). Lidt om landskabets udvikling i Midtjylland. *Hedeselskabets Tidsskrift,* **85** (5-6), 84-92, 102-7.

Olafsson, E. (1943). *Ferdabok Eggerts Olafssonar og Bjarna Palssonar um ferdir theirra* *a Islandi arin 1752-1757.* (2nd edition) Reykjavik.

Oldfield, F. (1970). Some aspects of scale and complexity in pollen-analytically based palaeoecology. *Pollen et Spores,* **12,** 163-71.

Olsen, M. (1926). *Ættrgård og helligdom. Norske stedsnavn sosialt og religionshistorisk belyst.* Institutt for sammenlignende kulturforskning, Oslo.

Olsson, I. U. (1979). A warning against radiocarbon dating of samples containing little carbon. *Boreas,* **8,** 203-7.

Olsson, I. U. (1986). A study of errors in ^{14}C-dates of peat and sediment. *Radiocarbon,* **28,** 429-35.

Orme, A. R. (1967). Drumlins and the Weichsel glaciation of Connemara. *Irish Geography,* **5,** 262-74.

OSA (1791-99). *Old Statistical Account of Scotland.* 20 volumes, Edinburgh.

Osvald, H. (1962). *Vallodling och växtföljder. Uppkomst och utveckling i Sverige.* Natur och Kultur, Uppsala.

Paasivirta, P. (1936). Piirteitä Hailuodon kulttuurimaantieteestä. *Terra,* **48,** 72-94.

Påhlsson, I. (1982). *Cannabis sativa* in Dalarna. *Striae,* **14,** 79-82.

Pals, J. P. (1984). Verkoolde plantenresten uit een 11e/12e eeuwse huisplaats te Oostzaan. *De Jol, Mededelingenblad van de Stichting Oudheidkamer Oostzaan,* **1** (2), 6-10.

Pals, J. P., van Geel, B. & Delfos, A. (1980). Palaeoecological studies in the Klokkeweel bog near Hoogkarspel (prov. of Noord-Holland). *Review of Palaeobotany and Palynology,* **30,** 317-418.

Parry, M. L. (1975). Secular climatic change and marginal agriculture. *Transactions of the Institute of British Geographers,* **64,** 1-13.

Parvela, A. A. (1923). Tietoja Kuusamon viljelykasvistosta. *Luonnon Ystävä,* **27,** 11-8.

Patterson, W. A. III, Edwards, K. J. & Maguire, D. J. (1987). Microscopic charcoal as a fossil indicator of fire. *Quaternary Science Reviews,* **6,** 3-23.

Paul, H. & Ruoff, S. (1927). Pollenstatistische und stratigraphische Mooruntersuchungen im südlichen Bayern I. *Berichte der bayerischen Botanischen Gesellschaft,* **29,** 1-84.

Paul, H. & Ruoff, S. (1932). Pollenstatistische und stratigraphische Mooruntersuchungen im südlichen Bayern. II. *Berichte der bayerischen Botanischen Gesellschaft,* **30,** 1-264.

Pearson, G. W. & Pilcher, J. R. (1975). Belfast radiocarbon dates VIII. *Radiocarbon,* **17,** 226-38.

Pearson, G. W. & Stuiver, M. (1986). High-precision calibration of the radiocarbon time scale, 500-2500 BC. *Radiocarbon,* **28,** 839-62.

Pécsi, M. (1968). Evolution of the flood-plain levels of the Danube and their principal bearings on the geography of agriculture. *Földrajzi Közlemények, Budapest,* **16,** 215-71.

Pécsi, M. (1977). Landscape types of Hungary. *Memorie della Societ Geografica Italiana, Roma,* **39,** 643-54.

Pécsi, M. & Somogyi, S. (1983). Magyarország tátípustérképe (Map of landscape types in Hungary). 1:1,250,000. *MTA Földrajztudományi Kutató Intézet,* Budapest (map).

Pécsi, M., Somogyi, S. & Jakucs, P. (1971). Landscape units and their types in Hungary. *Hungary. Geographical Studies, Budapest, IGU European Regional Conference 1971.*

Peglar, S. M., Fritz, S. C., Alapieti, T., Saarnisto, M. & Birks, H. J. B. (1984). Composition and formation of laminated sediments in Diss Mere, Norfolk, England. *Boreas,* **13,** 13-28.

Peglar, S. M., Fritz, S. C. & Birks, H. J. B. (1988). Vegetation and land-use history at Diss, Norfolk, England. *Journal of Ecology* (in press).

Pennington, W. (1975). The effect of Neolithic man on the environment of north-west England: the use of absolute pollen diagrams. *The Effect of Man on the Landscape: the Highland Zone* (Ed. by J. G. Evans, S. Limbrey and H. Cleere). *Council for British Archaeology Research Report,* **11,** London.

Pennington, W. (1979). The origin of pollen in lake sediments: an enclosed lake compared with one receiving inflow streams. *New Phytologist,* **83,** 189-213.

Persson, A. (1986a). *Villie socken ca. 1700. Bebyggelse - markanvändning - social struktur. Technical report. The rural landscape during 6000 years in South Sweden.* Lund University, Lund.

Persson, A. (1986b). *Agrar struktur i Bjäresjö by 1570 och 1699. Technical report. The* *rural landscape during 6000 years in South Sweden*. Lund University, Lund.

Persson, A. (1987). *Agrar struktur i Bjäresjö by 1570 och 1699. En trendanalys utförd med ADB-stöd - En arbetesrapport från kulturlandskapet under 6000 år.* Institute of Human Geography, Lund University, Lund.

Peterken, G. F. (1981). *Woodland Conservation and Management.* Chapman & Hall, London.

Peterken, G. F. (1983). Woodland conservation in Britain. *Conservation in Perspective* (Ed. by A. Warren and F. B. Goldsmith). J. Wiley & Sons, Chichester.

Peterken, G. F. & Game, M. (1984). Historical factors affecting the number and distribution of vascular plant species in the woodlands of central Lincolnshire. *Journal of Ecology,* **72,** 155-82.

Peterken, G. F. & Lloyd, P. S. (1967). Biological Flora of the British Isles: *Ilex aquifolium* L. *Journal of Ecology,* **55,** 841-58.

Peterkin, A. (1820). *Rentals of the Ancient Earldom and Bishoprick of Orkney.* Privately printed, Edinburgh.

Phillips, J. B. (1971). Effect of cutting technique on coppice regrowth. *Quarterly Journal of Forestry,* **65,** 220-3.

Pierce, T. Jones (1961). Pastoral and agricultural settlements in early Wales. *Geografiska Annaler,* **43,** 182-9.

Pigott, C. D. (1983). Regeneration of oak-birch woodland following the exclusion of sheep. *Journal of Ecology,* **71,** 629-46.

Pilcher, J. R. & Larmour, R. (1982). Late-glacial and post-glacial vegetational history of the Meenadoan nature reserve, County Tyrone. *Proceedings of the Royal Irish Academy B,* **82,** 277-95.

Pilcher, J. R., Smith, A. G., Pearson, G. W. & Crowder, A. (1971). Land clearances in the Irish Neolithic: New evidence and interpretation. *Science,* **172,** 560-2.

Pococke, R. (1986). *Tours in Scotland, 1747, 1750 and 1760 by Richard Pococke.* (Ed. by D. W. Kemp). Scottish History Society, Series 1, **1,** Edinburgh.

Polak, B. (1929). *Een onderzoek naar de botanische samenstelling van het Hollandsche veen.* Amsterdam.

Pons, L. J. & van Oosten, M. F. (1974). *De bodem van Noord-Holland.* Stiboka, Wageningen.

Pontoppidan (1749). *Reisebeskrivelse over Fillefjell. Forsög paa Norges Naturlige Historie.* Reprinted as Norges naturlige historie 1752-1753 (1977). Rosenkilde & Bagger, Copenhagen.

Popper, K. (1980). Evolution. *New Scientist,* **87,** 611.

Prentice, I. C. (1980). Multidimensional scaling as a research tool in Quaternary palynology: a review of theory and methods. *Review of Palaeobotany and Palynology,* **31,** 71-104.

Prentice, I. C. (1985). Pollen representation, source area and basin size: toward a unified theory of pollen analysis. *Quaternary Research,* **23,** 76-86.

Prentice, I. C., Berglund, B. E. & Olsson, T. (1987). Quantitative forest-composition sensing characteristics of pollen samples from Swedish lakes. *Boreas,* **16,** 43-54.

Prescott, C. (1986). *Chronological, Typological and Contextual Aspects of the Late Lithic Period. A Study based on Sites Excavated in the Nyset and Steggje Mountain Valleys, Årdal, Sogn, Norway.* Thesis, University of Bergen.

Punt, W. & Malotaux, M. (1984). Cannabaceae, Moraceae and Urticaceae. *The Northwest European Pollen Flora,* IV (Ed. by W. Punt and G. C. S. Clarke). Elsevier, Amsterdam.

Raatikainen, M. & Raatikainen, T. (1972). Weed colonization of cultivated fields in Finland. *Annales Agricales Fennici,* **11,** 100-10.

Rackham, O. (1972). Grundle House: on the quantities of timber in certain East Anglian buildings in relation to local supplies. *Vernacular Architecture,* **3,** 3-8.

Rackham, O. (1975). *Hayley Wood: its History and Ecology.* Cambridgeshire & Isle of Ely Naturalists' Trust, Cambridge.

Rackham, O. (1976). *Trees and Woodland in the British Landscape.* J. M. Dent & Sons, London.

Rackham, O. (1977). Neolithic woodland management in the Somerset Levels: Garvin's Walton Heath and Rowland's tracks. *Somerset Levels Papers,* **3,** 65-72.

Rackham, O. (1978). Archaeology and land-use history. *Epping Forest - the natural* *aspect?* (Ed. by D. Corke). *Essex Naturalist* N.S. **2**, 16-57.

Rackham, O. (1980). *Ancient Woodland; its History, Vegetation and Uses in England.* Edward Arnold, London.

Rackham, O. (1986a). *The History of the Countryside.* J. M. Dent & Sons, London.

Rackham, O. (1986b). The ancient woods of Norfolk. *Transactions of the Norfolk & Norwich Naturalists' Society,* **27,** 161-77.

Rackham, O. (1986c). *Ancient Woodland of England; the Woods of South-East Essex.* Rochford District Council, Rochford.

Rackham, O., Blair, W. J. & Munby, J. T. (1978). The thirteenth-century roofs and floor of the Blackfriars Monastery at Gloucester. *Medieval Archaeology,* **22,** 105-22.

Radley, J. (1961). Holly as a winter feed. *Agricultural History Review,* **9,** 89-93.

Ralph, E. K., Michael, H. N. & Han, M. C. (1973). Radiocarbon dates and reality. *MASCA Newsletter,* **9,** 1-20.

Ralska-Jasiewiczowa, M. (1986). *Palaeohydrological changes in the temperate zone in the last 15,000 years.* IGCP 158 B. Lake and mire environments Project Catalogue for Europe. Lund.

Rasmussen, K. & Reenberg, A. (1980). Ecological human geography - some considerations of concepts and methods. *Geografisk Tidsskrift,* **80,** 81-8.

Rasmussen, R. O. (1985). *Flyttemarksbrug.* Geografforlaget, Brenderup.

Raynor, G. S., Hayes, J. V. & Ogden, E. C. (1975). Particulate dispersion from sources within a forest. *Boundary-Layer Meteorology,* **9,** 257-77.

RCAHM (1964). *Royal Commission on Ancient and Historical Monuments, Inventory Caernarvonshire Vol. III, West.* Her Majesty's Stationery Office, London.

Reinton, L. (1955, 1957 & 1961). Sæterbruket i Noreg. I, II & III. *Institutt for Sammenlignende Kulturforskning, Serie B, Skrifter,* **XLVIII,** Oslo.

Reinton, L. (1976). *Til seters. Norsk seterbruk og seterstell.* Det norske samlaget, Oslo.

Risser, P. G., Karr, J. R. & Forman, R. T. T. (1984). Landscape ecology - directions and approaches. *Illinois Natural History Survey Special Publication,* **2,** 18 pp.

Ritchie, W. (1966). The postglacial rise in sea level and coastal changes in the Uists. *Transactions of the Institute of British Geographers,* **39,** 79-96.

Robinson, D. E. (1983). Possible Mesolithic activity in the west of Arran: evidence from peat deposits. *Glasgow Archaeological Journal,* **10,** 1-6.

Romme, W. H. & Knight, D. H. (1982). Landscape diversity: the concept applied to Yellowstone Park. *BioScience,* **32,** 664-70.

Ropeid, A. (1960). *Skav. En studie i eldre tids fôrproblem.* Universitetsforlaget, Oslo.

Rösch, M. (1983). Geschichte der Nussbaumer Seen (Kanton Thurgau) und ihrer Umgebung seit dem Ausgang der letzten Eiszeit aufgrund quartärbotanischer, stratigraphischer und sedimentologischer Untersuchungen. *Mitteilungen der Thurgauischen Naturforschenden Gesellschaft,* **45,** 1-110.

Rösch, M. (1985). Ein Pollenprofil aus dem Feuenried bei Ueberlingen am Ried: Stratigraphische und landschaftsgeschichtliche Bedeutung für das Holozän im Bodenseegebiet. *Materialhefte zur Vor- und Frühgeschichte in Baden-Württemberg,* **7,** 43-79.

Rose, F. & James, P. W. (1974). The corticolous and lignicolous (lichen) species of the New Forest, Hampshire. *Lichenologist,* **6,** 1-72.

Rostholm, H. (1982). *Oldtiden på Herning-egnen.* Herning museum, Herning.

Rowley, J. R. & Rowley, J. (1956). Vertical migration of spherical and aspherical pollen in a *Sphagnum* bog. *Proceedings of Minnesota Academy of Science,* **24,** 29-30.

Rowley-Conwy, P. (1981). Slash and burn in the temperate European neolithic. *Farming Practice in British Prehistory* (Ed. by R. J. Mercer). Edinburgh University Press, Edinburgh.

Rump, H.-U. (1977). Füssen. *Historischer Atlas von Bayern, Teil Schwaben,* **9,** 1-479.

Rygh, O. (1898). *Norske gaardnavne.* Forord og Indledning. W. C. Fabritius & Sønner A/S, Kristiania.

Rygh, O. (1905). *Norske gaardnavne. Vol. 16.* W. C. Fabritius & Sønner A/S, Kristiania.

Salamon, F. (1976). A Hortobágy mezögazdasági hasznosítása (Agricultural utilization of the Hortobágy). *Hortobágy a nomád Pusztától a Nemzeti Parkig* (Ed. by G. Kovács and F. Salamon). Natura, Budapest.

Salmi, M. (1962). Investigations on the distribution of pollens in an extensive raised bog. *Comptes Rendus Societé Géologique de Finland,* **34,** 159-93.

Salminen, S. (1973). Tulosten luotettavuus ja karttatulostus valtakunnan metsien V inventoinnissa. *Commentationes Institutes Forestalia Fennica*, **78** (6), 1-64.

Salskontrakt frå 1774. *Wendelbo (Øvre Ljøsne)*. Private paper.

Saunders, D. (1984). Small woods scheme - Sussex success. *Forestry and British Timber.* (February) 20-1.

Schlichtherle, H. (1985). Prähistorische Ufersiedlungen am Bodensee - eine Einführung in naturräumliche Gegebenheiten und archäologische Quellen. *Materialhefte zur Vor- und Frühgeschichte in Baden-Württemberg*, **7**, 9-42.

Schrøder, H. (1985). Nitrogen losses from Danish agriculture - trends and consequences. *Agriculture, Ecosystems and Environment*, **14**, 279-89.

Schweingruber, F. H. (1976). *Prähistorisches Holz. Die Bedeutung von Holzfunden aus Mitteleuropa für die Lösung archäologischer und vegetationskundlicher Probleme.* Paul Haupt, Bern & Stuttgart.

Seebohm, M. E. (1927). *The Evolution of the English Farm.* George Allen & Unwin, London.

Sernander, R. (1920). Den svenska hagens historia. *Svenska betes- og vallföreningens årsskrift*, **1920**, 5-13.

Simberloff, D. (1978). Use of rarefaction and related methods in ecology. *Biological Data in Water Pollution Assessment: Quantitative and Statistical Analyses* (Ed. by K. L. Dickson, J. Cairns and R. J. Livingston). American Society for Testing and Materials.

Simberloff, D. (1979). Rarefaction as a distribution-free method of expressing and estimating diversity. *Ecological Diversity in Theory and Practice* (Ed. by J. F. Grassle, G. P. Patil, W. Smith and C. Taillie). International Co-operative Publishing House, Fairland.

Simberloff, D. (1983). Biogeography: the unification and maturation of a science. *Perspectives in Ornithology* (Ed. by A. H. Brush and G. A. Clark). Cambridge University Press, New York.

Simmons, I. G. (1964). Pollen diagrams from Dartmoor. *New Phytologist*, **63**, 165-80.

Simmons, I. G. (1975). Towards an ecology of Mesolithic man in the uplands of Great Britian. *Journal of Archaeological Sciences*, **2**, 1-15.

Simmons, I. G., Dimbleby, G. W. & Grigson, C. (1981). The Mesolithic. *The Environment in British Prehistory* (Ed. by I. G. Simmons and M. J. Tooley). Duckworth, London.

Simon, T. (1975). A Györi-medence. Természetes növényzet (The Györ basin. Natural vegetation). *A Kisalföld és a Nyugat-magyarrországi peremvidék* (Ed. by L. Ádam and S. Marosi). Akadémiai Kiadó, Budapest.

Singh, G. (1970). Late-glacial vegetational history of Lecale, Co. Down. *Proceedings of the Royal Irish Academy B*, **69**, 189-216.

Skansjö, S. (1981). Landowners, tenant farmers and cottagers. Gårdlösa. An Iron age community in its natural and social setting, I (Ed. by B. Stjernquist). *Regiae Societatis Humaniorum Litterarum Lundensis* LXXV.

Skansjö, S. (1983). *Söderslätt genom 600 år. Bebyggelse och odling under äldre historisk tid. Skånsk senmedeltid och renässans.* Vetenskapssocieteten i Lund, Lund.

Skansjö, S. (1985). *Estate building and settlement changes in S. Scania c 1500-1650 in a European perspective. Technical report. The rural landscape during 6000 years in South Sweden.* University of Lund, Lund.

Skansjö, S. (1987a). Estate building and settlement changes in southern Scania c. 1500-1650 in a European perspective. *The Medieval and Early-Modern Rural landscape of Europe under the Impact of the Commercial Economy* (Ed. by H. J. Nitz). Department of Geography, University of Göttingen, Göttingen (in press).

Skansjö, S. (1987b). Från Vikingatida stormansgård till renässansslott. Några huvuddrag i Bjersjöholms äldre historia - En arbetesrapport från kulturlandskapet under 6000 år. Department of Human Geography, University of Lund, Lund.

Skene, W. F. (1876). *Celtic Scotland.* 3 volumes, Edmonston & Douglas, Edinburgh.

Skogen, A. (1984). Sunndalen. *Breheimen-Stryn. Konsesjonsavgjørende botaniske undersøkelser* (Ed. by O. B. Meyer). *Botanisk institutt, Universitetet i Bergen Rapport* **34.**

Skoug, S. E. (1975). *Kongeveien over Fillefjell. Vindhella & Galdane.* Gröndal & Sön Forlag A/S, Oslo.

Slicher van Bath, B. H. (1963). *The Agrarian History of Western Europe.* Edward Arnold, London.

Smith, A. G. (1970). The influence of Mesolithic and Neolithic man on British vegetation. *Studies in the Vegetational History of the British Isles* (Ed. by D. Walker and R. G. West). Cambridge University Press, London.

Smith, A. G. (1981). The Neolithic. *The Environment in British Prehistory* (Ed. by I. G. Simmons and M. J. Tooley). Duckworth, London.

Smith, A. G. (1984). Newferry and the Boreal-Atlantic transition. *New Phytologist*, **98**, 35-55.

Smith, C. A. (1974). A morphological analysis of late prehistoric and Romano-British settlements in North West Wales. *Proceedings of the Prehistoric Society*, **40**, 157-69.

Smith, C. A. (1985). Some evidence of early upland settlement from Wales. *Upland Settlement in Britain. The Second Millennium BC and After* (Ed. by D. Spratt and C. Burgess). British Archaeological Reports, **143**, Oxford.

Smith, C. T. (1967). *An Historical Geography of Western Europe Before 1800.* Longman, London.

Smith, R. S. (1983). *Northern Haymeadows.* Unpublished report to the Yorkshire Dales National Park Committee.

Solomon, A. M. & Kroener, D. F. (1971). Suburban replacement of rural land uses reflected in the pollen rain of northeastern New Jersey. *Bulletin of New Jersey Academy of Sciences*, **16**, 30-44.

Sølvberg, I. Ø. (1976). *Driftsformer i vestnorsk jordbruk ca. 600-1350.* Universitetsforlaget, Oslo.

Somogyi, S. (1975). Contribution to the Holocene history of Hungarian river valleys. *Biuletyn Geologiczny*, **19**, 185-93.

Somogyi, S. (1978). Regulated rivers in Hungary. *Geographica Polonica*, **41**, 39-53.

Somogyi, S. (1984). Történeti földrajzi bevezetö (Historical-geographical introduction). *Magyarország története I. Elözmények és magyar történet 1242-ig (History of Hungary I. Precedents and Hungarian history to 1242)* (Ed. by A. Bartha). Akadémiai Kiadó, Budapest.

Sørensen, I. (1982). Pollenundersøgelser i møddingen på Niaquassat. *Tidsskriftet Grønland*, **8-9**, 296-302.

SRO Manuscript collections, Scottish Record Office, Edinburgh.

Statistics of Annexed Estates (1973). *Statistics of the Annexed Estates 1755-1756.* Her Majesty's Stationery Office, Edinburgh.

Steele, R. C. & Peterken, G. F. (1982). Management objectives for broadleaved woodland conservation. *Broadleaves in Britain* (Ed. by D. C. Malcolm, J. Evans and P. N. Edwards). Institute of Chartered Foresters, Edinburgh.

Steineck, S. (1984). Stallgödselproduktion. *Kursmaterial vid kurser anordade av Statens Lantbrukskemiska Laboratorium* 1984.

Sterbetz, I. (1976). A Hortobágy természetvédelme (Nature conservation in the Hortobágy). *Hortobágy a nomád Pusztától a Nemzeti Parkig* (Ed. by G. Kovács and F. Salamon). Natura, Budapest.

Sterten, A. K. (1974). Hardangerviddas klima. *Hardangervidda natur-kulturhistorie-samfunnsliv* (Ed. by S. Øvstedal). Norges offentlige utredninger, **30B**, Universitetsforlaget, Oslo.

Stjernquist, B. (1981). Arkeologisk forskning om den agrara bebyggelsen i Skåne vid vikingatidens slut - källäge och problemställningar. *Bebyggelsehistorisk tidskrift*, **2**, 17-25.

Stockmarr, J. (1971). Tablets with spores used in absolute pollen analysis. *Pollen et Spores*, **13**, 615-21.

Stockmarr, J. (1975). Retrogressive forest development, as reflected in a mor pollen diagram. *Palaeohistoria*, **17**, 37-51.

Straka, H. (1970). *Pollenanalyse und Vegetationsgeschichte.* A. Zeimsen, Wittenberg Lutherstadt.

Stubbs, A. E. (1982). Conservation and the future for the field entomologist. *British Entomological and Field Society Transactions*, **15**, 55-67.

Stuiver, M. & Pearson, G. W. (1986). High-precision calibration of the radiocarbon time scale, AD 1950-500 BC. *Radiocarbon*, **28**, 805-38.

Stuiver, M. & Reimer, P. J. (1986). A computer program for radiocarbon age calibration. *Radiocarbon,* **28,** 1022-30.

Sturludottir, S. A. & Turner, J. (1985). The elm decline at Pawlaw Mire: an anthropogenic interpretation. *New Phytologist,* **99,** 323-9.

Suomela, J. L. (1967). *Hailuoto. Entisiä vaiheita.* Kaleva, Oulu.

Suominen, J. & Pohjakallio, K. (1964). Kuusamon Juuman kylän rikkaruohoista. *Molekyyli,* **4,** 60-3.

Sutherland Estate (1972). *Papers on Sutherland Estate Management 1801-1816,* volume 1 (Ed. by R. J. Adam). Scottish History Society, Series 4, **8,** Edinburgh.

Szabo, M. (1970). Herdar och husdjur. En etnologisk studie över Skandinaviens och Mellaneuropas beteskultur och vallningsorganisation. *Nordiska muséets Handlingar,* **73,** 389 pp.

Tanner, P. W. G. & Shackleton, R. M. (1979). Structure and stratigraphy of the Dalradian rocks of the Beannabeola area, Connemara, Eire. *Special Publication of the Geological Society of London,* **8,** 243-56.

Tansley, A. G. (1939). *The British Islands and their Vegetation.* Cambridge University Press, London.

Tardy, J. (1982). *Bevezetés a magyar östörténet kutatásának forrásaiba (Introduction to the Sources of Research into Hungarian Prehistory).* **IV.** Tankönyvkiadó, Budapest.

Tauber, H. (1967). Differential pollen dispersion and filtration. *Quaternary Paleoecology* (Ed. by E. J. Cushing and H. E. Wright). Yale University Press, New Haven.

Tauber, H. (1977). Investigations of aerial pollen transport in a forested area. *Dansk Botanisk Arkiv,* **32** (1), 1-121.

Taylor, J. A. (1973). Chronometers and chronicles: a study of palaeoenvironments in west-central Wales. *Progress in Geography,* **5,** 247-334.

ter Braak, C. J. F. (1981). The biplot in multivariate analysis. *Institute TNO for Mathematics, Information Processing and Statistics, Wageningen, Report A 81 ST 61 38,* 25 pp.

ter Braak, C. J. F. (1986). Canonical correspondence analysis: a new eigenvector technique for multivariate direct gradient analysis. *Ecology,* **67,** 1167-79.

ter Braak, C. J. F. (1987). *CANOCO - a FORTRAN program for canonical community ordination by (partial) (detrended) (canonical) correspondence analysis, principal components analysis and redundancy analysis (version 2.1).* TNO Institute of Applied Computer Science, Wageningen, 95 pp.

ter Braak, C. J. F. & Barendregt, L. G. (1986). Weighted averaging of species indicator values: its efficiency in environmental calibration. *Mathematical Biosciences,* **78,** 57-72.

Tesch, S. (1983). Ystad II. *Riksantikvarieämbedets och Statens historiska museer, Rapport Medeltidsstaden,* **45,** 1-134.

Teunissen, D. & Teunissen-van Oorschot, H. G. C. M. (1980). The history of the vegetation in S.W. Connemara (Ireland). *Acta Botanica Neerlandica,* **29,** 285-306.

Thomas, D. (1963). *Agriculture in Wales during the Napoleonic Wars.* University of Wales Press, Cardiff.

Thorsteinsson, I. (1983). *Undersøgelser af de naturlige græsgange i Syd-Grønland 1977-1981.* Landbrugets Forskningsinstitut, Island. Forsøgsstationen Upernaviarsuk, Greenland.

Timaffy, L. (1980). *Szigetköz.* (Magya Néprajz - Hungarian ethnography). Gondolat, Budapest.

Tipper, J.C. (1979). Rarefaction and rarefiction - the use and abuse of a method in paleoecology. *Paleobiology,* **5,** 423-34.

Tittensor, R. M. (1985). Conservation of our historic landscape heritage. *Folk Life,* **23,** 5-20.

Tolonen, M. (1978). Palaeoecological studies on a small lake, S. Finland, with special emphasis on the history of land use. *Annales Botanici Fennici,* **15,** 177-208, 209-22, 223-40.

Tolonen, M. (1981). An absolute and relative pollen analytic study on prehistoric agriculture in South Finland. *Annales Botanici Fennici,* **18,** 213-20.

Tómasson, P. (1973). Meltekja á Herjólfsstödum i Álftaveri. *Árbók hins islenzka Fornleifafélags,* 43-61.

Troels-Smith, J. (1953). Ertebølle Culture - Farmer Culture. Results of the past ten years excavations in Aamosen Bog, West Zealand. *Aarbøger for nordisk Oldkyndighed og Historie*, **1953**, 5-62.

Troels-Smith, J. (1955). Karakterisering av løse jordarter. Characterization of unconsolidated sediments. *Danmarks geologiske Undersøgelse*, Series IV, **3** (10), 73 pp.

Troels-Smith, J. (1960). Ivy, mistletoe and elm: climatic indicators - fodder plants. *Danmarks geologiske Undersøgelse*, Series IV, **4** (4), 32 pp.

Troels-Smith, J. (1981). Naturwissenschaftliche Beiträge zur Pfahlbauforschung. *Archäologie der Schweiz*, **4**, 98-111.

Turner, J. (1970). Post-neolithic disturbance of British vegetation. *Studies in the Vegetational History of the British Isles* (Ed. by D. Walker and R. G. West). Cambridge University Press, London.

Turner, J. (1986). Principal components analyses of pollen data with special reference to anthropogenic indicators. *Anthropogenic Indicators in Pollen Diagrams* (Ed. by K.-E. Behre). A. A. Balkema, Rotterdam.

Tusser, T. (1573). *Five Hundred Pointes of Good Husbandrie*. (Ed. by W. Payne and S. J. Herrtage). Trübner, 1878.

Uggla, E. (1958). Skogbrandfält i Muddus Nationalpark. *Acta Phytogeographica Suecica*, **41**, 1-115.

Uhlig, H. (1961). Old hamlets with infield and outfield systems in western and central Europe. *Geografiska Annaler*, **43**, 285-312.

Ulbert, G. (1975). Der Auerberg. *Ausgrabungen in Deutschland*, **1**, 409-33.

Ulén, B. (1985). Åkermarkens erosion. *Ekohydrologi*, **20**, 26-35.

Utaaker, K. (1978). Lokal og vekstklima i Sogn. *Norges landbruksvitskaplige forskningsråd. Forsking og forsøk i landbruket*, **30**, 114-68.

van der Maarel, E. (1971). Plant species diversity in relation to management. *The Scientific Management of Animal and Plant Communities for Conservation* (Ed. by E. Duffey and A. S. Watt). Blackwell Scientific Publications, Oxford.

van Geel, B. (1978). A palaeoecological study of Holocene peat bog sections in Germany and the Netherlands. *Review of Palaeobotany and Palynology*, **25**, 1-120.

van Geel, B. (1984). Palynologische aanwijzingen voor landbouw op pasontgonnen hoogveen in middeleeuws Oostzaan. *De Jol, Mededelingenblad van de Stichting Oudheidkamer Oostzaan*, **1** (2), 11-5.

van Geel, B., Bos, J. M. & Pals, J. P. (1983). Archaeological and palaeoecological aspects of a Medieval house terp in a reclaimed raised bog area in North Holland. *Berichten van de Rijksdienst voor het Oudheidkundig Bodemonderzoek*, **33**, 419-44.

Varjo, U. (1977). Finnish Farming. Typology and economics. *Geography of World Agriculture*, **6**, 146 pp.

Virrankoski, P. (1973). *Pohjois-Pohjanmaa ja Lappi 1600-luvulla (III)*. Pohjois-Pohjanmaan, Kainuun ja Lapin maakuntaliittojen yhteinen historiatoimikunta, Oulu.

Vitousek, P. (1982). Nutrient cycling and nutrient use efficiency. *American Naturalist*, **119**, 553-72.

von Känel, H.-M., Furger, A. R., Bürgi, Z. & Martin, M. (1980). Das Seeland in ur- und frühgeschichtlicher Zeit. *Jahrbuch der geographischen Gesellschaft von Bern*, **53**, 71-165.

Vorren, K.-D. (1979). Anthropogenic influence on the natural vegetation in coastal North-Norway during the Holocene. Development of farming and pastures. *Norwegian Archaeological Review*, **2**, 1-21.

Vorren, K.-D. (1986). The impact of early agriculture on the vegetation of Northern Norway. A discussion of anthropogenic indicators in biostratigraphical data. *Anthropogenic Indicators in Pollen Diagrams* (Ed. by K. E. Behre). A. A. Balkema, Rotterdam.

Vorren, K.-D. & Alm, T. (1985). An attempt at synthesizing the Holocene biostratigraphy of a "type area" in northern Norway by means of recommended methods for zonation and comparison of biostratigraphical data. *Ecologia Mediterranea*, **11**, 53-64.

Vorren, T. (1976). Siste istids isbevegelser, sedimenter og stratigrafi på Hardangervidda, Sør-Norge. *I-II*. Universitetet i Tromsø, Institutt for biologi og geologi. (Unpublished manuscript).

Vorren, T. (1979). Weichselian ice movements, sediments and stratigraphy on Hardanger-vidda, South Norway. *Norges Geologiske Undersøkelser,* **350,** bulletin 50. 117 pp.

Vuorela, I. (1973). Relative pollen rain around cultivated fields. *Acta Botanica Fennica,* **102,** 1-27.

Wade, W. (1802). Catalogus plantarum rariorum in Comitatu Gallovidiae, praecipue Cunnemara inventarum. *Scientific Transactions of the Royal Dublin Society,* **2,** 103-27.

Walker, G. J. & Kirby, K. J. (1987). An historical approach to woodland conservation in Scotland. *Scottish Forestry,* **41,** 87-98.

Walker, J. (1980). *The Rev. Dr. John Walker's REPORT ON THE HEBRIDES of 1764 and 1771* (Ed. by M. M. McKay). J. Donald, Edinburgh.

Walker, M. F. & Taylor, J. A. (1976). Post-Neolithic vegetation changes in the western Rhinogau, Gwynedd, north-west Wales. *Transactions of the Institute of British Geographers N.S.,* **1,** 323-45.

Walker, M. J. C. (1984a). Pollen analysis and Quaternary research in Scotland. *Quaternary Science Reviews,* **3,** 369-404.

Walker, M. J. C. (1984b). A pollen diagram from St. Kilda, Outer Hebrides, Scotland. *New Phytologist,* **97,** 99-113.

Wartenberg, D., Ferson, S. & Rohlf, F. J. (1987). Putting things in order: a critique of detrended correspondence analysis. *American Naturalist,* **129,** 434-48.

Watts, W. A. (1977). The Late Devensian vegetation of Ireland. *Philosophical Transactions of the Royal Society of London B,* **280,** 273-93.

Watts, W. A. (1984). The Holocene vegetation of the Burren, western Ireland. *Lake Sediments and Environmental History* (Ed. by E. Y. Haworth and J. W. G. Lund). Leicester University Press, Leicester.

Watts, W. A. (1985). Quaternary vegetation cycles. *The Quaternary History of Ireland* (Ed. by K. J. Edwards and W. P. Warren). Academic Press, London.

Webb, D. A. & Glanville, E. V. (1962). The vegetation and flora of some islands in the Connemara lakes. *Proceedings of the Royal Irish Academy B,* **62,** 31-54.

Webb, D. A. & Scannell, M. J. P. (1983). *Flora of Connemara and the Burren.* Royal Dublin Society, Dublin, and Cambridge University Press, Cambridge.

Welinder, S. (1975). Prehistoric agriculture in eastern Middle Sweden. *Acta Archaeologica Lundensia, Series in 8⁰ Minore,* **4,** 102 pp.

Welinder, S. (1977). Ekonomiska processer i förhistorisk expansion. *Acta Archaeologica Lundensia, Series in 8⁰ Minore,* **7,** 222 pp.

Welinder, S. (1983a). The ecology of long-term change. *Acta Archaeologica Lundensia, Series in 8⁰ Minore,* **9,** 115 pp.

Welinder, S. (1983b). Ecosystem change at the Neolithic transition. *Norwegian Archaeological Review,* **16,** 99-105.

Welinder, S. (1984). A systematic approach to understanding long-term changes in culture landscape. *Settlement and Economy in Later Scandinavian Prehistory* (Ed. by K. Kristiansen). British Archaeological Reports, S211, Oxford.

Wheeler, B. D. (1980). Plant communities of rich-fen systems in England and Wales. III. Fen meadow, fen grassland and fen woodland communities, and contact communities. *Journal of Ecology,* **68,** 761-88.

Whittow, J. B. (1974). *Geology and Scenery in Ireland.* Penguin Ltd., London.

Wickham-Jones, C. & Sharples, N. (1984). *An Interim Report on the Excavations at Farm Fields, Kinloch, Rhum 1984.* National Museum of Antiquities of Scotland, Edinburgh.

Widdowson, A., Penny, A. & Hewitt, M. V. (1982). Results from the Woburn reference experiment. III. Yields of the crops and recoveries of N, P, K and Mg from manures and soil, 1975-79. *Rothamsted Experimental Station, Report for 1981,* 5-21.

Widgren, M. (1983). Settlement and farming systems in the early Iron Age. *Stockholm Studies in Human Geography,* **3,** 1-132.

Widgren, M. (1986). Bebyggelsesform och markrättigheter under järnåldern. *Svenska Sällskapet för Antropologi och Geografi, årsbok Ymer,* **1986,** 18-26.

Wilkins, D. A. (1984). The Flandrian woods of Lewis (Scotland). *Journal of Ecology,* **72,** 251-58.

Williamson, T. (1986). Parish boundaries and early fields: continuity and discontinuity. *Journal of Historical Geography,* **12,** 241-8.

Witte, H. J. L. & van Geel, B. (1985). Vegetational and environmental succession and net organic production between 4500 and 800 BP reconstructed from a peat deposit in the western Dutch coastal area. *Review of Palaeobotany and Palynology,* **45,** 239-300.

Woodman, P. C. (1985). Prehistoric settlement and environment. *The Quaternary History of Ireland* (Ed. by K. J. Edwards and W. P. Warren). Academic Press, London.

Worsøe, E. (1980). Jyske egekrat. Oprindelser, anvendelse og bevaring. *Flora og Fauna,* **86,** 51-63.

Wymer, J. J. (1977). *Gazetteer of Mesolithic Sites in England and Wales. Council for British Archaeology, Report* **22,** Geo Abstracts, Norwich.

Zachrison, A. (1922). Nyodling, torrläggning och bevattning i Skåne. *Skrifter utgivna av de skånska hushållningssällskapen med anledning av deras hundraårsjubileum år 1914.* **II,** 4.

Zólyomi, B. (1980). Landwirschaftliche Kultur and Wandlung der Vegetation im Holozän am Balaton. *Phytocoenologia,* **7,** 121-6.

Zólyomi, B. & Simon, T. (1969). Közép-Tiszavidék. Természetes növényzet (Middle Tisza region. Natural vegetation). *A tiszai Alföld* (Ed. by S. Marosi and J. Szilárd). Akadémiai Kiadó, Budapest, 124-31.

Zvelebil, M. (1988). *Hunters in Transition.* Cambridge University Press, Cambridge (in press).

Index

Abies expansion, 183, 303-6
Abkær Bog, Denmark, 210-27
Absolute pollen frequency, *see also* pollen influx and concentrations, 179, 180, 209
Acer campestre, 65, 81
Aerial photography, 9, 63, 82, 89, 153-63, 327, 465
Agriculture, transition to, 181, 183, 256-66
Air-photo interpretation, 158, 171, 173
Alder, *see Alnus glutinosa, A. incana, Alnus*
Alnus, 15, 17, 18, 19, 20, 23, 39
Alnus pollen expansion, 257, 274, 278, 340, 341, 342, 347, 452
Alnus glutinosa, 13
Alnus incana, 13, 15, 25, 26, 33, 40, 41, 42, 44
Alps, 72, 120, 290
America, *see also* United States of America, 33, 68, 76, 466
Amsterdam, The Netherlands, 321, 322, 325, 327, 329-31
Ancient woodland, 59, 70, 79, 80, 81, 82, 83, 86, 88, 89
Anglo-Saxon, 8, 53, 54, 57, 60, 62, 67, 72, 74, 76, 234, 235
Animal droving, 435, 438, 440, 441, 443
Anthropogenic pollen indicators, 179, 180, 181, 182, 192-99, 206, 230, 239, 250, 259, 264, 289, 297, 307, 366, 417-22, 435, 470, 473, 478-79
Årstrand, Norway, 370, 371-75
Ash, *see Fraxinus excelsior*
Aspen, *see Populus tremula*
Auerberg, Germany 301-10
Avena spp. 36, 126, 127, 128, 129, 131, 132, 143, 145, 392, 425, 459, 472

Bacteria, 134
Bark peeling, 16, 20, 27, 39, 262
Barley, *see Hordeum*
Beavers, 340
Beech, *see Fagus sylvatica*
Belgium, 69
Bere, *see Hordeum*
Betula, 12, 13, 14, 15, 16, 17, 18, 19, 20, 22, 23, 25, 26, 27, 28, 31, 39, 43, 51, 57, 81, 353, 356, 362, 363, 449
Betula pendula, 12, 40, 42, 44, 56
Betula pubescens, 12, 56, 369, 382

Bielersee, Switzerland, 296, 297, 298
Biplots, 231
Birch, *see Betula pubescens, B. pendula, Betula*
Bjärsjöholmssjön, Sweden, 244, 250, 251
Bjäresjö, Sweden, 247, 248, 409-28
Black Death, 49, 377, 379, 386, 443
Blanket bogs, 182, 267, 272, 279, 283, 285, 286, 456, 475
Blanket peat, 59, 182, 475
Blue-green algae, 135
Boar, 79
Bog reclamation, 321-31
Boreal forest, 189, 190, 196, 199, 206, 207
Britain, *see* Great Britain
British Isles, 8, 53-77, 255-66
Bronze Age, 49, 57, 117, 170, 171, 184, 186, 234, 235, 237, 238, 244, 246, 249, 251, 252, 269, 285, 286, 292, 308, 310, 333, 340, 341, 342, 346, 347, 395, 406, 415, 419, 421, 422-23, 426, 427, 428, 441, 445, 460, 461, 471, 472, 478-79
Bryophytes, 87, 93, 95, 99
Buckwheat, *see Fagopyrum esculentum*
Bulgaria, 470-71, 483-84
Buried seed, 86
Butter production, 50
Byre, 144-47, 148, 387, 480

Callanish, Standing Stones of, 186, 187, 445-61
Calluna vulgaris, 11, 183, 313, 458
Calluna pollen expansion, 312, 313, 317, 458
Cannabis sativa cultivation, 418, 419, 423, 424, 426, 427, 459, 472
Canonical correspondence analysis, 7, 95, 105, 106, 465
Carpinus betulus, 65, 81
Carrying capacity, 111, 112, 124, 242, 389
Caschrom, 148, 149
Castanea sativa, 69, 70, 75, 289
Cattle, 11, 17, 18, 19, 20, 33, 34, 37, 39, 47, 50, 51, 60, 65, 91, 93, 105, 106, 126, 128, 129, 141, 144, 146, 148, 173, 174, 175, 189, 327, 345, 349, 350, 369, 385, 387, 389, 399, 402, 405, 441, 458, 472, 476
Cattle plague, 172
Celtic influences, 8, 53, 54, 76, 77, 171
Cereal pollen morphology, 259, 260, 261, 302, 336, 372, 375, 418, 448
Cereal production, 33, 35, 43, 47, 48, 50, 123, 125, 126, 134, 137, 149, 174,